# SCIENCE *of* CERAMICS

## Volume 4

738.105

# SCIENCE *of* CERAMICS

*Proceedings of the fourth International Conference on "Science of Ceramics," held under the auspices of the European Ceramic Association at Maastricht, Netherlands, 23-27 April 1967*

Edited by

## G. H. STEWART

*The British Ceramic Society*

Volume 4

*Published by*

## THE BRITISH CERAMIC SOCIETY

1968

*NoC*

*Papers classified according to the D.E.C. system
of the European Ceramic Association*

PRINTED IN GREAT BRITAIN BY
HENRY BLACKLOCK & CO. LIMITED
MANCHESTER

# Contents

## Reactions in Multiphase Ceramics during Firing

## Development of Ceramics for the Newest Applications

*Part One*

# THE EFFECT OF MECHANICAL FORCES DURING HEAT TREATMENT

# 1.—Dependence of the Mechanical Properties of Dense Porous Refractories on Thermal Cycling

By J. P. KIEHL and G. VALENTIN

*Centre de Recherches de la Société Générale
les Produits Réfractaires*

## ABSTRACT                                                    D45

*Reports on mechanical deterioration observed in refractory materials that have been submitted to cyclic temperature variations. The experimental set-up and method of testing are briefly described, and the results obtained both with dense and with porous refractories are analysed. Some correlations with the physical properties of the refractories are developed. A tentative theory which interprets the observed correlations and effects is presented.*

*Variations spécifiques de quelques propriétés mécaniques de produits réfractaires denses et poreux sous l'influence de différents cycles thermiques*

*Etude sur des phénomènes de désagrégation mécanique observés sur des matériaux réfractaires soumis à des variations de température cycliques. On décrit brièvement les dispositifs expérimentaux ainsi que les méthodes d'essai, puis on analyse les résultats obtenus avec des produits réfractaires aussi bien denses que poreux. On développe ensuite quelques corrélations avec certaines propriétés physiques des produits réfractaires examinés. Enfin, on présente une tentative d'interprétation plus théorique des effets et des corrélations observés.*

*Abhängigkeit der mechanischen Eigenschaften dichter und poröser Keramik bei der Wärmebehandlung*

*Es wird über die Verschlechterung der mechanischen Eigenschaften berichtet, die man an feuerfesten Materialien nach der Beeinflussung durch zyklische Temperaturänderung beobachtet. Die experimentelle Anordnung und die Prüfmethode werden kurz beschrieben. Die Resultate der Untersuchungen an dichten und an porösen Körpern werden analysiert, und es werden einige Beziehungen zu den physikalischen Eigenschaften der Keramik hervorgehoben. Ferne wird eine vorläufige*

3

*Theorie angegeben, die die beobachteten Beziehungen und Effekte beschreibt.*

## 1. PRINCIPLES OF THE METHOD

Most of the usual methods that are standardized or recommended for the determination of resistance to thermal shock suffer from a fair degree of uncertainty about the physical phenomena involved, as well as about their practical usefulness in industry.

The weak points of these methods are:

The poor reproduction of the real conditions.

The empirical or arbitrary values of the parameters considered.

The impossibility of fundamental research, owing to inability to describing quantitatively the behaviour of the materials.

After a number of experiments, we devised a method based on the following principles:

(1) Experiments were carried out on samples taken directly from the fabrication process, without prior reduction in size or cutting.

(2) One of the faces was brought to high temperature at a rate that was either proportional to the time or adapted to a heating cycle used in an actual industrial process.

(3) The cooling rate was slower than the heating rate so that only deterioration due to the latter needed to be taken into account.

(4) The deterioration of the mechanical properties of the refractory material was characterized by measuring the modulus of rupture in tension, deduced from a classical bending experiment.

## 2. EXPERIMENTAL METHOD

The resistance of the materials to thermal shock was investigated on samples which were all the same size.

The following procedure was used:

(1) Determination of the normal modulus of rupture $P$ at room temperature, by bending experiment.

(2) Determination of the critical heating rate ($V°C/min.$), at which cracking occurred, in the furnace described below.

(3) The leading face of the sample was subjected to a heat treatment consisting of $N$ cycles of $20°C–\theta–20°C$ at a rate of $x°C/min.$ ($x < V$) followed by measuring $P$ again. Several samples of the same material were treated in this way with different values of $x$, the heating rate in the range between $0°C$ and $V°C/min.$ (always for $N$ cycles for a given sample and rate).

(4) The different ratios $P/P_0$ were plotted as a function of $x$ (the heating rate).

## 3. SELECTION OF SAMPLES

If the method and the resulting curves are to be used correctly, and if doubts concerning the selection and the study of the effect of just a single parameter, e.g. density, manufacturing method, etc., are to be eliminated, it is necessary to work with a batch of samples as homogeneous as possible, particularly in respect of the mechanical properties.

### 3.1 Control by Density

A first selection was effected by measuring the apparent density $d$, which is related directly to the other characteristics. Only samples with densities within $\pm 5\%$ of the mean value were selected.

### 3.2 Ultrasonic Testing

A second selection was made according to a process of dynamic auscultation which measured the time $\tau$ (and thus the velocity $C$) of the propagation of ultrasonic waves (100 kHz) along the axis of samples (e.g. the length $L$ of a brick).

This velocity $C$ is related to the apparent density $d$, Young's modulus $E$, and the Poisson coefficient $\sigma$, by the formula:

$$C = \frac{L}{\tau} = \sqrt{\frac{1-\sigma}{(1+\sigma)(1-2\sigma)} \cdot \frac{E}{d}}$$

Experience had shown that samples with the same apparent density and the same value of Young's modulus had very similar mechanical properties.

In practice some 30 bricks, previously selected on the basis of apparent density, were tested ultrasonically, the value of $\tau$ being measured along and across the brick.

A dozen samples, selected on the basis of density and propagation time (within $5\%$ tolerance), were kept for the thermal-shock experiments as well as for measurements of the modulus of bending, according to ASTM C 93–54, before or after the heat treatment.

## 4. THE CRITICAL HEATING RATE FOR FRACTURE

### 4.1 Furnace

Figure 1 is a diagram of the furnace used for bricks of standard dimensions ($230 \times 115 \times 65$ mm). The bricks were heated from the front ($115 \times 65$). The rise in temperature was linearly proportional to

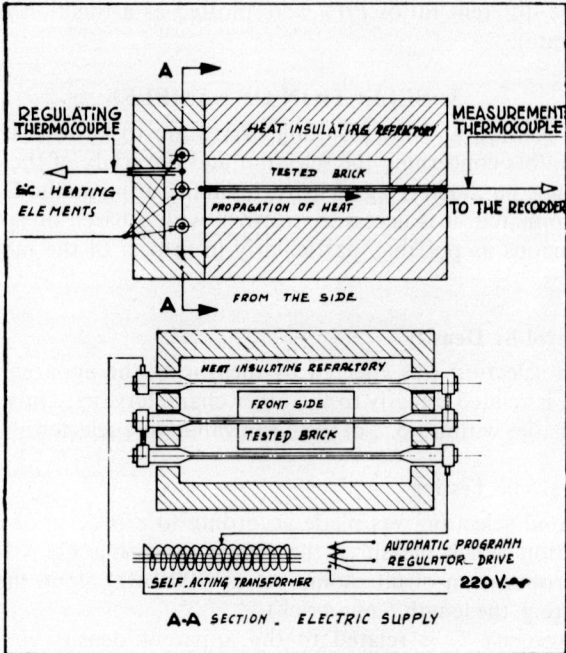

FIGURE 1.—Test furnace.

the time. Electric heating enabled heating and cooling to be accurately controlled. Each brick was laid biggest face down $(230 \times 115)$ on a refractory support of low thermal conductivity, and the sides and rear face of the brick were covered with the same thermally insulating material.

A Pt/Pt–10% Rh thermocouple was attached to the centre of the front face of the brick and in good thermal contact with it. A series of thermocouples was also placed along the sides of the brick (Figure 2). By recording their temperature it was possible to determine the temperature distribution along the sample. The front of the furnace was heated by several tubular heating-elements carried on a support; these elements were connected to an autotransformer, which provided a variable increasing voltage by manual control or by proportional control, or a fixed voltage adjustable to the rate of rise in temperature by an on–off switch.

## 4.2 Detection of the Moment of Fracture

Fracture of the brick was detected either by the ultrasonic method, which is very sensitive, or by the breaking of an electric circuit. A

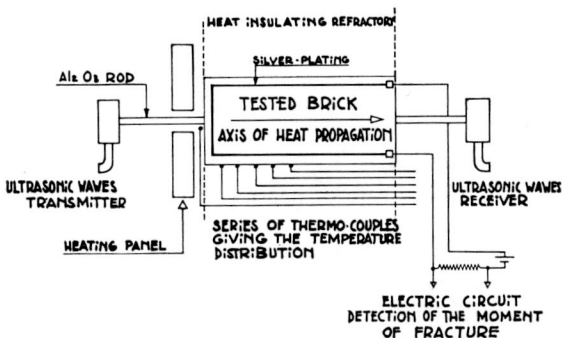

FIGURE 2.—Diagram of the lay-out of the apparatus.

strip of silver deposited along the edges of a face of the brick was included in an electric circuit (Figure 2). When the brick fractured, the circuit was broken and the potential across a series resistance dropped to zero.

### 4.3  Influence of Thermal Cycling on the Mechanical Properties

*Furnace*

Identical with that described above.

*Number of Cycles*

Each rate of heating experiment comprised five cycles. Experience showed that the material underwent little change after the third or fourth cycle.

*Maximum Temperature of the Front of the Brick*

This temperature was limited to 1000°C, most of the effects of thermal shock occurring in the range of 0°C–1000°C. It is possible, however, to use other temperatures according to the material and the conditions in which it is used.

*Cooling Rate*

In order to take mainly the rate of rise of the temperature into account (the nature of the phenomena being slightly different during the cooling phase), the rate of cooling was never allowed to exceed half the heating rate.

### 5.  EXPERIMENTAL RESULTS

The above methods were applied to some 60 different refractories. Not only dense materials, based on aluminium silicate, magnesia, chromite and zircon, as well as materials that had been sintered to zero porosity, but also a large series of thermally insulating refractories with densities between 0·4 and 1·2 and a limit of operation between 1000° and 1800°C were studied.

## 5.1  General Observation

In all cases a critical rate of heating $V$ (°C/min) can be defined, which causes "total fracture" (total separation or separation > 90 %) of that part of the brick that had been heated. This limiting rate of heating varied considerably for samples of the same size (from 2 or 3°C/min. up to 100°C/min.).

The zone in which total fracture occurred was always 6 to 8 cm behind the face (115 × 65 mm) that had been heated, whatever the nature of the material or its apparent density. The crack is essentially parallel to the heated front face. Thus the brick fractured in what is a low-temperature zone.

Fracture always occurred when the heated face was below 1000°C; the thermal shock is therefore a low-temperature phenomenon. The deterioration of mechanical properties due to variation in temperature occurs only in the range of temperature during which the refractory is considered to be rigid, and which we believe to be below 800°C. This observation supports the findings of SCHWIETE et al.[2] who showed that it was possible for refractories to creep from 700°— 800°C.

No cracks or mechanical deterioration were observed if a product known for its poor properties was brought to 1000°C at a rate of heating well below the critical rate for mechanical deterioration, followed by heating to 1500°C at a rate well above the critical value. The deterioration of mechanical properties therefore occurs in the temperature region in which the refractory can be considered to be a true solid (see below).

In some materials the temperature at which total fracture occurred varied very little with the rate of heating, even when this rate surpasses the critical value 4 or 5 times. This was observed specially with materials that had been sintered to zero porosity, i.e. materials that are very sensitive to temperature variation. It can therefore be concluded that, for rupture to occur, there must be a difference in temperature between the front face and the cold part.

When a sample was heated at a rate well below the critical value, the properties of the material changed. In particular, a rapid decrease of the modulus of rupture was observed. Sometimes internal cracks extending over bigger or smaller parts of the cross-section were found, though the brick looked perfectly all right from the outside.

These internal changes occurred if the rate of heating ($x$) remained smaller than $V$. The process continued for 2 or 3 cycles but after a total of 4 or 5 cycles no changes were observed.

This decrease in the modulus was a function only of the rate of heating at $x°C/min$. As the rate approached the critical value, this decrease became more important but varied greatly, depending on the product.

It was then possible to plot curves (still for the sizes under consideration) of:

$$\frac{P}{P_0} = \frac{\text{Modulus of fracture after 5 thermal cycles}}{\text{Normal modulus of fracture of the material}}$$

as a function of the rate of heating. These diagrams exhibited two limits:

When the ratio $(P/P_0)$ became unity on the ordinate (rate→0)

When the ratio $(P/P_0)$ became zero at the minimum critical rate for total (or for 90%) fracture.

## 5.2 Special Considerations

The curves of $(P/P_0)$ *versus* $f$ (rate of rise), plotted on linear co-ordinates and comparable for all materials, indirectly cover all the thermo-mechanical properties of refractories. Examination revealed three fundamental functions (A, B, C) (Figure 3).

### 5.21 Dense Refractories

Groups A and B covered the materials commonly known as having poor resistance to sudden changes in temperature, i.e. magnesia types sintered to zero porosity, acid-resisting refractories, etc. On the other hand, these diagrams show that those products that have a very mediocre general behaviour perform very differently in practice.

Thus the mechanical properties of a type A refractory, e.g. magnesia, gradually deteriorated from the lowest rates of temperature increase ($2·5°C/min.$) although total fracture did not occur until about $20°C/min$. The B-type (low-porosity sintered refractory,) however, retained its properties unaltered up to a certain value, and then suddenly fractured. Refractory B would have excellent resistance to thermal shock under temperature fluctuations below this critical limit.

Group C contained the vast majority of the alumino-silicate refractories with porosities of 20 to 24%, and mullite and corundum types. The differences between them, although still clear, were less striking than those of the products in groups A and B.

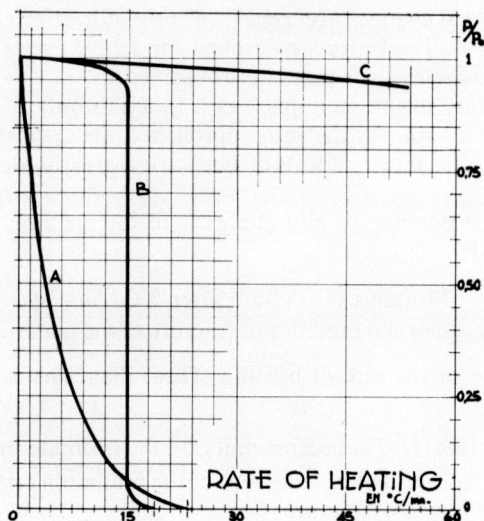

FIGURE 3.—Rate of rise in temperature plotted against $P/P_0$.

## 5.22 Insulating Refractories

Here again, we had the three functional groups but, unlike the
case with the dense refractories, most of the light porous materials
came into group B: i.e. beyond a certain critical rate ranging between 15 and 30°C/min., they showed total fracture. Most of the
other insulating products fell into group C and exhibit excellent
behaviour.

## 6. EFFECT OF SOME PHYSICAL PROPERTIES OF REFRACTORY MATERIALS

After the results referred to had been obtained, we attempted to
relate good or poor behaviour under temperature fluctuations (i.e.
the critical rate for fracture and the rate of reduction in Young's
modulus of elasticity to 75%) to certain physical properties which
should be significant in these respects.

### 6.1 Coefficient of Linear Expansion ($\alpha$)

Materials with a high coefficient were not generally very strong.
This was true of the basic products (with an $\alpha$ of between 9 and $13 \times 10^{-6}$) and the silicas (with a complex $\alpha$ ranging from $20 \times 10^{-6}$ to
$40 \times 10^{-6}$ within the 0°C to 500°C range). However, a material based
on diatomite was exceptional, with $\alpha = 18 \times 10^{-6}$ at 150° and 580°C,
which resisted temperature fluctuations remarkably well.

Unfortunately the other refractories with $\alpha$ values around 5 or $6 \times 10^{-6}$ also exhibited widely varying behaviour under thermal shock.

## 6.2 Diffusivity ($D$), Thermal Conductivity ($\lambda$), Specific Gravity ($\rho$), and Specific Heat ($c$)

Materials with a low diffusivity, i.e. insulators, should be more greatly affected. Actually, if diffusivity is of essential significance (when it is low) in the creation of thermal gradients varying from one point to another, and therefore in the very mechanism of thermal shock, differences in diffusivity seem to have only a minor influence.

In fact:

Diffusivity is generally low for all refractories except SiC, amounting at the most to about $10^{-2}$ cm$^2$/s;

Thermal fluctuations at a point are great in relation to this diffusivity, which represents the rate of heat transfer.

## 6.3 Mechanical and Elastic Properties

*6.31 Ratio* $\dfrac{P_0\,(\textit{Modulus of fracture in the cold})}{E\ \ (\textit{Young's modulus in the cold})}$

This ratio would, overall, constitute the relative terminal tensile deformation if the material were perfectly elastic. The behaviour should become better as $P_0/E$ increases. In fact it appears that this rule applies only to extreme cases:

$P_0/E$ is small (for basic products and materials with a very homogeneous structure) $P_0/E \simeq 2$ to $4 \times 10^{-4}$.

$P_0/E$ is high (diatomite product) 9 to $10 \times 10^{-4}$.

However, a knowledge of this ratio $P_0/E$ was often not enough for a proper appreciation of behaviour under thermal shock. This observation applied particularly to alumino-silicate insulating refractories with densities between 0·5 and 1·2, where the ratio $P_0/E$ varied slightly around $5 \times 10^{-4}$, while the differences between the behaviour at the critical rate and at the terminal fatigue were considerable.

*6.32 Deformation and Terminal Sag at Fracture*

The curves showing the sag as a function of the force of bending to the tested materials were studied. Such curves usually exhibit three sections (Figure 4):

(1) A purely elastic zone, often very small, in which the deformation and the force applied are proportional.

(2) A largely linear zone, but one where the deformation no longer disappears with the load (permanent deformation $d_0$ if $p$ becomes 0).

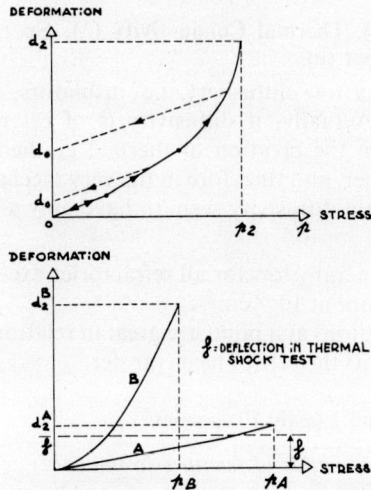

FIGURE 4.—Deformation as a function of bending force.

(3) A final zone in which the deformation rapidly increases to fracture, $(d_r)$, the permanent deformation becoming even greater. This final zone is extremely important. In fact, the performance of a material with a large final deformation $(d_r)$ or a relatively low breaking strain when exposed to thermal fluctuations may be better, particularly as regards the critical rate, than that of a product with a high breaking strain but poorer "elasticity".

Figure 5 shows the sag and tensile modulus curves for various products, and Figure 6 gives the very clear correlation between the sag of a flexed bar at fracture and the critical heating rate. This observation supports the opinion of KLASSE.[3]

## 7. THEORETICAL INTERPRETATION OF THE PHENOMENA OBSERVED

The following arguments apply to the experimental conditions under which our work was done, but the essential principles should nevertheless be susceptible to more general application.

### 7.1 Current Hypotheses

Most theories on fracture due to temperature fluctuations are based partly on the hypothesis of fracture by direct tension and partly

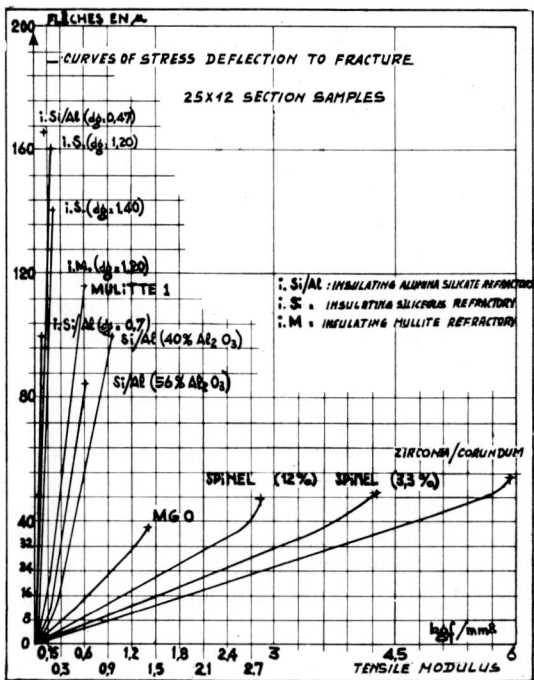

FIGURE 5.—Deformation against tensile modulus for various products (test-pieces 25 × 12 mm).

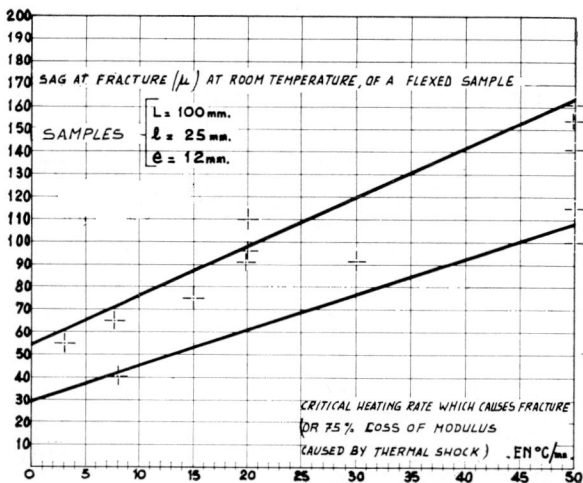

FIGURE 6.—Deformation of stressed bar at fracture and critical heating rate.

on the application of the general principles of elasticity, especially Hooke's law, the latter being applied to heat expansion. We might mention the equation of Winkelmann and Schott connecting unit linear elongation by expansion, tensile strength and Young's modulus modified by coefficients which take into account diffusivity, heat exchange with the environment (Biot number), the homogeneity of the material, and the probability of fracture.[4,5]

Reasoning is often based, too, on isothermal sections, assumed to be very thin, showing tensile forces perpendicular to the thermal flux which should thus cause superficial and successive peeling.

Unfortunately, these interpretations do not explain any of our experimental observations, and we should like to make the following remarks about them.

### 7.11 Existence of an Elastic Region

Mathematical propositions involving Young's modulus apply only to sintered ceramics with a fine and uniform structure. These products do have a certain reversible elasticity over the entire stress/strain curve. It does not, however, apply to refractories with a heterogeneous grain/binder structure, where the elastic zone is sometimes extremely limited, although the total acceptable deformation may be considerable.

### 7.12 Identification of Expansion Deformation with Mechanical Deformation

We are not satisfied with this identification and equalization. Let us, in fact, compare two experiments on prismatic samples subjected to:

Traction and flexion tests to ultimate fracture. We can deduce the final unit linear elongation to which a bar of length $l_0$ can still be subjected, i.e. $a_r = \Delta l / l_0$, from the stress/strain curves;

Heat expansion from 20°C to 1000°C or 1500°C. From this we find a unit linear elongation $a_d = \Delta l' / l_0$, such that:

$$a_d = \bar{\alpha}_{20}^{\theta°} \left(\theta - 20°C\right)$$

$\bar{\alpha}_{20}^{\theta°}$ being the mean coefficient of thermal expansion between ambient temperature and $\theta°C$ or, more generally,

$$a_d = \int_{20}^{\theta} \alpha(T) \, dT$$

These experiments, made on most of the materials tested within the scope of our study, are summarized in Table 1. Now it is found that the unit expansions from 20 to 1000°C are 10 to 30 times the linear elongation at fracture under tension. This observation allows us to deduce that:

**Table 1**

$a_t = \frac{\Delta\ell}{\ell}$ ; FINAL RELATIVE ELONGATION, AT FRACTURE, IN A MECHANICAL TEST (SAG OF A 100 x 25 x 12 mm PRISMATIC BAR).

$a_D = \frac{\Delta\ell}{\ell}$, $\sum_i \alpha_i \Delta\theta_i$ RELATIVE LENGHTHENING CAUSED BY UNIFORM THERMAL EXPANSION, FROM 20°C TO 1.000°C

| MATERIAL | $a_t = \frac{\Delta\ell}{\ell}$ CAUSED BY BENDING (OUTER FIBER UNDER TENSION) X $10^4$ | $a_D = \frac{\Delta\ell}{\ell}$ CAUSED BY THERMAL EXPANSION FROM 20°C TO 1.000°C X $10^4$ |
|---|---|---|
| I. Si/Al (d = 0,45 /cm³) | 6,9 | 50 |
| I. Si/Al (d = 0,5 /cm³) | 7,8 | 43 |
| I. Si/Al (d = 0,7 /cm³) | 5,8 | 44 |
| I. Si/Al (d = 0,9 /cm³) | 4,4 | 47 |
| I. Si/Al (d = 1 /cm³) | 4,92 | 45 |
| I.M. (d = 1,2 /cm³) | 5,86 | 55 |
| I.A. (d = 1,35 /cm³) | 4,1 | 79 |
| I.S. (d = 1,20 /cm³) | 7,7 | 63,7 |
| I.S. (d = 1,40 /cm³) | 6,7 | 63,7 |
| Si/Al (40% Al₂O₃)(d=2,20 /cm³) | 5 | 56 |
| Si/Al (50/55% Al₂O₃)(d=2,25 /cm³) | 4 | 57 |
| MULLITE 1 (d=2,30 /cm³) | 5,6 | 53 |
| MULLITE 2 (d=2,40 /cm³) | 3 | 53 |
| AL 90% (d=3,18 /cm³) | 3,9 | 64 |
| AL 100% (d=3,20 /cm³) | 1,9 | 83 |
| MGO (d= 2,90 /cm³) | 1,8 | 130 |
| SPINELLE Al₂O₃/MGO (d=2,85/cm³) | 2,2 | 95 |
| ZIRCONE/... (d=3,40/cm³) | 2,7 | 62 |

I.Si/Al: INSULATING ALUMINA SILICATE REFRACTORY.. IM = INSULATING MULLITE REFRACTORY.. IA: INSULATING ALUMINA REFRACTORY.
I.S. = INSULATING SILICEOUS REFRACTORY.

A refractory material, when uniformly heated, can undergo, without any danger, relative deformations very many times greater than the limit of linear elongation to fracture under mechanical load.

Thermal expansion as a phenomenon is basically different from a conventional mechanical effect. According to current ideas, expansion acts in a solid at the atomic level.[6,7] It is produced by outwardly directed internal pressures which are, in fact, radiation pressures exerted by the phonons associated with elastic waves liberated by thermal vibrations in the crystal lattice. Expansion lies outside the range of elasticity, departures from Hooke's law (non-zero mean displacement of the atoms) explaining expansion exactly, and the additive translations of the atoms in relation to one another explain why the elongations can be very much greater than those obtained by mechanical stress or strain. A force acts only within narrow ranges, limited by faults, defects and dislocations which determine the polycrystalline state of a real solid.

A distinction should therefore be made between internal expanding forces or pressures and mechanical stresses of thermal origin caused by unequal expansion. Now particularly with fracture by thermal shock, as our experiments show, the fracture zones at very low temperatures at some distance from the heated surface can be caused only by purely mechanical effects, to which we can, as a first approximation, apply Hooke's law, and not by the direct effect of the thermal expansion forces, which are of an entirely different nature.

Conventional theories, which identify the forces of expansion with mechanical stresses, and application of Hooke's law to them, lead to improbable results. Hence the equation:

$$\frac{\Delta l}{l} = \bar{\alpha}_{20^\circ}^{\theta^\circ}(\theta^\circ - 20) = \frac{P_0}{E}$$

applied to a zirconia–corundum type material in which

$$\alpha_{20}^{1000} = 6 \cdot 2 \times 10^{-6},\ P_0 = 55\ \text{kgf/cm}^2\ \text{and}\ E = 1 \cdot 6 \times 10^6\ \text{kgf/cm}^2$$

gives $\Delta\theta = \theta - 20$ of 56°C

Such a product would be in danger of breaking when its temperature is raised by 56°C. This, however, has never been observed. At the moment of complete fracture, a standard brick, for example, exhibits a difference of 540°C between the heated surface and the rear zones at ambient temperature.

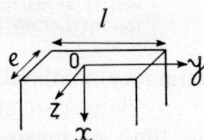

### 7.13 Hypotheses and Proposed Mechanism

Let us take for our example a parallelepiped of a given material, of infinite length, of width $l$ and thickness $e$, the centre of the upper surface forming the origin of a system of axes $\vec{Ox}, \vec{Oy}, \vec{Oz}$. This surface $x = 0$ is heated over a given temperature range $\theta(0,t)$ as a function of time. At a given moment $t$ there will be a decreasing temperature distribution $\theta(x, t)$ in the solid, and this will be assumed to be solely a function of $x$ (the unidimensional heat transmission). $\theta(x,t)$ is governed by the rate of heating $\theta(0,t)$, the diffusivity $D = \lambda/\rho.c$ of the material ($\lambda$ = thermal conductivity; $\rho$ = specific gravity; $c$ = specific heat) and must be the solution to the transmission equation:

$$\frac{\partial^2\theta}{\partial x^2} = \frac{1}{D}\frac{\partial\theta}{\partial t} \qquad\qquad (1)$$

Moreover, these temperatures decrease rapidly and are virtually imperceptible after 8 to 10 cm because of the low diffusivities in question of $0.1$ to $1 \times 10^{-2}$ cm²/sec.

We assume that the expansion effects actually take place in three dimensions, but, for simplicity, we shall work first in two dimensions only, $\overrightarrow{0x}$ and $\overrightarrow{0y}$ (plane $z = 0$).

Let us reduce our solid, plane $\overrightarrow{0x}$, $\overrightarrow{0y}$ (Figure 7) into longitudinal fibres of thickness $\Delta y$ parallel to $\overrightarrow{0x}$, and then reduce each fibre into zones of thickness $\Delta x$ parallel to $\overrightarrow{0y}$, e.g. 1, 2, 3, 4 . . . where the temperatures will be supposed to be uniform $\theta_1$, $\theta_2$, $\theta_3$, $\theta_4$, and decreasing. We thus reconstitute the sample into unit elements like $a$, $b$, $c$, $d$, which in the limit become infinitesimal (dimensions $dx$, $dy$).

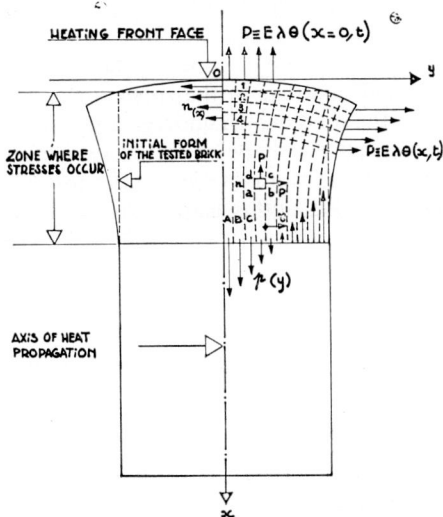

FIGURE 7.—Proposed mechanism.

### 7.14 Proper Transverse Deformation of Each Fibre

Let us consider what occurs in each elementary fibre, e.g. the first one of ordinate $y = 0$. We shall disregard, within the fibre, the effects of expansion along $\overrightarrow{0x}$, that exist only in the very first centimeter and cause nothing but minor stresses. Now, from our point of view, the transverse effects are essential here.   An elementary element of abscissa $x$, would, if it were alone, undergo an expansion deformation along $\overrightarrow{0y}$: $f(x) = \alpha . \theta(x,t) . \Delta y$. But, because of the unequal expansion in the elements above and below it (that are not at the same temperature) there will be a stress $n(x)$ along $\overrightarrow{Oy}$ in the given element. So that the algebraic resulting deformation of one element of a fibre may be written:

$$f(x) = \alpha . \theta(x, t) . \Delta y + \frac{n(x)}{E} \Delta y \quad . \qquad . \qquad . \quad (2)$$

(assuming that, as a first approximation, the elastic zone is defined by the mechanical stress $n(x)$).

Along the fibre, the stresses $n(x)$ will be compressive in the higher elements, tensile elsewhere, and will disappear fairly rapidly with $x$ because of the decrease in the temperatures. The effects of expansion and of its differences vanish beyond a few centimetres distance. The phenomena thereafter are as though the rest of the fibre, and hence the rest of the solid, from an abscissa $x = X$ was unable to deform and were to seem firmly embedded. $X$ corresponds exactly to the zone of fracture by thermal shock (6 to 8 cm from the heated surface). That is nearly a state of concentrated stress; it still subsists in the fibre local deformations in the zones under constraint, compared to the rest of the solid, which undergoes no alteration.

We shall assume that the shape of the proper deformation of this fibre is circular in view of the minuteness of the possible deformations $f(x)$. Now if $R_1$ is the radius of this shape and $C_1 = (1/R)$ the equivalent curvature, we may practically write:

$$C_1 = \frac{1}{R_1} \simeq \frac{2f(x)}{(X-x)^2} \quad . \quad \text{(3) (geometrical property)}$$

Taking Equations (2) and (3), equilibrium conditions of stresses

$$\int_0^X n(x)\,dx = 0 \qquad . \qquad . \qquad . \qquad . \qquad . \quad (4)$$

and limit conditions into account (continuity of the deformations in the embedding), there are in succession:

$$f(x) = \frac{C_1 \cdot (X-x)^2}{2} = \alpha\theta(x, t). \Delta y + \frac{n(x)}{E} \Delta y$$

$$\int_0^X \frac{C_1 \cdot (X-x)^2}{2}\,dx = \frac{C_1}{6} X^3 = \alpha\Delta y \int_0^X \theta(x, t)\,dx + \frac{\Delta y}{E} \int_0^X n(x)\,dx$$

$$\left( \int_0^X n(x)\,dx = 0 \right)$$

whence

$$C_1 = \frac{6 \cdot \alpha \cdot J(x) \cdot \Delta y}{X^3} \qquad \cdot \qquad \cdot \qquad \cdot \qquad \cdot \qquad (5)$$

where

$\alpha =$ Coefficient of thermal expansion

$X =$ The above-defined distance, beyond which there is no appreciable transverse expansion along $0y$.

$$J(x) = \int_0^x \theta\,(x,\ t)\ \mathrm{d}x;\ \text{integral of temperatures } C_1 \text{ is then the}$$

value of the circular shape that the edge of each fibre takes on.

### 7.15 Mechanism of Appearance of the Stresses Leading to Fracture

The result of the preceding line of argument is that the first fibre of ordinate $y=0$ undergoes a proper deformation, approximately circular, its curvature being $C_1$, which has repercussions on the whole part of the solid on its right and which causes a mechanical flexion and hence a curvature $C_2$ over the whole of that part. Because of this flexion, each fibre beyond the first then undergoes linear elongation (or contraction) along $\overrightarrow{0x}$ under the appearance of forces $p(y)$ parallel to $\overrightarrow{0x}$ and which are a function solely of $y$. This flexion appears solely in the lower zones of the solid that undergo no alteration, i.e. in the distance $X$.

This takes us back to the theory of embedded beams. We obtain a distribution of tensile and compressive forces that can be calculated if curvature $C_2$ is known; we will take it with a fair degree of accuracy to be the same as that of the shape of the sole fibre which undergoes a deformation, such as:

$$C_2 = C_1 = C = \frac{6 \cdot \alpha \cdot J(x)\, \Delta y}{X^3} \qquad \cdot \qquad \cdot \qquad \cdot \qquad (6)$$

It is then shown that the stresses per unit length $p(y)$ caused by the influence of the first fibre may be written:

$$p_1\,(y) = C \cdot E \cdot \left(\frac{l}{4} - y + \frac{\Delta y}{2}\right) \qquad \cdot \qquad \cdot \qquad \cdot \qquad (7)$$

($l =$ width of the sample.)

(If we still assume the elastic field; if we apply the standard results of embedded beams and take the actual position of the neutral fibre into account—change of origin). The stresses are tensile towards the centre of the solid, compressive towards the outside.

Now the second fibre also deforms in its own way and transmits a new curvature $C$ to the fibres to its right, which then undergo a fresh distribution of linear tensile and compressive forces in the same way as described above, distribution which is superimposed on the first ("equivalent beam" flexion to the right of the second fibre); moreover

$$p_2(y) = C \cdot E \cdot \left(\frac{l}{4} - y + \Delta y\right) \qquad . \qquad . \qquad . \qquad . \qquad (8)$$

the same applies to the third, fourth, etc. . . . . up to the $n$th fibre embedded at distance $X$.

Finally we add the partial distribution $p_1(y) + p_2(y)$ . . . After the limit has been reached and after integrating, we obtain the final distribution of longitudinal forces:

$$p(y) = \frac{C \cdot E \cdot}{4\Delta y} \cdot y\,(l - 3y) \qquad . \qquad . \qquad . \qquad . \qquad (9)$$

We then come to the deformation and the final formula, $C$ being replaced by its value:

$$\frac{\delta u}{\delta x} = \frac{3}{2} \frac{\alpha \cdot J(x)}{X^3} y\,(l - 3y) \qquad . \qquad . \qquad . \qquad . \qquad (10)$$

These deformations and stresses parallel to the axis of the solid will cause the fracture by thermal shock.

### 7.16 Summary of Main Results

Let us quite simply summarize the main results, without going deeply into the calculation:

Distribution of forces:

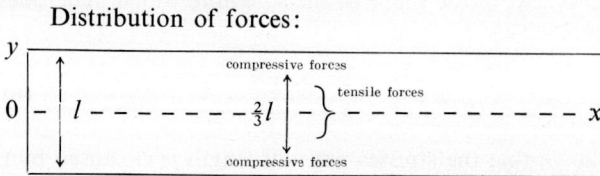

The central two-thirds of the section are under tension, while the one third remaining at the edge is in compression.

The maximum relative elongation per unit length takes place with $y = \pm \frac{l}{6}$ and is given by:

$$\frac{\partial u}{\partial x} = \frac{\alpha J(x)\,l^2}{8 X^3}$$

The maximum compression per unit length is four times the above value.

The mean elongation per unit length over the two thirds of the section (under tension) is:

$$\frac{\overline{\partial u}}{\partial x} = \frac{\alpha J(x) \, l^2}{12 X^3}$$

Fracture therefore starts in the central zones when either the maximum elongation or the mean elongation is exceeded, depending on the hypothesis adopted.

This distribution of stresses at break tallies well with many of our observations (shape of the fracture, internal fissures invisible from outside). This can explain why fracture occurs fairly far from the heated surface as a result of flexural forces.

### 7.17 Adoption of Three Dimensions and Experiments to Verify the Formulae

We should, on the one hand, modify the above formulae by the coefficient $(1 - \sigma)$, where $\sigma$ is Poisson's coefficient, and, on the other, we can use as a basis the mean linear elongations which may be taken as being additive, i.e. we shall have an elongation along $\overrightarrow{0x}$:

$$\left( \frac{\overline{\partial u}}{\partial x} \right)_1 = \frac{(1 - \sigma) \, \alpha J(x) \, l^2}{12 X^3}$$

due to the flexions along $0y$, and a mean elongation, always along $\overrightarrow{0x}$, but due to flexion along $\overrightarrow{0z}$:

$$\left( \frac{\overline{\partial u}}{\partial x^2} \right)_2 = \frac{(1 - \sigma) \, \alpha J(x)}{12 X^3} \, e^2$$

Since $(\overline{\partial u}/\partial x)_1$ and $(\overline{\partial u}/\Delta x)_2$ are colinear, we shall find

$$\left( \frac{\overline{\partial u}}{\partial x} \right)_1 + \left( \frac{\overline{\partial u}}{\partial x} \right)_2 = \frac{\overline{\partial u}}{\partial x} = \frac{(1 - \sigma) \, \alpha J(x)}{12 X^3} (l^2 + e^2)$$

With the purely theoretical hypothesis adopted, with the approximations in the calculation, and with the actual experimental conditions, the verification of such a formula can only lead to a comparison of orders of magnitude. Nevertheless, attempts can be made to check it for materials behaving badly under thermal fluctuations in the following manner:

$J(X)$ and then $\delta u / \delta x$ can be recalculated, using the temperature distribution and distance $X$ of fracture found in the experiments.

Furthermore, on the basis of our bending tests and the stress/ strain curves, we can deduce the unit elongation limit in the cold $a_r$ that a bar of length $l_0$ near $X$, which, as will be remembered, varies from 5 to 8 cm, can withstand.

The comparisons between $a_r$ and $\delta u/\delta x$ are these:

|  | $a_r$ | $\overline{\delta u/\delta x}$ | Fracture distance $X$ |
|---|---|---|---|
| Zero-porosity sintered zirconia-corundum product (10°C/min) | $2\cdot7\times10^{-4}$ | $2\cdot3\times10^{-4}$ | 6·2 cm |
| 10%-porosity aluminate– forsterite spinel (25°C/ min) | $2\cdot2\times10^{-4}$ | $3\cdot1\times10^{-4}$ | 5 cm |

These values seem to tie in well with our experiments and we do not think that under present circumstances, it is possible to obtain greater accuracy. Nevertheless, with the limiting characteristics of mechanical deformation known, it is possible, with the above formulae, to find heat cycles to which standard format refractories can be subjected without excessive risk.

## 8. CONCLUSION

This study of the resistance of refractory materials of heterogeneous structure to temperature fluctuations, which should not be considered as anything more than an approach to the problem, has led to the definition of a few points:

(1) Critical rates of increase in temperature can be expressed in figures, interpreted, and brought into line with practical data.

(2) A deterioration in the mechanical properties before total fracture can often be demonstrated, such a deterioration sometimes occurring even at the very lowest rates of temperature increase.

(3) Similarity between the behaviours of dense and porous materials under temperature fluctuations.

(4) The importance of some observations for their theoretical applications.

(5) Confirmation of the correlation between maximum deformation under flexion and resistance to temperature fluctuations.

(6) An interpretation of these phenomena, which shows good agreement with the experimental observations, and should make it possible to propound the first hypothesis on the problem of the shape

factor (the effect of the size and shape of sections whose temperature is increased), a problem which we are currently engaged upon.

## REFERENCES

1. FOURNEAU, T., "Comparative, Experimental and Statistical Study of Various European Methods of Thermal Shock Testing" 7th International Ceramic Congress, 1960.
2. EHRCKE, ., and SCHWIETE, H. E., *Arch. Eisenhütten.*, **34**, 795, 1963.
3. KLASSE, F., and HEINZ, A., *Mitteilung des Forschungsinstituts DIDIER (Wiesbaden) Biebrick*, (19/20), 296, 1955.
4. KINGERY, W. D., "Property Measurements at High Temperature." (John Wiley & Sons, Inc.: New York, 1959).
5. Symposium on Thermal Fracture, *J. Amer. Ceram. Soc.*, 1955.
6. KITTEL, C., "Introduction à la Physique de l'Etat Solide" (translated by E. L. Huguenin et R. Papoular) Dunod,–Paris, 1968).
7. BRILLOUIN L., "Les Tenseurs en Mécanique et en Elasticité" (Masson, Paris; 1960).
8. TIMOSHENKO, S., and GOODIER, J. N., "Theory of Elasticity" (McGraw Hill Book Company, Inc. 1951) (2nd Edition).

# 2.—The Viscosity of Pure Silica

## By G. HOFMAIER and G. URBAIN

*Institut de Recherches de la Sidérurgie Française*
*St. Germain-en-Laye, France*

**ABSTRACT**                                               A5256/D454

The viscosity of pure silica has been measured between 1600°
and 2480°C by means of a rotating-crucible viscometer and a tungsten
tube resistance furnace. All viscometer and furnace elements exposed
to high temperatures were made of molybdenum or tungsten. The
silica investigated was Brazil quartz of optical quality with an overall
impurity content of approximately 50 p.p.m. Several series of measurements gave consistent results. The energy of activation for viscous
flow E was deduced from the slope of the linear function log viscosity
versus $1/T$ (°K) as $E = 123 \cdot 1 \pm 1 \cdot 8$ kcal/mole. The pre-exponential
constant A was determined with $-\log_{10}A = 6 \cdot 24 \pm 0 \cdot 17$. Entropy of
activation $\Delta S^*_0$ was calculated as $11 \cdot 0$ cal/°K mole.

### La viscosité de la silice pure

La viscosité de la silice pure est mesurée entre 1600°C et 2480°C
à l'aide d'un viscosimètre à creuset rotatif et d'un four à résistances
de tungstène. Tous les éléments du viscosimètre et du four, destinés
à être exposés à des températures élevées, sont soit en molybdène,
soit en tungstène. La silice étudiée est du quartz du Brésil de qualité
optique avec une teneur totale en impuretés de l'orde de 50 ppm.
Plusieurs séries de mesures ont fourni des résultats concordants.
L'énergie d'activation correspondant à l'écoulement visqueux E est
déduite de la pente de la fonction linéaire log viscosité en fonction de
$1/T$ (°K); elle est de $E = 123,1 \pm 1,8$ kcal/mol. La constante pré-
exponentielle A a été déterminée avec $-\log_{10}A = 6,24 \pm 0,17$. L'en-
tropie d'activation $\Delta S^*_0$ obtenue par calcul est de 11,0 cal/°K mole.

### Die Viskosität reiner Kieselsäure

Die Viskosität reiner Kieselsäure wurde zwischen 1600 und 2480°C
mit einem Rotationstiegelviskosimeter in einem Wolframrohr-
widerstandsofen gemessen. Alle Viskosimeter- und Ofenteile, die
hohen Temperaturen ausgesetzt wurden, waren aus Molybdän oder
Wolfram gefertigt. Die untersuchte Kieselsäure war brasilianischer
Quarz von optischer Qualität mit einem Gesamtverunreinigungsgehalt

*von ungefähr 50 ppm. Mehrere Meßserien ergaben übereinstimmende Resultate. Die Aktivierungsenergie des viskosen Fließens wurde aus dem Anstieg der linearen Funktion Logarithmus der Viskosität von $1/T$ (°K) zu $E=123,1\pm1,8$ kcal/mol abgeleitet. Die Konstante A vor der e-Funktion wurde mit $-\log_{10}A=6,24\pm0,17$ bestimmt. Die Aktivierungsentropie $\Delta S^*_0$ wurde zu 11,0 cal/°K mol berechnet.*

## 1. INTRODUCTION

Exact knowledge of the viscosity/temperature relationship of pure silica is of interest because it might afford a better understanding of the glass structure. In theories concerning the vitreous state, pure silica with its relatively simple structure of tetrahedra connected at the edges plays an important part and it was the point of departure for the lattice theory. As the best way to discover changes in flow mechanism and structure is to make measurements over as wide a range as possible, a rotating-crucible viscometer measuring from $10^{-1}$ to $10^8$ poise was constructed, so that the viscous behaviour of a material could be examined from the very fluid to the nearly "solid" state.

## 2. APPARATUS

The apparatus was essentially the same as had been used in previous studies,[1] and so only its essential features will be described, together with some improvements, which enabled the range of viscosity and of temperature to be considerably extended.

For high-temperature measurements a Margules-type viscometer was chosen because of technological difficulties caused by the efflux method and the counter-balanced sphere procedure and of the restricted range of measurement of the oscillating hollow cylinder method.

Following equation of motion is valid:

$$C=4\pi\eta\, l\,\omega\, r^2{}_1 \cdot r^2{}_2/(r^2{}_1-r^2{}_2) \quad \cdots \quad (1)$$

in which $C$ is the momentum exerted by the liquid investigated (viscosity $\eta$) on the cylinder $A_5$ (radius $r_2$), when the crucible ($B_1$) (Figure 1) (radius $r_1$) is rotating. The depth of immersion of the cylinder is $l$ and $\omega$ is the angular velocity of the crucible. By means of an electro-magnetically controlled null-point method, an equal and opposite momentum $C^*$ is exerted on the cylinder ($A_5$) by the magnet coils ($A_3$). $C^*$ is proportional to a constant $k$ and to the square of the current $i$, according to the following equation:

$$\eta=k\, i^2\,(r^2{}_1-r^2{}_2)/(4\,\pi\, l\,\omega\, r^2{}_1 r^2{}_2) \quad \cdots \quad (2)$$

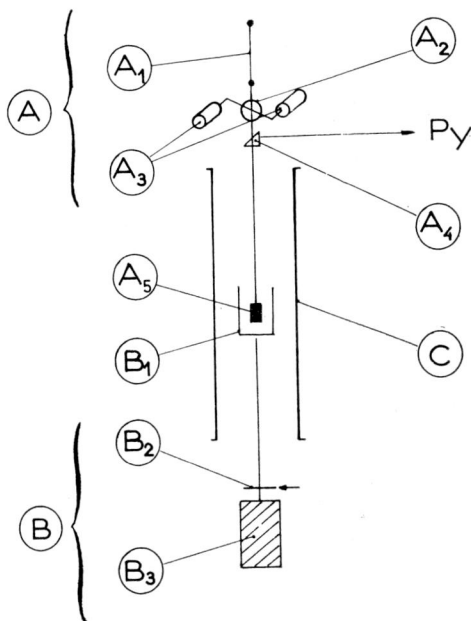

FIGURE 1.—Diagram of the viscometer

| A. SENSITIVE UNIT | B. ROTATING ELEMENTS |
|---|---|
| $A_1$, suspension wire | $B_1$, crucible |
| $A_2$, mirror | $B_2$, planetary reduction gear |
| $A_3$, magnet coils | $B_3$, synchronous motor |
| $A_4$, reflecting prism | |
| $A_5$, cylinder | |

C. TUNGSTEN TUBE RESISTANCE FURNACE

By eliminating all the parameters,

$$\eta = K\, i^2/\omega \quad . \quad . \quad . \quad . \quad . \quad . \quad . \quad . \quad . \quad (3)$$

The viscosity $\eta$ of the liquid being examined is now a function of $i^2$ and of the speed of rotation of the crucible, both easily measurable. A synchronous motor ($B_3$) connected to a planetary reduction gear ($B_2$) gives ten speeds between 10 r.p.m. and 0·01 r.p.m. The constant $K$, combining the electrical constant k with the geometry of the apparatus, was determined by means of oils the viscosity of which had been accurately determined.

For these experiments the heating-element of the furnace was a tungsten tube resistance. Because the viscosity of pure silica is very

3

sensitive to changes in temperature, exact measurement of temperature is very important. The temperature inside the hollow cylinder
($A_5$) was measured by optical pyrometer, the approximation to
ideal black body conditions being sufficiently close. The real temperature $T$ (°K) is obtained from the equation:

$$1/S - 1/T = A \quad \cdots \quad \cdots \quad (4)$$

$S$ (°K) is the measured temperature. By determining two secondary
reference points, the melting point of Pt (1769°C) and the solidification point of $Al_2O_3$ (2051°C), $A$ was determined as $10 \cdot 0 \pm 1$.

## 3. ERRORS

The accuracy of the values of absolute viscosity are about $\pm 3,5\%$,
owing to the dispersion observed by the determination of $K$, the
possible variation in the depth of immersion, and the eccentricity
of the cylinder and the crucible at high temperatures. Between 1600°
and 2500°C the accuracy of measurement of temperature is $\pm 10°$.
The following errors have to be allowed for: $\pm 4°$ for the 1948
temperature scale in connection with the solidification point of
$Al_2O_3 = 2{,}051 \pm 4°C$ (not yet official);[2] $\pm 2°$ resulting from the calibration and graduation of the pyrometer and $\pm 4°$ for the values of
$A$. To check the calibration of the apparatus, standard glass No.
710 was measured.[3] Results lie within the limits of error given by the
N.B.S.

## 4. EXPERIMENTAL

Selected pieces of Brazil quartz of optical quality were used for
experiments 1, 2 and 3, and a vacuum-fused silica block was used
for experiment 4. Table 1 gives the analyses of each sample.

All the samples were heated in an atmosphere of purified argon
to approximately 2200°C, at which temperature the cylinder was
immersed in the now liquid silica. After a stabilization period lasting

**Table 1**

| Sample No. | Impurity (p.p.m.) | | | | | | |
|---|---|---|---|---|---|---|---|
| | OH | Na | K | Ca | Al | Ti | Mo |
| 1 | 22–23 | 0 | <5 | 70 | 14 | 2·8 | 0·2 |
| 2 | 3–4 | 2 | <5 | 11 | 32 | 3·4 | 1 |
| 3 | 4 | 0 | <5 | 0 | 19 | 5·1 | 1 |
| 4 | | | | | 70 | | |

up to several hours, the viscosity was measured, and at the same time 10 readings were taken with the pyrometer and averaged. This procedure was repeated at progressively lower temperatures until the highest measurable viscosity was reached, and then at progressively higher temperatures up to the highest temperature obtainable. When plotted on a graph, the values obtained for decreasing and for increasing temperatures lie on the same straight line, which proves that they correspond to a state of equilibrium. Table 1 shows that practically no reaction occurred between the silica samples and the molybdenum crucibles. Evaporation was negligible (0·2% to 0·4%, depending on the duration and temperature of the experiment). As the vapour condensed in the cooler regions of the furnace, it did not interfere with the measuring of temperature. After the experiments, all the samples were entirely transparent and free of bubbles, an important improvement avoiding a frequent source of error.

A complete analysis of all the materials used is available. Two of them came from different sources; nevertheless, consistent results were obtained. By using the secondary reference point ($Al_2O_3$), absolute values of temperature above 2000°C could be determined with more reliability and the range of measurement could be extended to 2480°C and so a very large range could be tested.

## 5. RESULTS

Throughout the whole range of measurement, from 1600° to 2480°C, log $\eta$ versus $1/T$ (°K) was linear, and therefore the results could be described by the Arrhenius equation:

$$\eta = A \exp(E/RT)$$

where $\eta$ = Viscosity $(P)$
$\quad A$ = A pre-exponential constant
$\quad E$ = The energy of activation of viscous flow.

The following results within 95% confidence limits were obtained:

$$E = 123·1 \pm 1·8 \text{ kcal/mole}, \qquad -\log_{10}A = 6·24 \pm 0·17$$

They are in good agreement with results for high temperatures published previously (Table 2). Since the limits of the errors could be considerbly narrowed, the pre-exponential constant $A$ was obtained more exactly for the first time and the entropy of activation $\Delta S^*_0$ could be determined better. Assuming the $SiO_2$ unit as the unit of flow with a corresponding molar volume $V = 27$ cm$^3$/mole,[15] $\Delta S^*_0 = 11·0$ cal/°K mole follows from the equation;

$$\eta = \frac{Nh}{V} \exp(-\Delta S_0^*/R) \exp(\Delta H_0/RT). \quad . \quad . \quad . \quad (6)$$

**Table 2**

| Author | E(kcal/mole) | $-log_{10}A$ | Temperature range (°C) |
|---|---|---|---|
| INUZUKA[4] | $185 \pm 10$ | — | 1300–1430 |
| VOLAROVITCH[5] | $170 \pm 80$ | — | 1300–1450 |
| YOVANOVITCH[6] | — | — | 1000–1200 |
| HETHERINGTON[7] | 170* | — | 1000–1400 |
| FONTANA[8] | $127 \cdot 1 \pm 2 \cdot 3$ | — | 1236–1335 |
| SOLOMIN[9] | $151 \pm 10$ | — | 1710–2000 |
| BOCKRIS[10] | $134 \pm 9$ | $8 \cdot 2 \pm 0 \cdot 9$ | 1924–2060 |
| BACON[11] | $89 \cdot 2 \pm 23$ | — | 1935–2322 |
| ROSSIN[12] | $119 \cdot 7 \pm 9$ | $5 \cdot 8$ | 2100–2250 |
| ROSSIN[13] | $119 \cdot 4 \pm 6$ | $6 \cdot 56 \pm 0 \cdot 7$ | 2100–2250 |
| BRÜCKNER[14] | $122 \cdot 0 \pm 1 \cdot 7$ | — | 1686–2006 |
| HOFMAIER | $123 \cdot 1 \pm 1 \cdot 8$ | $6 \cdot 24 \pm 0 \cdot 17$ | 1600–2480 |

*I. R. Vitreosil

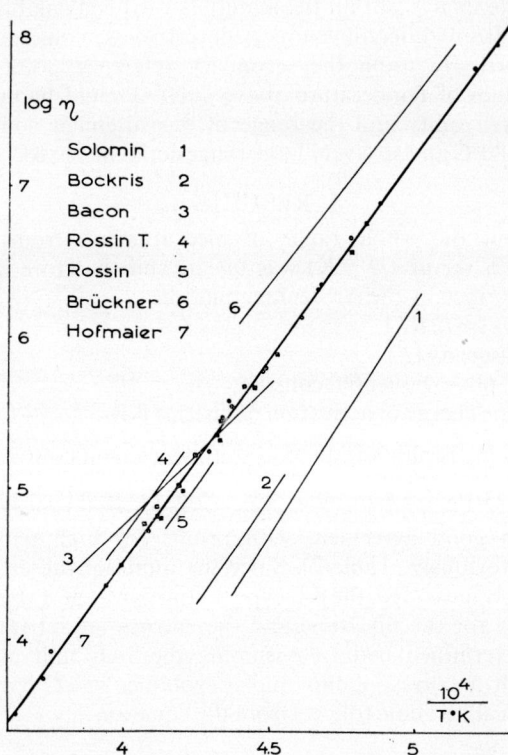

FIGURE 2.—Log $\eta$ versus $1/T$ over a range of 880°C.

where $N$ is Avogadro's number, $h$ is Planck's constant and $\Delta H^*_0$ is the heat of activation for the flow process, i.e. with the same assumptions BOCKRIS[10] determined $\Delta S^*_0$ as 20·0 cal/°K mole.

The strict linearity of log $\eta$ *versus* $1/T$ over a range of 880°C (Figure 2) is in contradiction to the deductions of BOCKRIS[16] and MACKENZIE,[17] who proposed a continuous increase in the energy of activation $E$ at decreasing temperatures.

In the low-temperature range most authors[4,5,7] (Figure 3) have measured the activation energies $E$ as 170–185 kcal/mole, but this is no proof of the temperature dependence of $E$ because of the inevitable crystallization[18] of silica when annealed to the experimental temperatures. The mixture of glass and crystallites thus formed definitely has a higher viscosity and energy of activation. This energy can be identified with energy necessary for 1 mole of silicate ions to form a mole, surmount the energy barrier from one position and fall into the adjacent hole. As flow properties of liquids depend strongly on the size of structural elements, the general

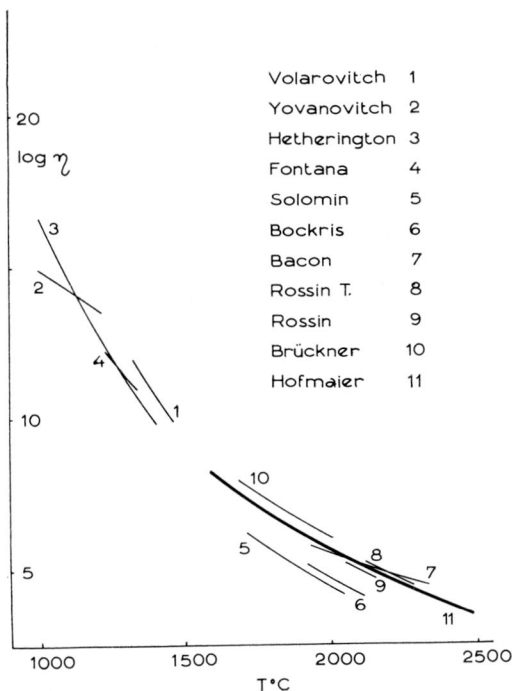

FIGURE 3.—Activation energies in the low-temperature range.

mobility decreases when crystallization occurs and forms very large rigid arrangements inside the liquid. Nevertheless recent data of FONTANA[8] at low temperatures show that, if devitrification is minimized, an energy of activation close to the high-temperature values can be obtained.

## ACKNOWLEDGMENT

The authors thank the Directors of IRSID for enabling this work to be carried out and Professor P. Kozakevitch for much helpful advice. They also thank M. Bruneau for technical help and I. Yovanovitch of "Quartz et Silice" for all the analyses and the preparation of sample No. 4.

## REFERENCES

1. ROSSIN, R., BERSAN, J. and URBAIN, G., *Rev. Hautes Temp. Réfract.*, **1**, 159, 1964.
2. SCHNEIDER, S. J. and McDANIEL, G. L., *Rev. Hautes Temp. Réfract.*, **3**, 351, 1966.
3. Department of Commerce, National Bureau of Standards Washington 25, D.C.: Standard Sample No. 710, Soda–Lime–Silica Glass.
4. INUZUKA, H., *J. Japan Ceram. Ass.*, **47**, 292, 1936.
5. VOLAROVITCH, M. P., and LEONTIEVA, A. A., *J. Soc. Glass Techn.*, **20**, 139, 1936.
6. YOVANOVITCH, J., *Bull. Soc. Franç. Céram.*, **55**, 25, 1962.
7. HETHERINGTON, G., JACK, K. H., and KENNEDY, J. C., *Phys. Chem. Glass.*, **5**, (5), 130, 1964.
8. FONTANA, E. H., and PLUMMER, W. A., *Phys. Chem. Glass*, **7**, (4), 139, 1966.
9. SOLOMIN, N. V., *Zh. Fiz. Khim.*, **14**, 235, 1940.
10. BOCKRIS, J. O'M., MACKENZIE, J. D., and KITCHENER, J. A., *Trans. Farad, Soc.*, **51**, 1734, 1955.
11. BACON, J. F., HASAPIS, A. A., and WHOLLEY Jr., J. W., *Phys. Chem. Glass*, **1**, (3), 90, 1960.
12. ROSSIN, R., Thesis, Rennes, 1963.
13. ROSSIN, R., BERSAN, J., and URBAIN, G., *C.R. Acad. Sci. Paris*, **258**, (7), 562, 1964.
14. BRÜCKNER, R., *Glastech. Ber.*, **37**, (9), 413, 1964.
15. EWELL, E., *J. Chem. Phys.*, **5**, 726, 1937.
16. BOCKRIS, J. O'M., and LOWE, D. C., *Proc. Roy. Soc.*, A, **226**, 423, 1954.
17. MACKENZIE, J. D., General Electric Research Lab. Rep. 61–RL2698–M, 1961.
18. WAGSTAFF, F. E., and RICHARDS, K. J., *J. Amer. Ceram. Soc.*, **49**, (3), 118, 1966.

# 3.—The Hot Pressing of Magnesium Hydroxide and Magnesium Carbonate

By T. A. Wheat and T. G. Carruthers

*Department of Ceramics, Houldsworth School of Applied Science,*
*The University of Leeds*

**ABSTRACT**                                                      C327/E141

*The behaviour during hot pressing of magnesium hydroxide and of magnesium carbonate have been studied at temperatures up to 850°C and pressures up to 20 t.s.i. A pressing die has been designed which enables differential thermal analysis, continuous monitoring of evolved gases and dilatometry to be carried out simultaneously throughout the hot pressing. When magnesium hydroxide is pressed at high pressures, sufficient water vapour is entrapped within the compact to raise the temperature at which it is completely decomposed by about 200°C. Much less entrapping of $CO_2$ occurs when pressing the carbonate and the decomposition temperature is raised only slightly. The magnesium oxide produced by pressing magnesium hydroxide at 7·5 t.s.i. to about 600°C was found to be strongly preferentially oriented in the die. [111] was parallel to the direction of pressing if the pressure was applied before dehydroxylation, and weakly oriented if the pressure was applied after dehydroxylation. Under no conditions of pressing was the formation of oriented grains detected when $MgCO_3$ was pressed. The possible mechanism of this behaviour is discussed.*

## Le pressage à chaud de l'hydroxyde et du carbonate de magnésium

*Le comportement de l'hydroxyde et du carbonate de magnésium pendant le pressage à chaud a été étudié à des températurés allant jusqu'à 850°C et à des pressions atteignant 3150 kg/cm². Un dispositif de pressage a été mis au point qui permet de procéder simultanément pendant toute la durée du pressage à chaud, à l'analyse thermique différentielle, au contrôle continu des gaz dégagés et aux mesures dilatométriques. Lorsque l'hydroxyde de magnésium est traité à des pressions élevées, la quantité de vapeur d'eau emprisonnée à l'intérieur du matériau comprimé est suffisamment grande pour que la température de décomposition complète s'élève d'environ 200°C. La quantité de $CO_2$ emprisonné est bien plus faible dans le cas du pressage du carbonate et la température de décomposition n'augmente que légèrement. Il a été constaté que l'oxyde de magnésium produit en pressant* *l'hydroxyde de magnésium à 1180 kg/cm² à 600°C, présente une orientation préférentielle prononcée dans le moule. [111] était parallèle au sens du pressage, si la pression est appliquée avant la déshydroxylation, alors que l'orientation n'est que faible si la pression est appliquée après la déshydroxylation. Dans le cas du pressage de MgCO₃ on n'a jamais pu déceler dans aucun cas la formation de grains orientés. Le mécanisme possible de ce comportement est discuté.*

### Heisspressen von Magnesiumhydroxid und Magnesiumcarbonat

*Das Heißpreßverhalten von Magnesiumhydroxid und Magnesiumcarbonat wurde bei Temperaturen bis 850°C und Drücken bis 3000kg/cm² untersucht. Es wurde eine Preßform entworfen, die gleichzeitig die Durchführung von DTA, kontinuierliche Überwachung der abgegebenen Gase und dilatometrische Messungen während des Heißpressens erlaubt. Wenn Magnesiumhydroxid bei hohen Drücken gepreßt wird, ist genügend Wasserdampf im Pressling eingefangen, um die Temperatur, bei der es sich zersetzt, um etwa 200°C ansteigen zu lassen. Wesentlich weniger CO₂ wird beim Pressen der Carbonate eingeschlossen, die Zersetzungstemperatur wird nur wenig angehoben. Das Magnesiumoxid, hergestellt durch Pressen von Magnesiumhydroxid bei etwa 1180 kg/cm² bis 600°C, erwies sich als stark bevorzugt orientiert in der Preßform, mit [111] Richtung parallel zur Druckrichtung, wenn vor der Dehydroxilierung Druck angelegt wurde, und nur wengi orientiert, wenn der Druck nach der Dehydroxilierung angelegt wurde. Unter keinen Pressbedingungen wurde die Bildung orientierter Körner gefunden, wenn MgCO₃ gepresst wurde. Der mögliche Mechanismus dieses Verhaltens wird diskutiert.*

## 1. INTRODUCTION

Previous work on the behaviour, during hot-pressing, of kaolinite and alumina–silica gels [1] suggested that the gaseous water produced during their dehydroxylation or dehydration under pressure played a significant role in accelerating densification and in promoting crystallization. The present work is an attempt to compare the hot-pressing behaviour of two compounds, one of them releasing gaseous water and the other releasing carbon dioxide during decomposition to the oxide. Brucite and magnesite were chosen for comparison because they give the same oxide, they both decompose within the working temperature range of the die materials available and they are obtainable in reasonable purity. Furthermore the behaviour of both substances during calcination and sintering have previously been investigated widely at atmospheric pressure.

## 2. MATERIALS

### 2.1 Magnesium Hydroxide

The magnesium hydroxide (brucite) was stated by the suppliers (May & Baker Ltd) to contain not less than 97% $Mg(OH)_2$. The chief impurities were 1·26% CaO and 1·29% $SiO_2$. The loss on ignition to 1000°C was 31·40% (theoretical 30·4% for 97% $Mg(OH)_2$). The material gave a normal differential thermogram with a main endothermic peak at 444°C. Examination by optical microscopy showed the material to consist of aggregates 5–10 $\mu$m diam. The ultimate crystallite size was estimated from electron micrographs to be 0·5 $\mu$m or less.

### 2.2 Magnesium Carbonate

Pure synthetic magnesium carbonate could not be obtained; it is extremely difficult to prepare free from the basic carbonate. The material used was therefore a natural Grecian magnesite, selected for its purity and supplied in lump form by The Steetley Co. Ltd. After crushing and grinding, the chief impurities were 1·2% CaO, 0·2% $SiO_2$ and 0·1% $Fe_2O_3$. The loss on ignition to 1000°C was 52·04 % (theoretical 51·20% for 98% $MgCO_3$). The differential thermogram was normal for magnesite, with a main endothermic peak at 635°C. The aggregate size was 10–20 $\mu$m and the crystallite size was about 1 $\mu$m.

## 3. APPARATUS AND TECHNIQUE

The press and the general procedures for hot-pressing were closely similar to those described previously by the present authors.[1] However, an improved method of measuring the shrinkage of the compact during pressing has been used, the design of the die has been modified to permit sampling of the gases evolved, and means of simultaneously recording all the measured variables has been added.

### 3.1 Measurement of Shrinkage of Samples

Shrinkage of the compact during pressing was obtained by measuring the relative movement, by means of a displacement transducer (Philips Model No. 9301/01), between the upper water-cooled bolster and the end of a nickel–chrome alloy rod (Brightray B, Henry Wiggin & Co. Ltd) which rested on the lower bolster (Figure 1). This rod was lagged with refractory cement so that its heating rate and thermal expansion closely matched that of the heated parts of the die assembly. Trial runs with an empty die enabled a small

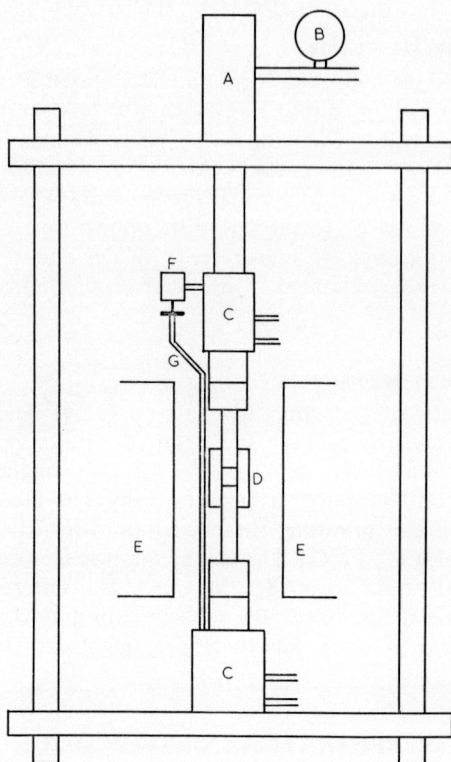

FIGURE 1.—Press and die assembly (diagrammatic):

A. Hydraulic jack                D. Nimonic die and plungers
B. Pressure gauge                E. Split electric furnace
C. Water-cooled bolsters         F. Displacement transducer
              G. Expansion reference rod

correction factor to be determined, which allowed for the residual thermal expansion mismatch.

## 3.2  Design of the Die and the Detection of Gases Evolved

The die and plungers were made of Nimonic 105 alloy (Henry Wiggin & Co. Ltd). As shown in Figure 2, two annular grooves, (A) connected to one another and to the outside of the die by drilled passages, (B) were machined above and below the central part of the die. The top and bottom caps fitted closely round the plungers so that pure argon (Purargon, Air Products Ltd) at 3 p.s.i. could be passed through the two chambers formed by the grooves. Gas connections to the die were made by forcing stainless steel hypo-dermic tubing (Accles & Pollock Ltd) into the drilled holes. The clearance between the faces of the plunger and the central portion of the die was made large (0·003 in.) to prevent seizure and to allow evolved gases to escape readily into the chambers.

A flow diagram of the gas detector system is shown in Figure 3. Most of the argon flow is vented to waste at the T-piece but a small quantity is drawn into the vacuum system of a commercial argon-purity meter (Colstone Electrical Co. Ltd, Bristol). The principle of this instrument has been described by LOVELOCK[2] and its practical use by RILEY.[3] It has the advantage, over detectors normally used in gas chromatography, of being sensitive to both water vapour and $CO_2$ over a concentration range of 10–1500 p.p.m. The connecting tubes from the die were heated to prevent condensation of water vapour: their volume was sufficiently small that a time lag of only 15 sec. occurred between the evolution of water vapour in the die and its detection.

### 3.3 Recording System

A multi-channel ultra-violet recorder (Southern Electrics Ltd–S.E. 2005/12) whose mirror galvanometers were chosen to match the electrical outputs of the thermocouples, the argon-purity meter and the displacement transducer, was used to record the measured variables.

### 3.4 Pressing-procedure

In all cases 0·3 g of material were used, resulting in compacts 0·020–0·050 in. thick. Two methods of pressing were used. In the first, pressure was applied at room temperature and maintained dur-ing heating (at a rate of 30°C per min.) until the chosen maximum desired temperature was reached. In the second method, the sample was lightly compacted at about 30 p.s.i. and heated at atmospheric pressure until the desired temperature was reached. After soaking for 5 min. at the working temperature, the full pressure was applied for a further 5 min. After pressing by either method the compact was extracted whilst hot into the upper die chamber, from which it was removed by dismantling the die when cold. Some samples pressed by the first method were "quenched", whilst still under pressure, by rapid cooling to 400°C before being extracted from the die.

### 3.5 Assessment of Orientation

When anisotropic minerals were present, preferential orientation could be detected by microscopical examination of thin sections

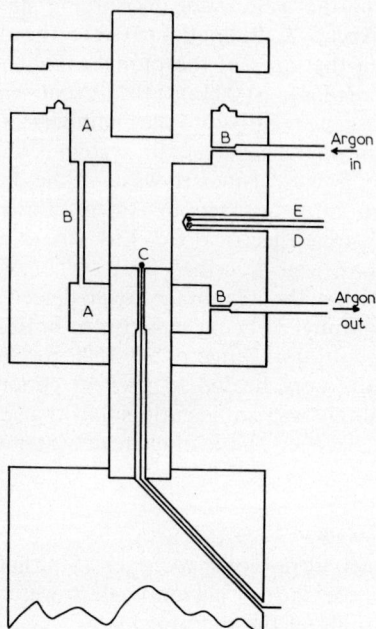

FIGURE 2.—Detail of die assembly:

A. Annular grooves    C. and D. Differential thermocouple
B. Argon passages     E. Die thermocouple

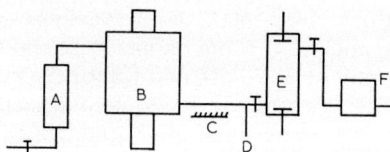

FIGURE 3.—Schematic diagram of gas-sampling and detector system:

A. Flowmeter      D. Vent
B. Die assembly   E. Gas detector
C. Heater         F. Vacuum pump

between crossed polars. Orientation in isotropic MgO was detected by mounting whole pellets, whose surfaces had been cleaned on silicon carbide paper, directly in an X-ray diffractometer. The ratio of intensities of the reflections from relevant planes relative to those found in a randomly oriented powder preparation was taken as an indication of preferential orientation.

## 3.6 Assessment of Crystallite Size

Increases in crystallite sizes were detected by X-ray diffraction: a size increase causes a decrease in peak widths in diffractometer scans and increases the resolution of the Cukα doublet in back-reflection photographs. Examination of fracture-surface replicas by electron microscopy was also used.

# 4. RESULTS

## 4.1 Variation of Bulk Density with Temperature and Pressure

### 4.11 Cold Pressing and Firing at Atmospheric Pressure

For comparison with the results obtained by hot pressing, the bulk densities of compacts of brucite and magnesite after cold pressing, and after subsequent firing to 750°C and 850°C respectively, are shown as a function of pressure in Figure 4. The much larger decrease in bulk density of the magnesite after firing is largely accounted for by its greater loss in weight.

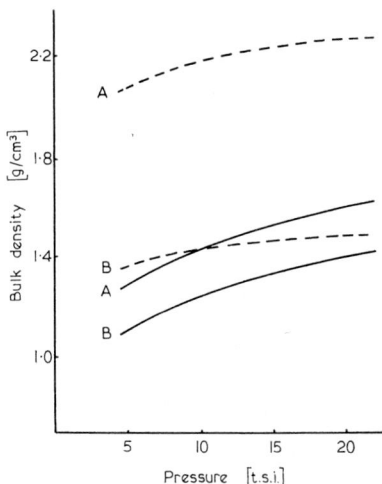

FIGURE 4.—Green and fired densities of brucite and magnesite as a function of pressing pressure:

Solid line. Brucite:
A. Unfired
B. Fired at 750°C for 24 h.
Dashed line. Magnesite:
A. Unfired
B. Fired at 850°C for 24 h.

## 4.12  Hot Pressing: Pressure applied at Room Temperature

The effect of pressure on the bulk density of compacts made from brucite and magnesite is shown in Figure 5. The behaviour of brucite (curves A, B and C) is characterized by the rapid increase in density with pressure below 5 t.s.i. for all three working temperatures used. Increasing the pressure to 5 t.s.i. produced little further increase in density at either 750°C or 800°C. In the special case of the compacts heated to 670°C and then "quenched" under pressure to 400°C before extraction, an increase of pressure resulted in a decrease in density.

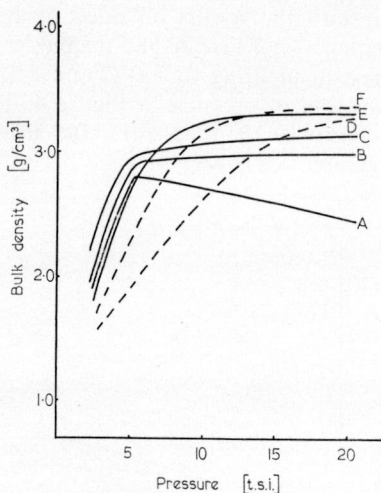

FIGURE 5.—Variation of hot-pressed density with applied pressure:

A.  Brucite hot-pressed to 670°C followed by cooling to 400°C under pressure prior to extraction from the die.
B.  Brucite hot-pressed to 750°C.
C.  Brucite hot-pressed to 800°C.
D.  Magnesite hot-pressed to 850°C.
E.  Magnesium oxide hot-pressed to 750°C (obtained from brucite calcined at 645°C).
F.  Magnesium oxide hot-pressed to 750°C (obtained from magnesite calcined at 850°C).

Magnesite under comparable conditions (curve D) showed a steady increase in bulk density with pressure. Although the maximum temperature used was 850°C for the magnesite, the bulk density it attained was lower than that of brucite at all pressures below 10 t.s.i.

Curves E and F are for brucite and magnesite which had been previously calcined in air for 1 h at 645°C and 850°C respectively to convert them to MgO before being pressed to 750°C. The bulk density of the calcined brucite is the higher up to 10 t.s.i., though at 20 t.s.i. the calcined magnesite attains a marginally higher value.

### 4.13 Hot Pressing: Pressure applied at Maximum Temperature

Figure 6 shows the variation of bulk density with pressure for compacts made by heating brucite and magnesite in the die for 5 min. at the stated temperature before applying the pressure for a

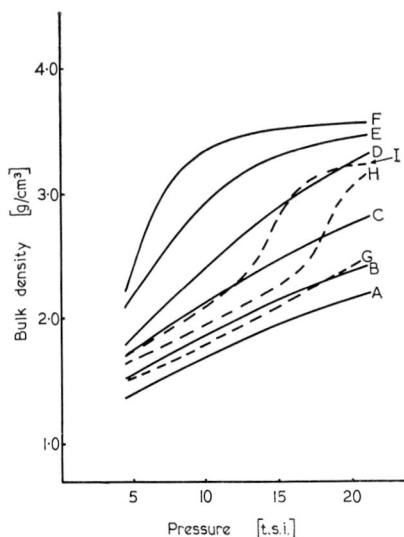

FIGURE 6.—Variation of hot-pressed density with applied pressure. Brucite precalcined in the die for 5 min. and subsequently pressed for 5 min. at:

A. 450°C
B. 500°C
C. 430°C
D. 600°C
E. 700°C
F. 800°C

Magnesite similarly treated:

G. 740°C
H. 760°C
I.  800°C

further 5 min. The lowest temperature used for each material was that found, by X-ray diffraction, to give complete conversion to the oxide in the die within 5 min. The densities reached by brucite at 450°C and 500°C were lower than those reached at 430°C. At 800°C, a density close to the theoretical limit (determined by X-rays) was reached at 20 t.s.i. Magnesite showed a progressive increase in density with temperature but theoretical density was not reached at 800°C and a pressure of 20 t.s.i.

## 4.2 Behaviour during Pressing

### 4.21 Brucite

Simultaneous records of d.t.a., compact shrinkage and water-vapour evolution obtained when brucite was heated to 750°C under different pressures are shown in Figure 7. At pressures below 5 t.s.i. the dehydroxylation, as indicated by the d.t.a. peak at about 500°C, coincides closely with rapid shrinkage of the compact. Further shrinkage occurs as the temperature increases, but at a much slower rate. At very low pressures, the gaseous water escapes from the compact at the dehydroxylation temperature but at increasingly

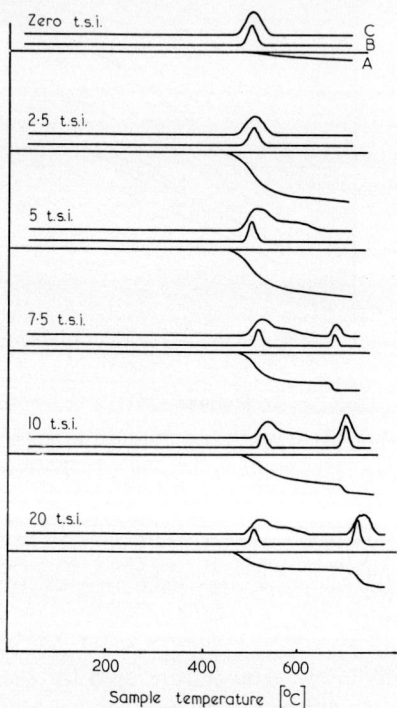

FIGURE 7.—Hot-pressing characteristics of brucite:
A. Compact shrinkage trace
B. D.t.a. trace
C. Evolved water trace

higher pressures its escape becomes increasingly delayed until at 5 t.s.i. it is not all removed until 600°C is reached. The slow delayed evolution of water is not associated with additional thermal activity or enhanced shrinkage and is thus the escape of physically trapped gaseous water.

At pressures above 5 t.s.i a second endothermic peak in the d.t.a. curve appears and is associated with a second peak in the evolved-water curve and with a rapid increase in the shrinkage rate. The temperature of the second d.t.a. peak increases with pressure as does the delay in the evolution of water from the compact.

## 4.22 Magnesite

The behaviour of magnesite during hot pressing is shown in Figure 8. At very low pressures, the thermal decomposition is associated

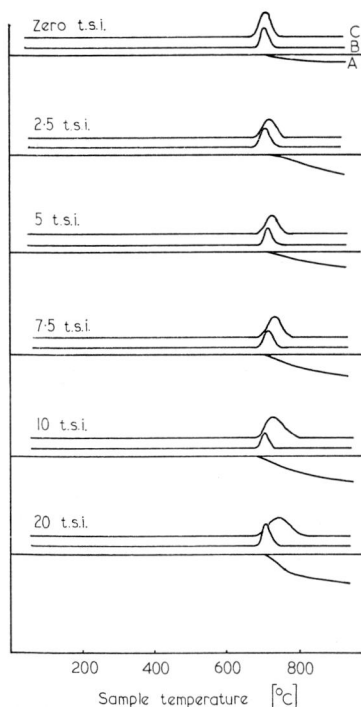

FIGURE 8.—Hot-pressing characteristics of magnesite:
    A. Compact shrinkage trace
    B. D.t.a. trace
    C. Evolved carbon dioxide trace

4

with marked shrinkage of the compact and the simultaneous evolution of $CO_2$. As the pressure is increased, the maximum rates of
compact shrinkage and $CO_2$ evolution occurred at slightly higher
temperatures whilst the temperature of decarboxylation remains
unchanged. No additional peaks in the d.t.a., $CO_2$-evolution or
shrinkage curves were observed in the pressure/temperature range
covered.

### 4.3 Appearance and Constitution of Hot-pressed Compacts

#### 4.31 Brucite

The general appearance of the compacts was found to depend on
the conditions of pressing. Those heated from room temperature to
750°C while continuously under pressures below 5 t.s.i. were white
and opaque and were shown by X-ray diffraction and optical microscopy to be composed entirely of MgO. X-ray diffractometer scans
of the compacts showed the MgO to be randomly orientated. Compacts pressed at 5 t.s.i. consisted of a slightly translucent rim about
6 mm wide surrounding an opaque core. Optical microscopy of such
samples revealed the presence of undecomposed brucite crystals
whose size was about 10 times that of the raw material. The amount
present was too small to be detected by X-rays. Diffractometer
scans showed a three-fold increase in the intensity of the (111)
reflections, indicating some orientation of the MgO crystallites so
that the [111] was parallel to the direction of pressing. At pressures
over 5 t.s.i the appearance was generally similar although the outer
rim, which still contained some undecomposed brucite, was increasingly translucent. The degree of preferential orientation of the
MgO increased to a maximum at 7·5 and 10 t.s.i. and then decreased
to nearly random at 20 t.s.i.

Compacts which were heated to 670°C whilst under pressures o
7·5 t.s.i. or more, and then "quenched", possessed a much larger
opaque central region; the width of the translucent rim was reduced
to about 100 $\mu$m at 15 t.s.i. Optical microscopy and X-ray diffraction
showed the central region to contain large amounts of brucite, the
majority of which was orientated so that its $c$ axis was parallel to the
pressing direction. The proportion which was so orientated decreased
with increasing pressure. Compacts of brucite which had been merely
cold-pressed, or cold-pressed and subsequently fired at atmospheric
pressure to 1600°C, showed no preferential orientation and it is
clear therefore that orientation is produced by some mechanism
during hot-pressing. The amount of brucite present was calculated
from the weight loss suffered on re-firing the compacts in air to 700°C
for 1 h, and the concentration is shown in Figure 9.

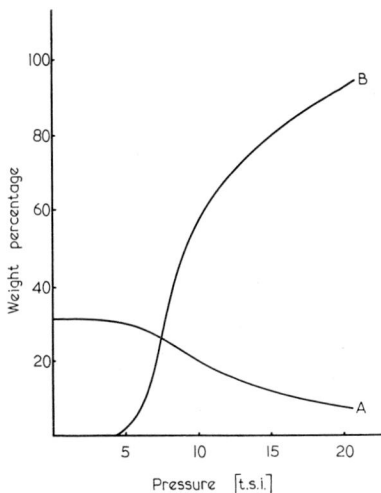

FIGURE 9

A. Loss in weight of brucite samples during hot-pressing to 670°C
followed by cooling under pressure to 400°C prior to removal
from die.

B. Total amount of brucite contained in samples prepared as in A.

The appearance of compacts made by pressing brucite pre-heated
in the die for 5 min. before pressing again depended on the conditions
of pressing and particularly on the temperature. Those pressed below 500°C were white and opaque whilst those pressed at higher
temperatures were translucent, the degree of translucency increasing
with temperature at a given pressure. All compacts were composed
entirely of MgO which was randomly orientated below 500°C,
with increasing preferential orientation at higher temperatures.

### 4.32 Magnesite

All compacts pressed from magnesite were white and opaque irrespective of the conditions of pressing. They were found to be composed entirely of MgO, which was randomly orientated, with no
trace of undecomposed magnesite.

### 4.4 Crystallite Size

Narrowing of the widths of X-ray diffractometer peaks and increasing resolution of the Cukα doublet showed that crystal growth
occurred in brucite pressed at temperatures above 600°C, irrespective of whether it had been pre-calcined in the die or not. The crystal
growth increased with both temperature and pressure. Electron

micrographs of fracture-surface replicas confirmed this. After pressing to 20 t.s.i. for 5 min. at 800°C, the crystallite size had reached about 0·5 μm.

No crystal growth could be detected in hot-pressed magnesite except at the highest temperatures and pressures used. Rapid crystal growth appeared to commence at about 800°C at a pressure of 15 t.s.i. and at 760°C at a pressure of 20 t.s.i.

## 5. DISCUSSION

### 5.1 Effect of Entrapment of Evolved Gases

The results of the present work indicate that, during the hot pressing of brucite and magnesite, considerable trapping of evolved gases can occur when the decomposition occurs under moderate pressure. In the brucite compacts some trapping occurs even at an applied pressure of only 5 t.s.i. This was shown by the presence of the discrete randomly orientated brucite crystals which were found in samples hot-pressed to 750°C. These crystals were approximately an order of magnitude larger than the crystallites in the starting material, indicating that isolated primary brucite crystals had been retained in random orientation within the bulk of the compact during pressing and that they had grown in the presence of MgO and trapped gaseous water. The metastable nature of this primary brucite was shown by its decomposition in hot-pressed samples refired to 700°C for 1 h at atmospheric pressure, a temperature lower than that to which the compact was originally hot-pressed. No evidence was found for the trapping of $CO_2$ in the magnesite compacts when pressed at pressures lower than 5 t.s.i.

At pressures of 7·5 t.s.i. and above, the trapping of both gaseous phases became more marked. In compacts of brucite the gaseous water pressure, developed by the decomposition of some of the brucite at about 500°C, was evidently sufficient to suppress the decomposition of the remainder until much higher temperatures. Calculation of the gas pressure developed is difficult, since it requires a knowledge of the porosity of the compact, of the fraction of the brucite decomposed, of the amount of gaseous water escaping from the die and of the departure of gaseous water from ideality. A trial calculation, involving the simplifying assumptions that all the brucite is decomposed and that all the gaseous water is trapped in a compact of 20% prosity, indicates that a pressure of about 200 t.s.i. could be generated, which is an order of magnitude greater than any externally applied pressure used in the present work. The maximum internal gaseous pressure which a compact can sustain might well be approximately equal to the externally applied pressure. Evidence that this is the case was provided by the explosive violence with which the final decomposition of retained brucite sometimes occurred. This caused the centre of the compact to be blown out and it was clear that the plungers had been forced back against the applied load. The agreement found between the start of the decomposition of the retained brucite at different applied pressures (as indicated by the d.t.a. curves of Figure 7) and the thermal equilibrium data of ROY and ROY[4] is additional supporting evidence.

The increase, at increasing pressures, of shrinkage of the compact, associated with the final decomposition and the decrease in that associated with the initial decomposition, as shown in Figure 7, may be taken as evidence that the amount of brucite retained is greater at higher pressures.

Some trapping of $CO_2$ in magnesite compacts is indicated in Figure 8 by its evolution, coincident with sample shrinkage, after the completion of decomposition to the oxide. At higher pressures the gas evolution, and concomitant shrinkage, occurred at progressively higher temperatures, whereas the decomposition temperature, as indicated by the d.t.a. curve, remained unchanged. Evidently insufficient $CO_2$ was trapped to cause the retention of undecomposed magnesite, since no second decomposition was observed nor was its presence detected by optical or X-ray examination. Trial calculation again shows that more than adequate gas pressures for the retention of magnesite to much higher temperatures are potentially possible. The $CO_2$ must therefore have escaped fairly readily from the compact. It is clear that gas must have been trapped within the compact and not merely in the die, because the die was the same for both brucite and magnesite and the clearance between the die and plungers (0·003 in.) was adequate for the easy passage of gases at high pressures. It is suggested that the difference in the degree of gas entrapment between the compacts of brucite and of magnesite was due to the much higher permeability of the latter. This can arise from two causes. Firstly, the size and the strength of the aggregates of the crushed natural magnesite rock would result in a more permeable compact. Secondly, at a given applied pressure the density of the oxide compact resulting from the decomposition of the magnesite would be lower than that from brucite, again resulting in greater permeability.

Comparison of the behaviour during pressing of the brucite and the magnesite (Figures 7 and 8) at pressures of less than 5 t.s.i. shows that the brucite underwent about twice as much shrinkage as the magnesite. This is confirmed by the final density data (Figure 5), which shows that the brucite attained a density of 2·8 g/cm³ at 750°C whilst the magnesite only reached 2·00 g/cm³ at 850°C. The higher densities achieved by the brucite may well be because the oxide produced had a higher surface area than that produced from the magnesite. Confirmation of this may be seen in Figure 6, which shows the final densities obtained after pressing for 5 min. brucite and magnesite which had been previously calcined in the die for 5 min. at the same temperature. In the case of brucite, higher final densities were reached at 430°C than at any other temperature below 600°C. It has been reported[5] that the surface area of MgO produced from $Mg(OH)_2$ falls rapidly with increasing calcination temperature in this range of temperatures and hence the ease of sintering and densification is reduced. At 600°C and above, the effect of increased temperature on the speed of densification under pressure presumably outweighs the retarding effect of decreasing surface area and hence higher densities are obtainable. The decrease in surface area with increasing calcination temperature of MgO produced from magnesite has been shown to be relatively slow.[6] The final densities obtained therefore show a continuous increase in density with temperature.

The failure of pressures in excess of 5 t.s.i. to cause further densi-fication of compacts of brucite pressed to 670°, 750° and 800°C is undoubtedly due to the trapping of gaseous water at pressures ap-proaching the applied pressure, leading to the retention of unde-composed brucite. The compacts pressed to 670°C were quenched and showed large amounts of brucite (Figure 9), much of which was orientated. It is believed that the orientated brucite was formed on cooling by back reaction between entrapped gaseous water and MgO. At the prevailing pressure much of the MgO is itself orient-ated with the [111] parallel to the direction of pressing and on re-hydration, since the reaction is topotactic,[7] this will lead to the formation of brucite with its $c$ axis parallel to the direction of press-ing. The decrease in the amount of orientated brucite at higher pressures indicates that increasing quantities of primary undecom-posed brucite must have been present. This is confirmed by the in-creasing magnitude of the compact shrinkage associated with the second d.t.a. peak (Figure 7). The decrease in density, with increasing pressure, of compacts pressed to 670°C is thus due to the increasing quantities of primary and secondary brucite, which is of lower density than MgO.

At 750°C very little undecomposed brucite remained and it might therefore be expected that a high density would be attained at this temperature. Figure 5 shows that this was not so, and that raising the final temperature to 800°C had little effect. A probable reason is that, at the pressure used, the dense MgO rim formed around the outside of the compact would not allow further compression of the pellet. The centre portion, containing the porous mass of MgO formed by the breakdown of the retained brucite, could not therefore be densified.

## 5.2 Development of Preferential Orientation

The development of preferentially orientated "secondary" brucite in quenched specimens has been satisfactorily explained on the basis of a topotactic conversion of the previously orientated MgO crystallites which were always found when brucite was hot-pressed at over 5 t.s.i.

Two possible mechanisms leading to the preferential orientation of MgO will be considered: plastic flow, and growth of crystal nuclei of favourable orientation. In the temperature range of the present work the (110) slip system is the only one likely to be operative. This would give rise to a (110) orientated texture. The development of such a texture has been reported in hot-pressed MgO.[8] At higher temperatures other slip systems become operative but the (111) system does not operate below about 1700°C.[9] It therefore seems unlikely that the (111) orientated texture found in the present work results from a plastic-flow mechanism.

The growth of crystal nuclei to produce the observed preferential orientation requires some discriminating factor favouring those lying with their [111] parallel to the direction of pressing. Such a factor is the strain energy of the growing crystal. The minimum strain energy stored in any crystal for a given stress level is developed when the stress axis coincides with the direction in which Young's modulus, $E$, is a maximum. In the case of MgO, $E$ has a minimum value along [100] and a maximum along [111]. Assuming a simple stress distribution, the strain energy developed in a cube of MgO under a stress of 7 t.s.i. (near the minimum needed for orientation to develop) is about 0·039 cal/mole when stressed along [111] and 0·055 cal/mole when stressed along [100]. These values are very small but are markedly different from one another and could possibly favour the growth of crystals already favourably orientated by chance to give a (111) texture of the type found in the present work. A similar explanation has been given for the development of orientated textures in hot-pressed BeO compacts.[10]

Magnesite pressed under any conditions never developed a preferentially orientated structure and unlike brucite did not show any detectable crystallite growth except under extreme conditions of temperature and pressure.

The formation of an oxide by decomposition of a salt may be divided into two stages. The first involves decomposition and the production of a metastable phase having the chemical composition of the oxide but structurally different from the stable oxide phase. In the second stage the structure changes to that of the stable oxide. These two processes need not occur simultaneously and HORLOCK, MORGAN and ANDERSON [11] have shown that in the case of MgO the time lag between them is almost eliminated by a water vapour pressure of 0·2 torr. Thus in the present work the MgO formed from brucite was found to have lattice parameters close to the correct values for periclase, whereas those for MgO formed from magnesite were slightly larger until crystallite growth occurred. CHEKURI and ROBERTS [12] have observed a similar effect when calcining large pieces of natural magnesite at atmospheric pressure, the metastable pseudo-lattice persisting in this case to be about 1700°C. At this temperature they observed rapid crystallization, coincident with the development of a true periclase lattice and with the appearance of a calcium silicate phase formed from small amounts of impurities which had been in solid solution in the pseudo-lattice. A similar phenomenon was seen in the present work; sudden crystallization occurred, with the appearance of a small amount of an unidentified second phase, at 760°C under a pressure of 20 t.s.i. and at 800°C under 15 t.s.i. This change in structure resulted in the upward inflections in the density/pressure curves of Figure 6. It is believed that the rapidity of the crystallization which accompanies this change overrides any tendency to develop a preferentially orientated structure.

# 6. CONCLUSIONS

(1) The behaviour of brucite during hot pressing is markedly different from that of magensite

(2) Gaseous water is readily entrapped within brucite compacts, at applied pressures greater than 5 t.s.i., causing the retention of brucite to temperatures much higher than its normal decomposition temperature. Little entrapping of $CO_2$ occurs in magnesite compacts under similar conditions so that magnesite is not retained to higher temperatures.

(3) Entrapment of gaseous water is avoided and consequently higher bulk densities are achieved if the brucite is first decomposed to the oxide in the die before pressure is applied.

(4) Detectable growth of the MgO crystallites does not occur in decomposed brucite below 600°C. Above this temperature, and at pressures greater than 5 t.s.i., the MgO crystallites are preferentially orientated with [111] parallel to the direction of the applied pressure. A mechanism is proposed which is based on the preferential growth

of nuclei so orientated that the growing crystallites have a minimum strain energy.

(5) Detectable crystallite growth in magnesite does not occur below 760°C and 800°C under applied pressures of 20 t.s.i. and 15 t.s.i. respectively. Above these temperatures and pressures crystallite growth is rapid but no preferential orientation is produced. This behaviour is believed to be due to the initial formation from the magnesite of a slightly expanded MgO lattice containing impurities in solid solution. Under the conditions mentioned above, the impurities are precipitated as a second phase with the formation of a true periclase lattice. Subsequent crystallite growth is so rapid that the mechanism, believed to be responsible for preferential orientated growth in brucite, is not effective.

## ACKNOWLEDGMENTS

The authors wish to express their indebtedness to the Science Research Council for a Post-Graduate Assistantship awarded to T.A. Wheat, during the tenure of which the work was done, and for a special grant for the purchase of the multi-channel recorder. They also wish to thank The Steetley Co. Ltd for donating the sample of Grecian magnesite.

## REFERENCES

1. CARRUTHERS, T. G. and WHEAT, T. A., *Proc. Brit. Ceram. Soc.*, (3), 259, 1965.
2. LOVELOCK, J. E., *J. Chromatography*, **1**, 35, 1958.
3. RILEY, B., "Gas Chromatography" (Ed. Scott) (Butterworths, 1960) p. 81.
4. ROY, D. M. and ROY, R., *Amer. J. Sci.*, **255**, 574, 1957.
5. LIVEY, D. T., WANKLYN, B. M., HEWITT, M. and MURRAY, P., *Trans. Brit. Ceram. Soc.*, **56**, 217, 1957.
6. GLASSON, D. R., *J. Appl. Chem.*, **13**, 111, 1963.
7. GARRIDO, J., *Amer. Min.*, **36**, 773, 1951.
8. VASILOS, T. and SPRIGGS, R. M., *Proc. Brit. Ceram. Soc.*, (3), 202, 1965.
9. DAY, R. B. and STOKES, R. J., *J. Amer. Ceram. Soc.*, **49**, 345, 1966.
10. KELLY, J. W., *J. Nuclear Mater.*, **8**, 227, 1963.
11. HORLOCK, R. F., MORGAN, P. L. and ANDERSON, P. J., *Trans. Farad. Soc.*, **59**, 721, 1963.
12. CHEKURI, N. S. P. V. and ROBERTS, A. L., *Trans. Brit. Ceram. Soc.*, **61**, 212, 1962.

# 4.—The Hot Pressing of Zirconium Diboride

J. CHOWN

Doulton Research Limited, Chertsey, Surrey

*ABSTRACT* C327/E18

*The final stage of densification of $ZrB_2$ during hot pressing follows accurately a logarithmic relationship between the fractional porosity and time at constant temperature, and fits the rate equation proposed by Murray* et al. *to describe densification according to a mechanism based on plastic or viscous flow. However, doubt is cast on the concept of such a mechanism as it implies that the high-temperature mechanical properties are inferior to these observed with this and related materials. The results have therefore been compared also with the work of Rossi and Fulrath on $Al_2O_3$, who derived an alternative, but similar, densification equation based on a mechanism of diffusion. It is shown that this leads to an activation energy of 140 kcal. $mole^{-1}$, corresponding to sintering of $ZrB_2$ via vacancy diffusion. It is believed that the initial rapid densification which began at c. 1400°C was caused by particle fragmentation and rearrangement, but this was later overtaken by the main final sintering mechanism.*

*Le pressage à chaud du borure de zirconium $ZrB_2$*

*Le stade final du compactage de $ZrB_2$ pendant le pressage à chaud suit à température constante une corrélation logarithmique entre la porosité et le temps. Il satisfait à l'equation proposée par Murray et al. pour décrire le compactage suivant un mécanisme basé sur l'écoulement plastique ou visqueux. Des doutes ont toutefois surgi quant à la validité du concept d'un tel mécanisme; en effet, d'après lui, les propriétés mécaniques aux hautes températures sont inférieures à celles observées sur ce matériau et d'autres composés semblables. Pour cette raison les résultats ont été comparés avec le travail de Rossi et de Fulrath sur $Al_2O_3$. Ces auteurs ont déduit une équation similaire pour le compactage, basée sur un mécanisme de diffusion. On montre que l'énergie d'activation est de 140 kcal.$mole^{-1}$, correspondant au frittage de $ZrB_2$ par diffusion des vacances. On pense que le compactage rapide initial qui a commencé à environ 1400°C était dû à la fragmentation des particules et leur réarrangement, mais que ce processus était ensuite dépassé par le mécanisme de frittage final.*

### Das Heisspressen von Zirkondiborid

*Das Endstadium der Verdichtung von $ZrB_2$ während des Heiß-pressens folgt gut einer logarithmischen Beziehung zwischen der Porosität und der Zeit bei konstanter Temperatur und erfüllt die von Murray und anderen aufgestellte Gleichung zur Beschreibung der Verdichtung auf Grund eines Mechanismus des plastischen oder viskosen Fließens. Jedoch treten Zweifel an einem solchen Mechanismus auf, da er bedeutet, daß die mechanischen Eigenschaften bei hohen Temperaturen geringer seien, als sie bei diesem und ähnlichen Stoffen beobachtet wurden. Deshalb wurden die Ergebnisse mit der Arbeit von Rossi und Fulrath an $Al_2O_3$ verglichen, der eine andere aber ähnliche Verdichtungsgleichung abgeleitet hat, die auf einem Diffusionsprozeß basiert. Es wird gezeigt, daß dies zu einer Aktivierungsenergie von 140 kcal/mol führt entsprechend dem Sintern von $ZrB_2$ auf dem Wege von Leerstellendiffusion. Man vermutet, daß die anfängliche schnelle Verdichtung, die bei 1400°C begann, durch Zerbrechen und Neuordnung der Teilchen bedingt wurde. Dies wurde später vom hauptsächlichen Endsintermechanismus eingeholt.*

## 1. INTRODUCTION

Zirconium diboride is a most promising material for long-term loadbearing applications at temperatures above which metals prove unsatisfactory because of their low strength and creep resistance. This compound is reported to have a melting point in excess of 3000°C and is stable in inert and reducing atmospheres and under vacuum at temperatures above 2000°C. As a constructional material, however, it suffers from two main disadvantages, namely lack of oxidation resistance at high temperatures, and the fact that fabrication has to be by hot-pressing.

Additions of $MoSi_2$ have been proved to confer a great increase in resistance to oxidation, increasing further the attraction of this material. However, even this blended material has to be hot pressed to obtain the optimum properties required for most applications.

This paper describes work on the hot pressing of $ZrB_2$, the first stage of an investigation of the high-temperature chemical, thermal and mechanical properties of $ZrB_2$ and related materials.

## 2. MECHANISMS OF HOT PRESSING

It is generally accepted that, at the start of a hot-pressing operation, initial densification takes place via a grain-boundary sliding and fragmentation process described by FELTEN [1] and CHANG and RHODES.[2] Thus, as densification progresses, this mechanism is overtaken by another which probably continues to near theoretical density. For most materials this final densification process can be described by the equation :

$$\frac{d\rho'}{dt} = m \ (1 - \rho') \qquad . \qquad . \qquad . \qquad . \qquad (1)$$

where $\rho'$ is the fractional relative density and $m$ is a constant for a given set of temperature and pressure conditions.

The first derivation of Equation (1) was by MURRAY, ROGERS and WILLIAMS,[3] who proposed that the densification of their oxides during hot pressing proceeded by a plastic or viscous flow mechanism. They added a term to the Mackenzie–Shuttleworth sintering equation [4] to represent the applied pressure and derived the equation

$$\frac{d\rho'}{dt} = \frac{3\sigma}{4\eta} (1 - \rho') \qquad . \qquad . \qquad . \qquad . \qquad (2)$$

where $\sigma$ is the applied stress (dyne cm$^{-2}$), and $\eta$ is the viscosity (poise), for high relative densities. However, the Murray equation has become suspect because it has been applied successfully to the densification behaviour of compounds at temperatures where little plastic or viscous flow was to be expected. This point was made by JAEGER and EGERTON [5] when applying the equation to the results of their study of the densification of potassium–sodium niobates.

Two papers frequently quoted as having conclusively confirmed the Murray equation and the plastic flow mechanism are those by VASILOS [6] on the hot pressing of fused silica and by MANGSEN et al.[7] on alumina. A mechanism of plastic flow is perfectly credible for the densification of fused silica, especially since Vasilos pointed out that the values of viscosity, $\eta$, calculated from his experiments, compared well with values obtained by the standard fibre elongation method.

However, although Mangsen et al. produced evidence which indicated that their alumina probably densified according to Equation (2), the calculated viscosity data ranged from $10^{12}$ to $10^{10}$ poise at 1300° to 1700°C. These values indicate that, for instance, the viscosity of $Al_2O_3$ at 1300°C is lower than that of fused $SiO_2$ at 1200°C, which is most improbable. Thus, although silica and alumina may densify according to the same rate equation, the mechanisms of transport are almost certainly different.

The first attempts to interpret densification results with models based on diffusion were made by COBLE and ELLIS [8] and by VASILOS and SPRIGGS.[9] Coble and Ellis calculated the contribution of plastic flow to the densification of hot-pressed $Al_2O_3$ and showed it to be small, concluding that the final stage of densification occurred by enhanced diffusion under the influence of stress. Vasilos and

Spriggs obtained consistent diffusion coefficients for hot-pressed alumina and their values of activation energy agreed well with those derived by other related processes. More recently, ROSSI and FULRATH [10] amplified the diffusion theory and reported good agreement between the activation energy for the final stages of the densification of alumina and the activation energy for the diffusion of aluminium ions in that oxide. They based their rate equation on the NABARRO–HERRING [11,12] creep model in which the strain rate is expressed by

$$\dot{\epsilon} = \frac{40\,D\,\Omega\,\sigma_e}{kTd^2} \qquad \qquad . \qquad . \qquad . \qquad (3)$$

where
$\dot{\epsilon}$ = Strain rate (sec$^{-1}$)
$D$ = Diffusion coefficient (cm$^2$ sec$^{-1}$)
$\Omega$ = Volume of a vacancy (cm$^3$)
$\sigma_e$ = Effective stress (dyne cm$^{-2}$)
$k$ = Boltzmann's constant
$T$ = Absolute temperature
$d$ = Mean grain diameter (cm).

In order to apply this model to densification during hot pressing, a stress concentration factor was introduced to allow for the effects of the presence of pores (at low values of total porosity) on the effective stress and the constraint in creep rate exerted by the die wall, resulting in the reduction of Equation (3) to

$$\dot{\epsilon} = \frac{40\,D\,\Omega}{kT} \cdot \frac{\sigma\,bQ}{d^2} \qquad . \qquad . \qquad . \qquad (4)$$

where
$b$ = Stress concentration factor
$\sigma$ = Applied stress
$Q$ = Porosity.

Consequently it was shown that for low porosities

$$\frac{dp'}{dt} = \frac{40\,D}{kT} \cdot \frac{b\,\sigma\,(1-\rho')}{d^2} \qquad . \qquad . \qquad . \qquad (5)$$

The similarity between Equations (1), (2) and (5) is readily apparent.

A completely different mechanism and rate equation for the sintering of Al$_2$O$_3$ during hot pressing has been proposed recently by FRYER.[13] Basing his treatment on a model involving bulk diffusion of vacancies from vacancy sources at the surfaces of pores to vacancy sinks at grain boundaries, Fryer derived the rate equation

$$\frac{1}{V_s}\frac{dV}{dt} = \frac{-Z\,\sigma}{l^2} \cdot \frac{D}{kT} \left(\frac{Q}{\rho}\right)^{\frac{5}{3}} . \qquad . \qquad . \qquad (6)$$

where

$V_s$ = Total volume of solid
$Z$ = Constant dependent on pore geometry
$l$ = Mean pore separation.

As for the reasoning of Rossi and Fulrath, which was based on the Nabarro-Herring creep model, the reasoning of Fryer also appears sound. It would therefore appear that $Al_2O_3$ can densify by two separate mechanisms, depending on experimental conditions. Unfortunately, Rossi and Fulrath have apparantly made at least one error in their calculations on their results since, if one examines their Figure 4 and thence plots $\log(1 - \rho')$ against time, it will be found that the plots are non-linear, suggesting that possibly their results do not fit their own theory. Furthermore, Fryer showed that the replotted data from this figure gave linear plots when tested against his equation.

## 3. EXPERIMENTAL PROCEDURE

### 3.1 Apparatus

The hot-press assembly used is shown in Figure 1. The dies, plungers, spacers and support blocks were of EY9 or EY110 grade graphite (Morganite Carbon Co.) and were surrounded by zirconia

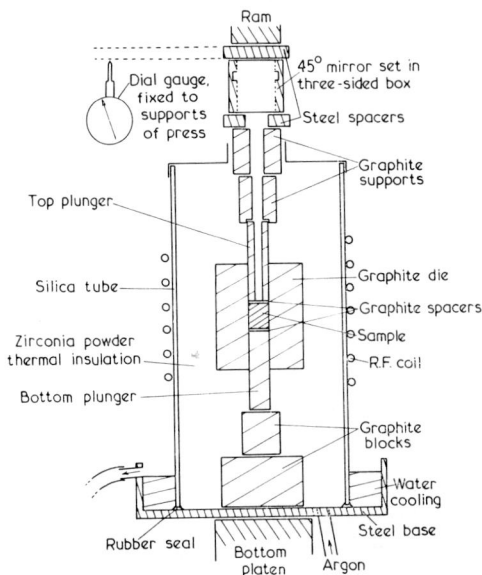

FIGURE 1.—Hot-pressing apparatus.

powder insulation contained in a silica tube. An argon atmosphere was maintained inside the silica tube in order to prevent the graphite being oxidized. Temperature was measured by sighting an optical pyrometer on the back face of the upper graphite spacer via an inclined chromium-plated mirror situated between the die and the ram of the press. Ram movement was measured with a dial gauge mounted on one of the pillars of the press. The dial gauge was calibrated in ten-thousandths of an inch. Power was provided by a 30 kVA radio frequency unit (Raydyne) operating at 450 kc . sec$^{-1}$.

### 3.2 ZrB$_2$ Powder

The powder used was the chemically pure grade of ZrB$_2$ produced by Borax Consolidated Limited. Typical analysis figures quoted by the manufacturers are:

| | | |
|---|---|---|
| Zr | minimum | 80·2% |
| B | minimum | 19·0% |
| C | maximum | 0·1% |
| Fe | maximum | 0·1% |
| O | maximum | 0·3% |
| N | maximum | 0·05% |

A technique similar to that reported by JONES [14] was developed for the examination of polished sections of this material in which the powder and Araldite mounting resin were stirred inside a metal container tube. After evacuation and curing, the mount was cut in half and then remounted in such a manner that the cross-section through the first powder-resin mount could be polished in the normal manner for examination under reflected light. A photo-micrograph of a typical area is shown in Figure 2. Microscopic examination has revealed the presence of trace quantities of at least one unidentified impurity phase in the ZrB$_2$ particles.

The mean particle diameter of this powder has been estimated to be c. 7 $\mu$m.

### 3.3 Hot-pressing Procedure

A pressure of 4000 lb.in$^{-2}$ was used for all experiments unless otherwise specified. Full pressure has been applied throughout each experiment, and was maintained until the temperature fell to c. 1000°C, when the pressure was allowed to decrease as the sample and supports contracted.

The rate of rise of temperature was varied, the time to reach peak temperature being between 15 and 60 min., but it had no effect on the final densification rate at constant temperature. The soak at constant temperature was always terminated when the rate of

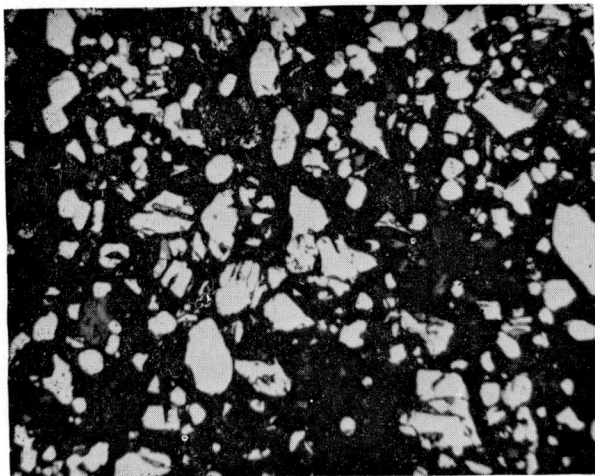

FIGURE 2.—Photomicrograph of ZrB$_2$ mounted in resin ($\times$ 560): White, ZrB$_2$. Black, resin and pull-out.

ram travel for samples of 0·5 in. diam. $\times$ 0·5 in. fell below 0·0010 in.min$^{-1}$, namely, when the rate of densification fell below $c$. 0·012 g. cm$^{-3}$.min$^{-1}$.

Cooling was rapid, a typical rate being from 2000° to 1200°C in 10 min.

## 3.4 Calibration Experiments

### 3.41 Mirror Calibration

The correction factor for the temperature when measured via the chromium-plated mirror was obtained by comparing a series of measurements taken by direct viewing and with the mirror in position and was found to increase linearly from 50°C at 1700°C to 100°C at 2000°C. (This correction factor has been applied to all temperatures given in this paper.)

### 3.42 Time to Equilibrium

A series of experiments was carried out to determine the expansion of the system as a whole and to determine the time for the system to reach an equilibrium position when no further ram movement could be detected at constant temperature and pressure. No powder sample was included in these runs.

It was found that the expansion was approximately linear between 1100°C and 2100°C. The effect of pressure (within the range 3000–5000 lb.in$^{-2}$) was negligible at these temperatures.

5

The time taken to reach equilibrium was determined by repeat runs in which the temperature was raised to the final temperature in 15–20 min., i.e., at approximately the rate of rise of temperature employed in the fastest runs. It was found that detectable expansion ceased within 3–4 min. at constant temperature.

### 3.5 Determination of Instantaneous Density

The calibration experiments described above indicated that, once the temperature had been constant for approximately 5 min., all observed shrinkage was then representative of the sintering of the sample. By assuming that no sintering took place during cooling and that the thermal expansion coefficient of the compact was independent of porosity during the small time interval under consideration, it can be said that the "equivalent room-temperature density" of the sample at the end of the temperature soak was equal to the measured density of the sample after extraction from the die. The equivalent room-temperature density of the sample at any earlier time during the soak at constant temperature could then be calculated from the final measured density and the contraction observed during the experiments. Thus the first step in tracing back the densification was to determine the shrinkage that took place at the constant-soak temperature. The equivalent room-temperature length at any time during the soak period was then calculated by adding the shrinkage that took place between that time and the end of the soak period to the final measured length of the cold sample. In this step the thermal contraction of the sample from 1800°–2000°C to room temperature is ignored (error approximately 2%). The instantaneous equivalent room-temperature density was then calculated from the instantaneous equivalent room-temperature length and the final bulk-density measured on a mercury balance and checked by calculation from the dimensions. This step involves the safe assumption that all porosity in the final sample is closed. The porosity $Q$ was then calculated from the theoretical X-ray density and $\log_e Q$ plotted against time according to Equation (7), which was obtained from Equation (1) by integration

$$\log_e Q = -mt + \log_e Q_0 \qquad . \qquad . \qquad . \qquad (7)$$

### 4. RESULTS

### 4.1 Densification Rate

When the density-histories of the powder compacts were determined, it was found that the relative density of the initial cold-pressed powder was generally 55–60%. When the R.F. power was applied and the temperature started to rise, there was at first a slow expansion of the system, but as the temperature reached 1350°–1400°C, contraction commenced, governed by the densification of the sample. By the time the temperature had been stabilized the period of maximum rate of densification had been passed. An interesting point is that, although the final soak temperature was approached slowly from below, the rate of densification invariably decreased to a steady rate when plotted in the logarithmic form of Figure 3. In other words, the initial densification mechanism was

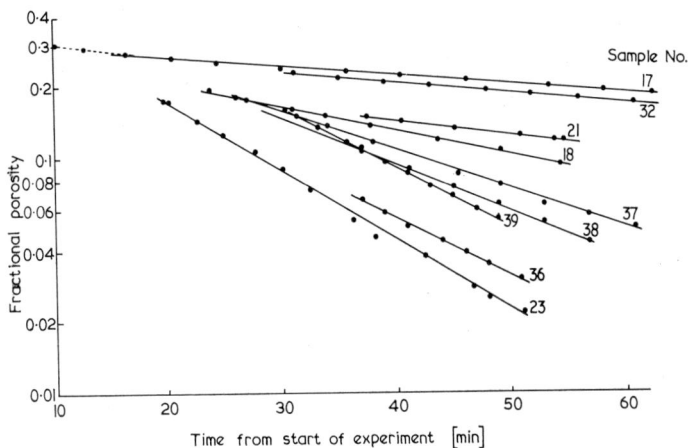

FIGURE 3.—Typical densification plots for ZrB$_2$

faster than that observed for the main controlled part of the experiment. It has been assumed that this was caused by a grain-boundary sliding and fragmentation process, as has been observed by other workers. It has not been possible to carry out any examination of this part of the densification, as it occurs simultaneously with the expansion of the system, owing to the uncontrolled rate of increase of temperature.

A selection of the resultant plots of $\log_e Q$ against time, $t$, are shown in Figure 3. All these plots show excellent agreement with Equation (1) and in themselves indicate that temperature control and measurement of shrinkage were reasonably accurate. As mentioned earlier, the runs have been terminated when the densification rate fell to 0·012 g.cm$^{-3}$.min$^{-1}$. This has caused the final density of the sample to be dependent on temperature. Generally, relative densities of c. 90% and c. 97% have been obtained for samples pressed at 1865°C and 2100°C respectively. It will be seen that there is no indication of any end-point density beyond which densification ceases.

Figure 4 shows the results of two experiments in which the temperature was held constant and the pressure was varied and vice versa. It can be seen that, within the range investigated, the rate of densification is more dependent on temperature than on pressure. In the so-called constant-temperature run, No. 90, the temperature actually increased from 1865° to 1885°C.

This experiment was the only run carried out to test the pressure dependence of Equations (2) and (5) when, as predicted by these equations, the ratio of each individual slope on the densification plot of Figure 4 to its corresponding pressure was found to be constant after allowance had been made for the temperature increase which took place. Many other workers have demonstrated the validity of this aspect of the rate equation.

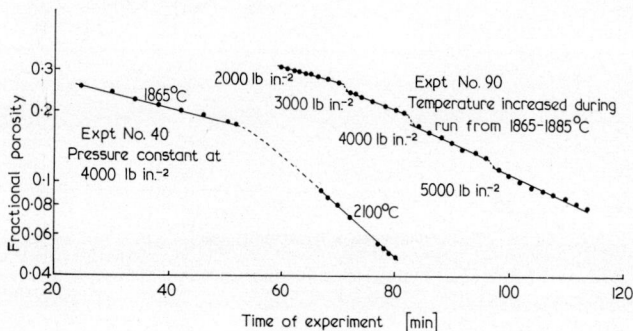

FIGURE 4.—Densification plots to show effects of temperature and pressure on sintering rate.

## 4.2 Metallographic Examination

It has not been possible to section many samples, as they are required for other parts of the programme, which is an investigation of the high-temperature chemical and physical properties of these materials. However, a number were mounted and polished, and where necessary, etched with 3N NaOH solution. Photomicrographs of the structure before and after etching are given in Figures 5 and 6. All porosity has been found to occur at grain boundaries. The optical properties of the impurity phases corresponded exactly to those of the impurities found in polished sections of the raw material powder.

This work has shown that, on hot-pressing $ZrB_2$, the observed densification rate can be accurately described by Equation (1). It

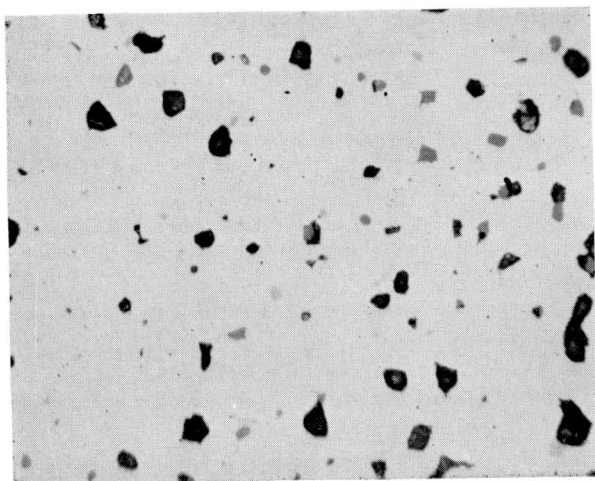

FIGURE 5.—Photomicrograph of hot-pressed $ZrB_2$ ($\times 560$): White, $ZrB_2$. Light and dark grey, unknown impurity phases. Black, porosity.

FIGURE 6.—Hot-pressed $ZrB_2$ after etching ($\times 560$). Porosity and impurity phases can be seen sited at grain boundaries.

## 5. DISCUSSION

was indicated earlier that there were alternative derivations of this equation, a plastic flow or viscous flow mechanism which, under the conditions prevailing in this and most other hot-pressing experiments, could be described by Equation (2), and a diffusion-controlled model system which could be described by Equation (5). It is therefore reasonable to deduce that $ZrB_2$ behaves closely like one or both of these model systems at some stage during its densification, and this point will be discussed further. It is certainly clear that $ZrB_2$ does not densify via the mechanism proposed by Fryer for his results on $Al_2O_3$.

According to the model system described by Equation (2), the slopes of the plots of $\log(1 - \rho')$ against time of sintering at constant temperature should enable calculation of the viscosity of $ZrB_2$ in the temperature range under examination. Results calculated in this manner are shown in Figure 7 and indicate a decrease in viscosity

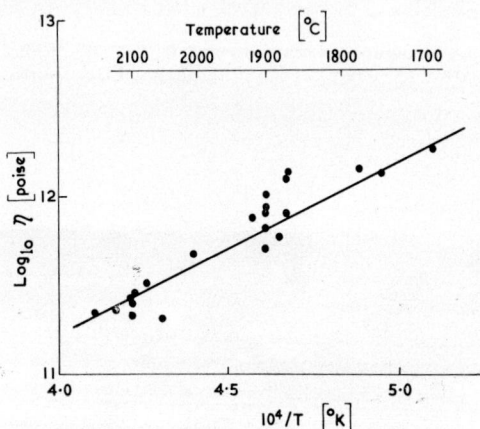

FIGURE 7.—Calculated viscosity data for $ZrB_2$.

of $ZrB_2$ from $2 \times 10^{12}$ poise at 1700°C to $2 \times 10^{11}$ poise at 2150°K. As mentioned earlier, when discussing in Section 2 the results of Mangsen, whose calculated viscosity results are in the same range, these values are very low, and represent values which are lower than that of, say, fused silica at 1200°C, and are lower than the viscosity at which fused silica is annealed. One would therefore conclude that, at 1700°C and above, $ZrB_2$ should exhibit low strength and poor creep properties. Unfortunately there is little in the literature on the high-temperature mechanical properties of $ZrB_2$ and it is an aim of the present study to obtain this information. However the high temperature properties of an 86 $^w/_o$ $ZrB_2$ : 12·8 $^w/_o$ $MoSi_2$ : 1·1 $^w/_o$ BN material have been well documented.[15] Examination of the hot pressing of this type of debased $ZrB_2$ material has been carried out concurrently with the work being described and has indicated that sintering proceeds some 100°–200°C below that of $ZrB_2$ alone, and even then the sintering rates are such that the slope of the viscosity/temperature plot is steeper, with a lower calculated viscosity range than is given here. Hence it could be concluded that this debased material should have an even lower high-temperature strength than $ZrB_2$ itself. However, this $ZrB_2$ alloy has been reported to exhibit a modulus of rupture of > 20,000 lb.in⁻² at 1700°C, decreasing to c. 12,000 lb.in⁻² at 1915°C. Creep rates of this material were reported to increase rapidly above 1800°C. It must be pointed out that the melting point of this alloy is probably in the range 2100°–2300°C. In general it is believed that it is unlikely that the high temperature mechanical properties of $ZrB_2$ support the plastic flow–viscous flow sintering mechanism.

Considering next the treatment of Rossi and Fulrath to the present results, it is firstly possible to determine an activation energy for diffusion by calculating diffusion coefficients (Table 1).    in the absence of better data, the value of $b$ was taken equal to 6, and the volume of a vacancy as $5 \times 10^{-23}$ cm³. These values do not of course affect the value of the activation energy, which was calculated from Figure 8 to be 140 kcal.mole⁻¹.

It is interesting also to attempt to predict the types of micro-structure that would be expected to result from plastic and viscous

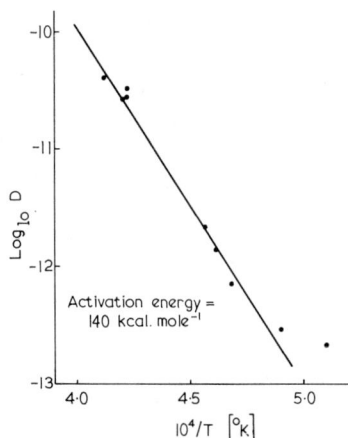

FIGURE 8.—Plot of log (diffusion coefficient) against reciprocal temperature.

**Table 1**
**Calculated Diffusion Coefficients**

| Experiment No. | Temperature ($^{\circ}C$) | Applied pressure ($lb.in^{-2}$) | Mean grain diam. $\times 10^4$ ($cm$) | $m$ ($sec^{-1}$) | $D \times 10^{13}$ ($cm^2.sec^{-1}$) |
|---|---|---|---|---|---|
| 86 | 1690 | 4000 | 2·95 | 0·47 | 3·34 |
| 17 | 1770 | 4000 | 3·45 | 0·60 | 6·07 |
| 32 | 1865 | 4000 | 3·33 | 0·69 | 6·80 |
| 30 | 1900 | 4000 | 3·70 | 1·00 | 12·4 |
| 85 | 1925 | 4000 | 4·00 | 1·15 | 16·8 |
| 37 | 2100 | 4000 | 10·00 | 2·80 | 277 |
| 38 | 2100 | 4000 | 10·00 | 3·33 | 329 |
| 41 | 2110 | 4000 | 9·10 | 3·30 | 272 |
| 36 | 2160 | 4000 | 11·00 | 4·03 | 408 |

flow, and from diffusional mechanisms. Although it is important to appreciate that more than one transport process may occur simultaneously, it is to be expected that when one examined the microstructure of a material undergoing densification predominantly by a plastic or viscous flow process one would be able to observe grains that had been extruded into accompanying voids during compaction. Further, with a porous and irregularly shaped powder, one would expect to find substantial amounts of residual porosity within the grains. On the other hand, with a vacancy diffusion-based mechanism, one would expect to find a microstructure indistinguishable from those produced on normal diffusion-controlled sintering without applied pressure. Further, if vacancy diffusion rates are high, the short duration of hot-pressing experiments would make it probable that residual porosity would be concentrated at grain boundaries but grain size would remain small.

In fact, we find little evidence of porosity within grains, or of "extruded" grains. On the contrary, the microstructure resembles that of most sintered ceramics, with all porosity concentrated at grain boundaries.

### 6. SUMMARY

Taking all the experimental evidence together, it is concluded that, without confirmatory creep rate data, diffusion coefficients, etc., it is at present impossible to distinguish the predominant sintering mechanism which occurs during the hot pressing of $ZrB_2$. If one assumes a viscous or plastic flow (Murray) mechanism to occur, then the calculated viscosity values are suspiciously low by at least 2–3 orders of magnitude. On the other hand, the diffusional interpretation of Rossi and Fulrath enables calculation of a reasonable activation energy for vacancy diffusion in $ZrB_2$, and this mechanism is supported by the metallographic evidence.

## ACKNOWLEDGMENTS

The author thanks Mr. P. B. Noakes and Mr. P. S. Brass for their assistance with many hot-pressing experiments, and Doulton Research Limited and Bristol Siddeley Engines Limited for permission to publish this paper.

## REFERENCES

1. FELTEN, E. J., *J. Amer. Ceram. Soc.*, **44**, (8), 381, 1961.
2. CHANG, R., and RHODES, C. G., *J. Amer. Ceram. Soc.*, **45**, (8), 379, 1962.
3. MURRAY, P., RODGERS, E. P., and WILLIAMS, A. E., *Trans. Brit. Ceram. Soc.*, **53**, (8), 476, 1054.
4. MACKENZIE, J. K., and SHUTTLEWORTH, R., *Proc. Phys. Soc. (London)*, **62B**, 833, 1949.
5. JAEGER, R. E, and EGERTON, L., *J. Amer. Ceram. Soc.*, **45**, (5), 109, 1962,
6. VASILOS, T., *J.Amer. Ceram. Soc.*, **43**, (10), 517, 1960.
7. MANGSEN, G. E., LAMBERTSON, W. A., and BEST, B., *J. Amer. Ceram. Soc.*, **43**, (2), 55, 1960.
8. COBLE, R. L., and ELLIS, J. S., *J. Amer. Ceram. Soc.*, **46**, (9), 438, 1963.
9. VASILOS, T., and SPRIGGS, R. M., *J.Amer. Ceram. Soc.*, **46**, (10), 493, 1963.
10. ROSSI, R. C., and FULRATH, R. M., *J. Amer. Ceram. Soc.*, **48**, (11), 558, 1965.
11. NEBARRO, F. R. N., Rept. Conf. Strength of Solids (Univ. of Bristol), July 1947 (published 1948).
12. HERRING, C., *J. Applied Phys.*, **21**, (5), 437, 1950.
13. FRYER, G. M., *Trans. Brit. Ceram. Soc.*, **66**, (3), 127, 1967.
14. JONES, T. I., New Types of Metal Pdrs., Met. Soc. Conferences, Vol. 23, *Trans. A.I.M.E.* Oct. 1963, pp. 139–67.
15. LOGAN, I. M., and NEISSE, J. E., U.S. Dept. of Commerce, Technical Report ASD-TDR-62-1055, 1962.

# 5.—The Hot Pressing of Aluminium–Magnesium Spinel (MgAl₂O₄)

By P. P. BUDNIKOV, F. KERBE and F. J. CHARITONOV

*Mendeleev Institute for Chemical Technology, Moscow*

**ABSTRACT**                                             C327/E212

*Incompletely crystallized $MgAl_2O_4$ with many defects, but of high purity, was hot-pressed in a hydraulic die assembly made of very dense graphite, between 1200° and 1500°C (0·56–0·7 of the melting point of the spinel) at 60 to 300 $kgf/cm^2$ for times ranging from 5 to 30 min. The influence of temperature, time and pressure of the linear shrinkage, density, porosity, and microstructure of the hot-pressed spinels was investigated. The shrinkage curves show a steep fall which, depending on the sintering pressure, occurs between 950° and 1000°C, but slight densification occurs at 1300°C. At 1300°–1400°C and 300 $kgf/cm^2$, a density of 92–98% of the theoretical (3·58 $kg/cm^3$) is obtained within 10 min. At optimal sintering conditions at 1450°C, spinel samples of theoretical density are obtained. Dense samples examined by means of an etching technique using $H_3PO_4$ (conc. 85%) had a regular fine-grained structure. The refractive index of the spinels increases with increasing temperature, pressure and duration of hot pressing, which confirms a decrease of defects in the spinel structure; the grain size increases from 1–2 μm up to 10 μm, converting into octahedra. Recrystallization is accompanied by an increase in micro-hardness from 642 to 1100–1530 $kgf/mm^2$.*

## Le pressage à chaud de spinelles d'aluminate de magnésium pur

*De l'aluminate de magnésium ($MgAl_2O_4$) présentant des défauts nombreux, mais de pureté élevée, est pressé à chaud entre 1200 et 1500°C (0,56 à 0, 7 du point de fusion du spinelle) sous 60 à 300 $kgf/cm^2$ et pendant des durées de 5 à 30 min., dans des moules de graphite très dense, à l'aide d'une presse hydraulique. On étudie l'influence de la température, du temps et de la pression sur le retrait linéaire, la densité, la porosité et la microstructure des spinelles pressés à chaud. Les courbes de retrait accusent une chute brutale qui, suivant la pression de frittage, apparaît entre 950 et 1000°C, mais une densification faible se produit à 1300°C. A 1300–1400°C et sous une pression de 300 $kgf/cm^2$, on obtient au bout de 10 min. une densité de*

# 70    BUDNIKOV, KERBE AND CHARITONOV:

*92–98 % du chiffre théorique (3,58 kg/cm³). Des échantillons de spinel-
les à la densité théorique sont obtenus à 1450°C dans des conditions
optimales de frittage. Une attaque à H₃PO₄ (concentré à 85 %) montre
que les échantillons denses ont une structure régulière à grains fins.
L'indice de réfraction des spinelles augmente avec l'accroissement de
la température, de la pression et de la durée du pressage à chaud,
ce qui confirme une diminution des défauts dans la structure des spinelles.
La dimension des grains passe de 1–2 µm à 10 µm, ceux-ci prennent
la forme d'octaèdres. La recristallisation s'accompagne d'une augmentation de la microdureté de 642 à 1100–1530 kgf/mm².*

### Das Heisspressen reiner Magnesium–Aluminiumoxidspinelle

*Ein fehlstellenreiches aber hochreines $MgAl_2O_4$ wurde in einer
hydraulischen Presse aus sehr dichtem Graphit bei Temperaturen
zwischen 1200 und 1500°C (0·56–0·7 des Schmelzpunktes von Spinell)
und Drücken von 60 bis 300 kgf/cm² in Zeiten zwischen 5 und 30 Min.
heiß gepreßt. Der Einfluß von Temperatur, Zeit und Preßdruck auf
lineare Schwindung, Dichte, Porosität und Mikrostruktur des heiß-
gepreßten Spinells wurde untersucht. Die Schwindungskurven zeigen
einen steilen Abfall zwischen 950 und 1000°C, der vom Sinterdruck
abhängt; unbemerkte Verdichtung tritt bei 1300°C ein. Bei 1300 bis
1400°C und 300 kgf/cm² wird eine Dichte von 92–98 % der theoretischen
Dichte (3,58 g/cm³) in 10 min. erreicht. Bei optimalen Sinterungsbed-
ingungen bei 1450°C werden Spinellkörper theoretischer Dichte
erreicht. Dichte Körper, die nach der Ätzmethode mit 85 % konz.
$H_3PO_4$ untersucht wurden, besaßen eine regelmäßige, feinkörnige
Struktur. Der Brechungsindex der Spinelle nimmt mit steigender
Temperatur, steigender Preßdauer und steigendem Druck zu, was
eine Abnahme der Fehlstellen in der Spinellstruktur anzeigt. Die
Korngröße wächst von 1–2 µm bis zu 10 µm. Die Rekristallisation
wird von einer Zunahme der Mikrohärte von 642 auf 1100 bis 1530
kgf/mm² begleitet.*

## 1. INTRODUCTION

Magnesium spinel ($MgAl_2O_4$) has a dense and stable structure for
which a relatively high firing-temperature is required to obtain
complete sintering. Detailed investigations[1–3] have shown that in
mixtures of powdered MgO and $Al_2O_3$ the spinel forms at a temperature below those of sintering and recrystallization. In magnesium
spinel—depending on the nature of the starting materials and the
technique—sintering is perceptible above 1400°–1500°C and at
higher temperatures it accelerates. The densities so obtained are
appreciably lower than the theoretical value for the spinel.

Hot pressing was considered to be a very effective means of obtaining densely sintered spinels of fine crystal structure at relatively low firing-temperatures. In 1910, HAMANO[4] reported on his hotpressing experiments on spinels; and extensive and systematic research on this subject has been carried out by American investigators,[5-10] who showed that, over a wide range of grain size, a spinel of theoretical density can be produced by hot pressing. The purity of the starting spinel powder, its synthesis and the method of hot pressing used were studied particularly, since these factors influence the mechanical and structural properties of the sintered product.

The main objectives of this investigation have been to obtain more detailed information about the sintering mechanism of extremely pure fine-grained magnesia spinel during hot pressing and to assess the structural and physical properties of the sintered products obtained under different hot-pressing conditions.

## 2. EXPERIMENTAL

### 2.1 Synthesis of Very Pure Spinels of Defective Structure

As starting components for the synthesis of a very pure $MgAl_2O_4$ we used $MgCO_3$ and $Al_2(SO_4)_3.18H_2O$, both spectrally pure. The two compounds, in 1:1 molar proportions, were dissolved in a water–ethyl alcohol mixture in a platinum dish by gentle heating. Complete dissolution gave a clear solution that changed into a brittle mass after partial evaporation of the solvent. X-ray analysis showed the residue to be wholly amorphous. After heating at 800°C for 1 h this product turned with bloating into a soft, biscuit-like material that could be easily mortared to an extremely fine powder with a specific surface of 39 m²/g. X-rays showed this product to consist of an incompletely synthesized defective $MgAl_2O_4$ spinel. No contamination could be detected by spectral analysis.

### 2.2 Method of Hot Pressing

Hot pressing was by means of an hydraulic lever press (Figure 1). Dies were of very dense graphite and were heated direct by electric current. The press has a recorder that registers the sintering curve. Temperatures were measured by means of an optical pyrometer through an opening in the graphite shield; conditions were close to black body conditions. Readings were checked by a wolfram–rhenium thermocouple, the hot junction being placed very close to the pressed material in the die. Experiments were carried out at temperatures between 1200° and 1500°C ($0.56-0.7$ $T_m$ of the spinel) using pressures between 60 and 300 kgf/cm² and with pressing times from 10 to 30 min. Preliminary experiments showed the importance of the pressing conditions for obtaining maximum density. Rapid heating with continuously rising pressure did not cause sufficient densification.

For our starting material the programme in Figure 2 proved to be optimal. The heating-rate was 45°C/min. for a minimum pressure

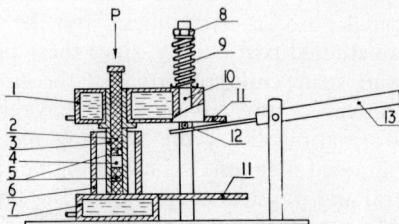

FIGURE 1.—Diagram of lever press:

1. Upper cooled contact
2. Press die
3. Graphite matrix
4. Test-piece
5. Stopper
6. Graphite casing
7. Lower cooled contact

8. Nut
9. Spring
10. Guide brace
11. Power supply bar
12. Sliding hinge
13. Hand lever to raise
    upper contact

(60 kgf/cm²); after the pressing-temperature had been attained, the pressure was raised to the required value in a few seconds. The hot-pressed test-pieces were cylindrical—10 mm diameter, 10–15 mm high. Preliminary experiments had shown that, with these dimensions, pressure was evenly distributed throughout the cylinder, irrespective of the method of pressing (uni- or bilateral).

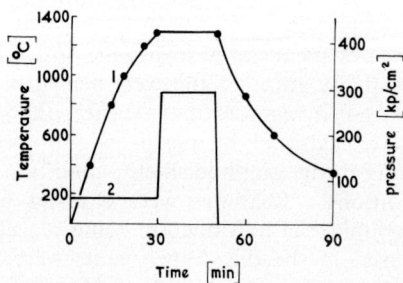

FIGURE 2.—Control of isothermal and isobaric pressure-sintering conditions:

1. Temperature          2. Pressure

## 3. EXPERIMENTAL RESULTS AND THEIR INTERPRETATION

The influence of temperature, pressure and sintering-time on the linear shrinkage, apparent specific gravity, porosity, water absorption index of refraction, micrahardness and microstructure of spinels was investigated.

### 3.1 Densification of Magnesia Spinel

Densification of the finely dispersed spinel powder is accompanied by an appreciable linear shrinkage which we can describe as a simultaneous process of sintering and flow under pressure. Figure 3 gives curves for the relative changes in length under pressures of 60, 100 and 200 kgf/cm²; the first detectable shrinkage begins at 950° to 1000°C. Shrinkage is complete at 1450° to 1500°C. The apparent specific gravities of the hot-pressed cylinders (measured by weighing under petroleum) show that, above 1300°C, important densification takes place.

After 10 min. at 1300°–1400°C at a pressure of 300 kgf/cm², densities of 92 to 98% of the theoretical, (3·58 g/cm³) were obtained. Under optimum conditions (300 kgf/cm²; 10 min. at 1400°–1500°C) we succeeded in attaining theoretical density. Figures 4A and B indicate the densities obtained under 200 and 300 kgf/cm² (the most effective pressures) as a function $\ln t = f(1/T)$.

The curves connect points of equal density, and comparison shows that the slope depends on the applied pressure. The straight part of each curve indicates that to extend the sintering time considerably enhances the degree of densification. When theoretical density is approached, however, the curves change slope, and the higher the

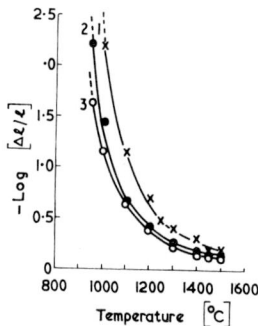

FIGURE 3.—Relative change in length of test-piece under constant pressure.
1. 60 kgf/cm²     2. 100 kgf/cm²     3. 200 kgf/cm²

pressure the greater the change. Theoretical density can only be obtained by prolonging the sintering time above a certain temperature. This temperature is about 1450°C. It can be concluded that an energy barrier has to be overcome before theoretical density can be attained.

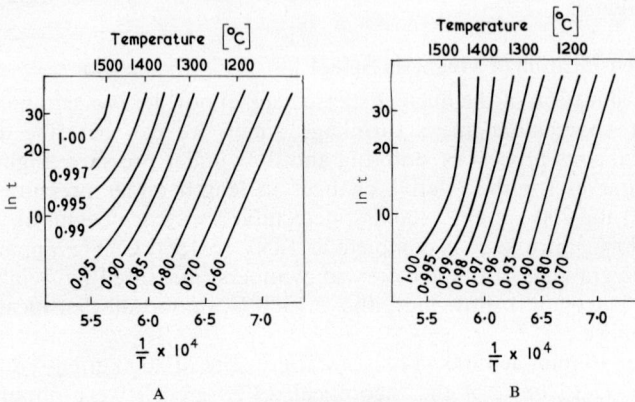

FIGURE 4.—Change in relative density of hot-pressed in the co-ordinate system ln t − 1/T for pressing pressures of:

A. 200 kgf/cm²              B. 300 kgf/cm²

## 3.2 Microstructure

Preparation of the hot-pressed samples for petrographic examination proved to be difficult. Etching with the normal etching-fluids proved unsuccessful because of the very fine structure of the product. Etching with 85% phosphoric acid proved to be effective.

Polished surfaces were immersed for 1 to 3 min. in the boiling acid and subsequently rinsed with distilled water. In the surface so obtained the spinel crystals were sufficiently distinct to be suitable for microscope examination.

Table 1 contains the most important physical properties of the spinel products. Microhardness was measured by the indentation method using a diamond pyramid under a load of 200 lb. The dense sintered products have a homogeneous and extremely fine structure, and are practically non-porous. The refractive index rises in parallel with the density, both being caused by the high degree of perfection of the spinel crystals. At the same time the dimensions of the crystals increase from about 2 to 10 μm. The development of a homogeneous structure and the beginning of crystal growth are associated with a pronounced increase in microhardness (from 642 kgf/mm² to 1100–1530 kgf/cm²). The microhardness value proved to be an extremely sensitive indication of density changes.

Table 1
Physical Properties of Hot-pressed Spinel

| Temperature (°C) | Pressure (kgf/cm²) | Sintering period (min.) | Relative density | Total porosity (%) | Water absorption | Microstructure | Crystal shape of spinels | Mean grain diameter (μm) | Refractive index | Micro-hardness (kp/mm²) |
|---|---|---|---|---|---|---|---|---|---|---|
| 1300 | 200 | 20 | 0·87 | 13·4 | 0·07 | Heterogeneous | Irregular Prisms, | 1–2 | 1·714 | 824 |
|  | 300 | 10 | 0·92 | 8·1 | 0·06 | Heterogeneous | squares | 4–6 | 1·719 | 926 |
|  | 300 | 20 | 0·96 | 3·9 | 0·05 | Uniform | Squares, hexagons | 4–6 | 1·726 | 1018 |
| 1400 | 300 | 10 | 0·98 | 2·2 | 0·03 | Heterogeneous | Irregular Prisms, | 4–6 | 1·718 | 946 |
|  | 300 | 20 | 0·99 | 1·0 | 0·02 | Uniform | squares | 4–6 | 1·718 | 1070 |
|  | 300 | 30 | 0·995 | 0·5 | 0·02 | Uniform | Squares, hexagons | 5–10 | 1·724 | 1100 |
| 1450 | 300 | 10 | 0·997 | 0·3 | 0 | Uniform | Squares, hexagons | 4–10 | 1·718 | 1240 |
| 1500 | 60 | 10 | 0·78 | 21·8 | 2·42 | Heterogeneous | Irregular Prisms, | 2–4 | 1·716 | 642 |
|  | 100 | 10 | 0·92 | 7·8 | 0·20 | Heterogeneous | hexagons | 4–6 | 1·720 | 944 |
|  | 200 | 10 | 0·99 | 0·8 | 0·02 | Uniform | Squares, hexagons | 4–10 | 1·724 | 1100 |
|  | 300 | 10 | 1·00 | 0 | 0 | Uniform | Squares, hexagons | 6–10 | 1·726 | 1530 |

## REFERENCES

1. BUDNIKOV, P. P., and ZLOCHEVSKAYA, K. M., *Ogneupory*, **23**, (3), 111, 1958.
2. BRON, V. A., and DISPEREVA, M. I., in "Silicates & Oxides in High-Temperature Chemistry" (Moscow 1963), p. 137.
3. DEGTYAREVA, E. V., KAINARSKIJ, I. S., and TOTSENKO, S. B., *Ogneupory*, **31**, (8), 47, 1966.
4. HAMANO, Y., *Osaka Kogyo Gijutsu Shikensho Kiho*, **11**, (1), 30, 1960.
5. PALMOUR, H., III, CHOI, D. M., and KRIEGEL, W. W., *Amer. Ceram. Soc. Bull.*, **41**, (4), 311, 1962.
6. CHOI, D. M., and PALMOUR, H., III, *Amer. Ceram. Soc. Bull.*, **42**, (4), 261, 1963.
7. CHOI, D. M., PALMOUR, H., III, and KRIEGEL, W. W., *NASA Tech. Pub. Announcements*, **2**, (17), 1012, 1962. *Chem. Abstr.*, **60**, (9), 10, 359, b/d, 1964.
8. PALMOUR, H., III, CHOI, D. M., and BARNES, L. D., "Materials Science Research" (New York: Plenum Press, 1963) Vol. 1, p. 158.
9. KRIEGEL, W. W., PALMOUR, H., III and CHOI, D. M., "Special Ceramics 1964" (London–New York: Academic Press, 1965), p. 167.
10. CHOI, D. M., and PALMOUR, H., III. "Materials Science Research" (New York: Plenum Press, 1966), Vol. 3, p. 473.
11. LEBEDEVA, S. I., "Determination of the Microhardness of Minerals". (Moscow: Akad. Nauk, U.S.S.R., 1963).

*Part Two*

# FABRICATION

# 6.—A Cryochemical Method for Preparing Ceramic Materials

By F. J. SCHNETTLER, F. R. MONFORTE and W. W. RHODES

*Bell Telephone Laboratories, Incorporated, Murray Hill, New Jersey*

## ABSTRACT
C20/B9

*The development of microstructure in sintered ceramic bodies is highly dependent on the nature of the starting materials. A method is described for producing uniformly fine powders of alumina, magnesia, and spinel. No added reagents or mechanical comminuting is required. The product is thus as pure as the raw material from which it is processed. The method is particularly useful for studies of the effect of additives on densification and grain growth, since a uniform distribution of such additives, regardless of amount, can be assured. Features of the method are given by the results of d.t.a., t.g.a., and X-ray studies. Process parameters for controlling particle size from 100 to 5000 Å are discussed. The preparation of free-flowing granules containing a binder for dry pressing is also described.*

### Méthode cryochimique de préparation des matériaux céramiques

*Le developpement de la microstructure dans les produits céramiques frittés dépend fortement de la nature des matériaux initiaux. La méthode décrite ici a pour objet la production de poudres d'alumine, de magnésie et de spinelle uniformément fines. Elle ne nécessite ni ajout de réactif ni opération de broyage fin. Elle est particulièrement utile pour des études concernant l'action des ajouts sur la densification et la croissance des grains puisqu'une distribution uniforme des ajouts peut être assurée indépendamment de leur quantité. Les résultats d'analyses thermiques différentielles, d'analyses thermo-gravimétriques et d'études aux rayons X montrent les caractéristiques de la méthode. Les paramètres à considérer pour obtenir des dimensions de particules entre 100 et 5000 Å sont examinés. La préparation, pour le pressage à sec, de granules s'écoulant librement et contenant un liant est également décrite.*

### Eine gefrierchemische Methode zur Herstellung keramischer Materialien.

*Die Entwicklung von Mikrostrukturen in gesinzerten keramischen Körpern hängt in großem Maß von der Art der Ausgangsmaterialien*

80     SCHNETTLER, MONFORTE AND RHODES:

*ab. Eine Methode zur Herstellung gleichmäßig feiner Pulver aus Al₂O₃, MgO und Spinell wird beschrieben. Bei dieser Methode werden keine Zusätze benötigt. Das Produkt ist infolgedessen ebenso rein wie das Rohmaterial aus dem es hergestellt wird. Diese Methode ist nützlich zur Untersuchung von Zusätzen auf die Verdichtung und das Kornwachstum, da eine gleichmäßige Verteilung der Zusätze unabhängig von der Menge gewährleistet ist. Gekennzeichnet wird die Methode durch Ergebnisse der DTA, der TGA und von Röntgenuntersuchungen. Herstellungsparameter zur kontrollierten von 100 bis 5000 Å großen Teilchen werden diskutiert. Die Herstellung von frei fließenden Granulaten mit Binder zum Trockenpressen wird ebenfalls beschrieben.*

## 1. INTRODUCTION

Studies concerned with the structure-sensitive properties of ceramic materials are often limited by inability to synthesize the desired microstructures. Long strides toward this end have been taken by the various techniques of applying mechanical force during heat treatment. These give some measure of independent control of grain size and densification. As a supplement to such advanced forming methods we wish to describe a technique for preparing controlled starting materials—controlled, that is, with respect to composition, particle size and size range, chemical homogeneity, and, if desired, the uniform distribution of additives.

## 2. PROCESS

The method requires four basic operations:
1. Mixing
2. Freezing
3. Sublimation
4. Decomposition.

### 2.1 Mixing

To achieve chemical homogeneity we have resorted to solution chemistry, which assures mixing on an atomic scale. It does, however, limit the process to the preparation of those materials for which mutually soluble salts are available. The method is amenable to a high degree of precision in formulation. Analysed stock solutions of suitable salts of the various metal components are prepared in advance and these solutions can then be apportioned either by weight or by volume to give the desired composition. As no other reagents are required and as no material is lost in the process, the composition of the final product is that of the starting mixture. Furthermore, no mechanical attrition is required, which might contaminate the product.

## 2.2 Freezing

A principal objective of the method is to achieve chemical homogeneity. To this end the solution is first frozen to immobilize the ions and then the water is removed via sublimation. In general, salt solutions on freezing form hydrates which are immiscible in ice. A typical temperature/concentration diagram—that of the aluminium sulphate–water system,[1] is reproduced in Figure 1.

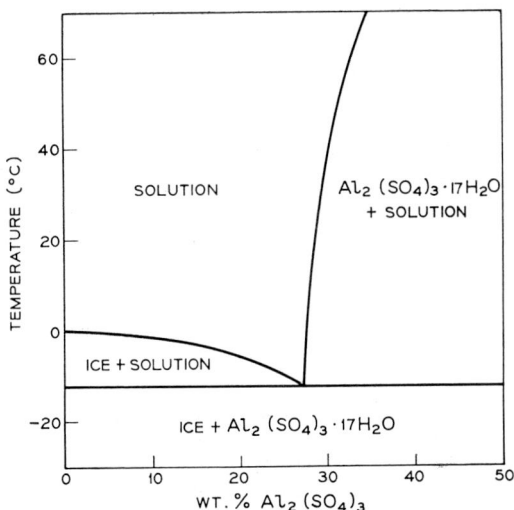

FIGURE 1.—Temperature/concentration relations in the system $Al_2(SO_4)_3$–$H_2O$.

When a solution in this system is cooled, a temperature will be reached at which either ice or hydrated aluminium sulphate will separate. With further cooling the separation will continue and the composition of the solution will change in accordance with the equilibrium diagram. Eventually the solution will assume a composition such that any further cooling will completely solidify it. The success of the cryochemical method for preparing uniformly fine homogeneous powder rests on the ability to pass from the liquidus to the solidus as quickly as possible so as to minimize any changes in salt concentration. Such changes will affect the particle size distribution, as will be shown later. Furthermore, in multisalt systems, composition gradients can be similarly induced if freezing is not too rapid.

A simple laboratory method for quick freezing is shown in Figure 2. A liquid such as hexane, which is immiscible with water

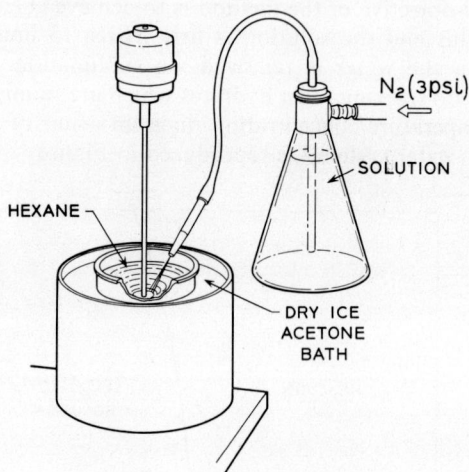

FIGURE 2.—Freezing-apparatus.

and which has a sufficiently low freezing point, is chilled with dry ice and acetone. The solution to be frozen is then introduced into the organic bath as a fine stream, where it breaks into droplets. The droplets are frozen as spheres, the diameter of which depends on that of the nozzle and on the stream velocity. By using a stirrer, a vortex can be formed that helps disperse the droplets and disseminate the heat evolved. The frozen product is then separated from the hexane by screening and poured into pre-chilled trays prior to sublimation.

## 2.3 Sublimation

A pressure/temperature diagram of water and a salt such as aluminium sulphate is shown in Figure 3. The dotted lines define the equilibrium between the three states of pure water. When salt is added to the system the freezing point is lowered, as was shown in Figure 1, and the vapour pressure of the water is diminished. At saturation an invariant point Q is defined where four phases can co-exist in equilibrium; ice, anhydrous salt, saturated salt solution, and water vapour. At pressures below this point, which is sometimes

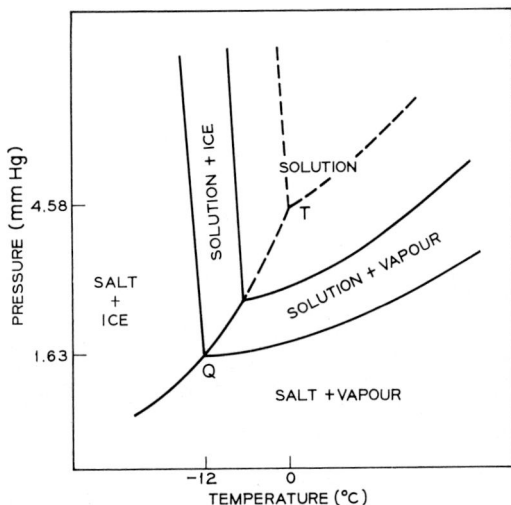

FIGURE 3.—Temperature/pressure relations in a salt–water system.

called the cryohydric point, the water can be removed by sublimation until only the anhydrous salt remains. Freeze dryers, as the equipment designed for this operation are sometimes called, are available in various sizes from bench-top laboratory models up to large-scale food processors that dry several tons per load. A freeze dryer is shown schematically in Figure 4. It consists of a chamber and a means to evacuate it. Within the chamber is a product shelf that can be either chilled to receive the frozen ice–salt mixture or,

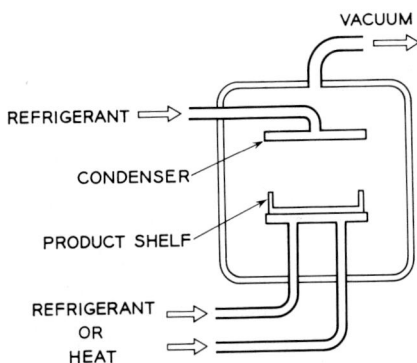

FIGURE 4.—Schematic diagram of a freeze dryer.

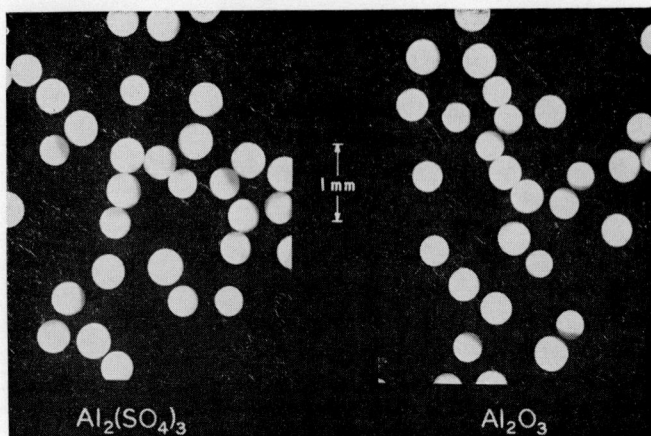

FIGURE 5.—Left: $Al_2(SO_4)_3$ prepared by cryochemical method. Right: $Al_2O_3$ prepared from sulphate by calcining for 10 h at 1000°C.

after the chamber is evacuated, heated to supply the necessary heat of sublimation. A means for recondensing the water vapour after its removal from the product is also included. An example of aluminium sulphate after sublimation is shown in Figure 5. It was prepared from a $10^w/_o$ solution. The size and uniformity of the spheres, as already noted, depends on the freezing technique. In this example they are about 0·4 mm in diameter. The spheres have sufficient strength to be transferred to a furnace for thermal decomposition.

## 2.4 Decomposition and Binder Addition

After calcination the spherical granules are ideal for handling because they are free-flowing and dust-free. As is common with freeze-dried materials, the spheres are completely permeated with fine channels through which the water vapour escaped. These same channels can be impregnated with a temporary organic plasticizer and binder, making the spheres suitable for dry pressing. To impregnate the spheres they are simply immersed in a solution of the organic material, drained and dried. While they have sufficient strength to be handled without disintegration, the spheres at the same time are quite friable and easily converted by the application of slight compressive or shearing forces to fine powders suitable for other processes. Thus ceramics can also easily be formed by extrusion, casting and other conventional ceramic manufacturing methods. Aluminium, oxide as obtained from the sulphate salt after calcining for 10 hours at 1000°C is shown in Figure 5.

## 3. RESULTS AND DISCUSSIONS

### 3.1 Thermal Analysis of MgSO₄

We have used differential thermal analysis as a tool for investigating the decomposition of salts prepared by the cryochemical technique described. In Figure 6 an analysis of freeze-dried magnesium sulphate is compared with that of the hydrated crystals from which it was prepared. The series of endothermic peaks corresponding to the

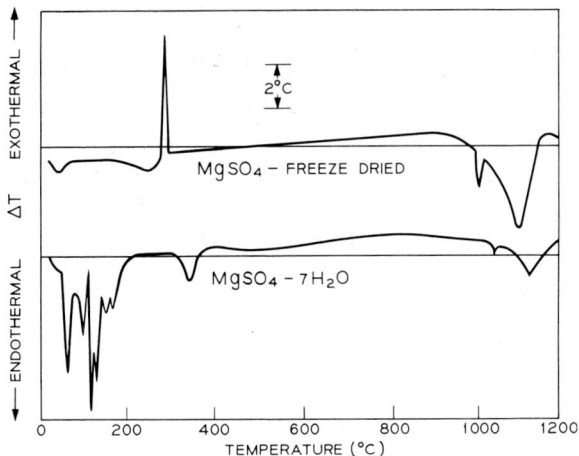

FIGURE 6.—D.t.a. of magnesium sulphate.

removal of the water of crystallization in the hydrated salt are not found in its freeze-dried counterpart. Such water as is present in the latter is either held by absorption or in a zeolite-type solution. Anhydrous magnesium sulphate undergoes a structural transformation about 350°C.[2] An endothermic peak at a corresponding temperature is found in the hydrated salt. The freeze-dried salt, on the other hand, has an exothermic peak at a somewhat lower temperature. To check this, two samples were prepared from the freeze-dried salt by heating for 2 h: one below the peak, at 200°C; the other above it, at 400°C. An X-ray diffraction pattern from the 200°C sample revealed only a few broad lines, which were at the expected positions for reflections from the low-temperature modification of anhydrous magnesium sulphate.[3] The 400°C sample gave a sharply defined pattern of the high-temperature modification. Apparently the heat generated as the near-amorphous material crystallized into the high-temperature modification was greater than that required by the transformation. The decomposition of the sulphate to the oxide starts about 900°C. Again, the temperature is somewhat lower for the freeze-dried material. We cannot account for the second peak found in both salts in this region. Perhaps some melting precedes the decomposition.

## 3.2 Formation of Spinel from Magnesium Aluminium Sulphate

A solution was prepared of magnesium and aluminium sulphates, the proportion of the metal ions corresponding to that of spinel. After processing the salt the decomposition was followed by d.t.a. as shown in Figure 7. For comparison, thermograms of the pure salts of magnesium and of aluminium sulphate prepared by freeze drying are included in the figure. We do not know at present if the salt mixture is in the form of a solid solution or of an intimate mixture of the amorphous salts. In either case the decomposition occurs at a temperature lower than that of either of the parent salts. A physical mixture of the same proportions was also made from the component freeze-dried sulphates of magnesium and aluminium. Its thermogram was a superposition of the thermograms of the individual salts, with no evidence of interaction.

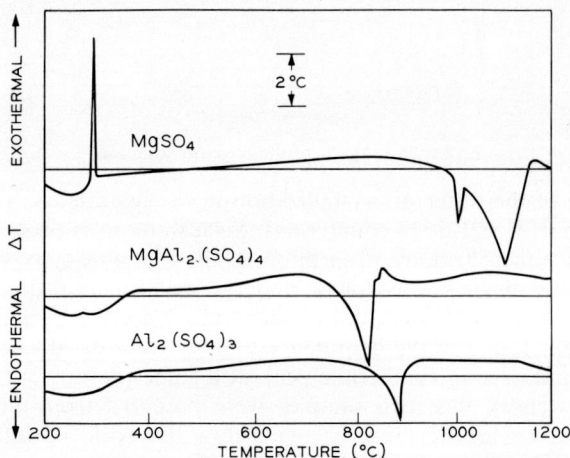

FIGURE 7.—D.t.a. of sulphates prepared by cryochemical method.

The d.t.a. of the freeze-dried magnesium aluminium sulphate has been reproduced as curve (a) in Figure 8. On the same temperature scale the results of a thermogravimetric analysis of this material is shown in curve (b). The weight scale has been normalized at the

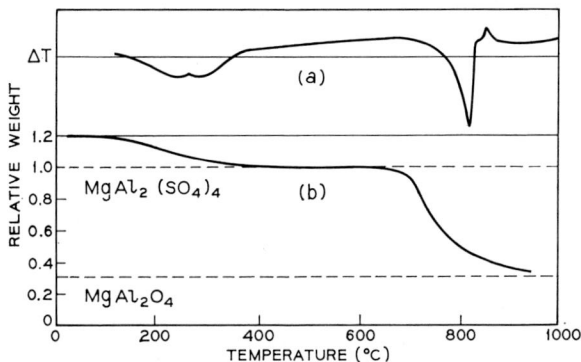

FIGURE 8.—Thermal decomposition of $MgAl_2(SO_4)_4$
Curve (a). Differential thermal analysis
Curve (b). Thermogravimetric analysis.

amorphous salt. This analysis was made at a heating rate of 3 C°min. It indicates complete conversion to the oxide at 1000°C. Actually, an X-ray diffraction pattern of a sample held for 1 h at 870°C contained the lines of the spinel structure only. The initial weight change of about 20% may be attributed to water which may have been either residual or absorbed after drying; no special precaution having been taken to avoid the latter.

### 3.3 Particle Size Analysis of $Al_2O_3$

Two series of aluminium oxides were prepared by calcining two different freeze-dried aluminium sulphates at various temperatures. One of the sulphates was made from a solution containing the equivalent of 2·6 g of the anhydrous salt per 100 ml of solution; the other contained 26·0 g/100 ml. The surface area of the many samples was determined by nitrogen gas adsorption using the B.E.T. technique. A value of 16·2 Å² was used as the area covered by a nitrogen molecule at its normal boiling point, 77·4°K. The results are shown in Figure 9 as a function of the temperature at which the sample was held for 2 h. A scale of equivalent diameters calculated from the measured surface area and the known density of aluminium oxide is also given. Here we have a demonstration of the "Hedvall Effect."[4] X-ray analysis of the sample calcined at 1000°C revealed the structure to be that of $\gamma$-$Al_2O_3$. The diffraction lines were broad, as would be expected of 100 Å particles. After a 1200°C treatment the structure was found to be fully converted to $\alpha$-$Al_2O_3$. Although a thorough kinetic study has not yet been made, it is apparent that there is a marked increase in the rate of particle growth during the period of the structure transformation. In materials such as $Fe_2O_3$ in which there is no structure transformation, the change of surface area with temperature is much more gradual. The change of surface area with time at temperature and solution concentration was also found to be gradual if no structural transformation was encountered. An electron photomicrograph and electron diffraction pattern of each of two aluminium oxide samples are shown in Figure 10. The sample on the left was calcined for 2 h at 1000°C. The particles are

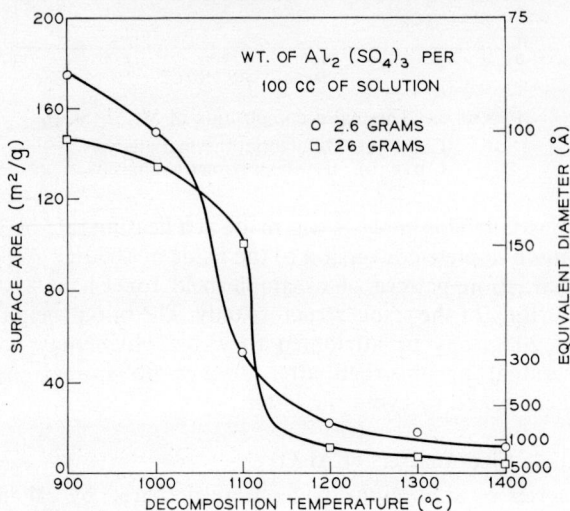

FIGURE 9.—Variation of surface area with calcination temperature for two aluminium oxides.

hardly resolved at a magnification of × 40,000. The sharp rings of the diffraction pattern indicate a particle size less than 200Å, which is in agreement with the surface area measurements. The sample shown on the right was treated for 2 h at 1200°C. The particles, though of uniform size, appear to be fused into chain-like clusters, which can probably be attributed again to the increased activity during transformation. Iron oxide with a similar treatment appeared as uniform but discrete particles. The diffraction pattern is spotty, as would be expected from particles a thousand Å or more in diameter.

## 4. SUMMARY

The cryochemical method of preparing ceramic materials has many novel features of interest to the research worker. It is both

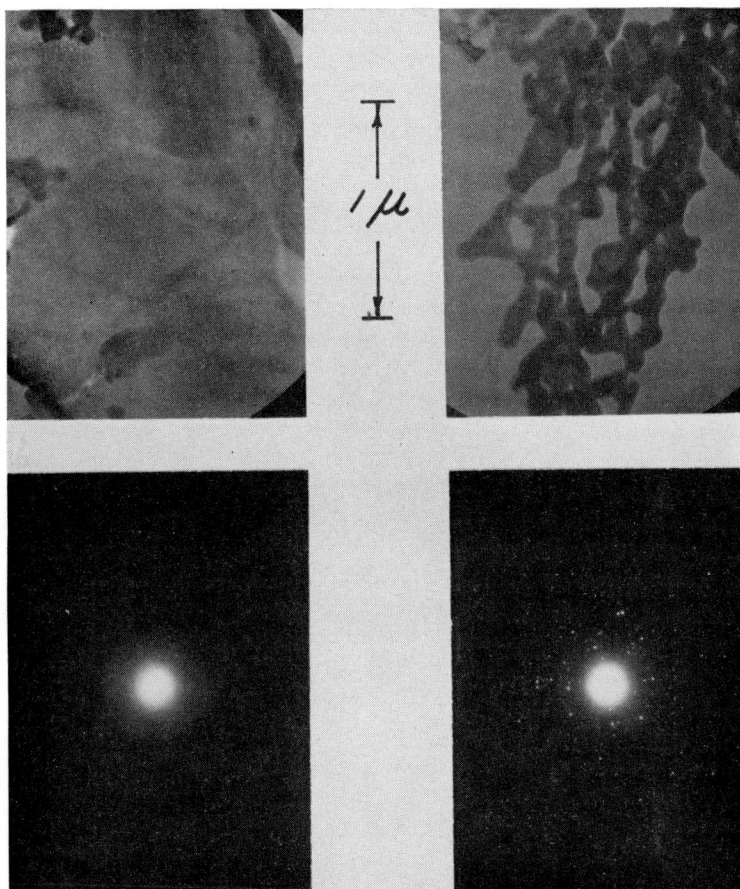

FIGURE 10.—Electron photomicrographs and electron diffraction patterns of the two aluminium oxides. Left: Calcined 2 h at 1000°C. Right: Calcined 2 h at 1200°C.

simple and clean. Compositions can be precisely formulated from high-purity reagents without danger of contamination. Minor components can be added with the assurance of a uniform distribution. The powder particles prepared cryochemically are uniform in size and homogeneous in composition. Using the freezing method described, the powders are naturally agglomerated as near perfect spheres. This too is of advantage. Spheres are free-flowing and easy to handle. As they are dust-free, there is little or no hazard toward health or danger of contaminating other materials.

90     SCHNETTLER, MONFORTE AND RHODES:

## ACKNOWLEDGMENTS

The authors extend their gratitude to their colleagues Dr P. K. Gallagher, Dr H. M. O'Bryan, and Dr M. D. Rigterink for their help and valuable discussions.

## REFERENCES

1. SMITH, N. O., and WALSH, P. N., *J. Amer. Chem. Soc.*, **76**, 2054, 1954.
2. COING-BOYAT, J., *Compt. Rend.*, **255**, 1962, 1962.
3. RENTZEPERIS, P. J., and SOLDATOS, C. T., *Acta Cryst.*, **11**, 1958.
4. HEDVALL, J. A., "Reactivity of Solids" (Verlag: Johann Ambrosius Barth, Leipzig, 1938).

# 7.—Application of Isostatic Pressing Techniques to the Production of Dense Ceramic Bodies

R. M. GILL and J. BYRNE

*Morganite Research and Development Limited*

## ABSTRACT

C322/E13

*A 10-in. and a 30-in. diameter isostatic press and tooling for the manufacture of simple specimens and complicated finished pieces are described. The merits of basic types of tooling such as "wet bag" and "dry bag" are considered. The manufacture of bags in the laboratory by dipping or casting methods and the advantages of natural rubber and synthetic materials are discussed. Curves illustrate the variation of density with forming pressure for alumina materials and a markedly anisotropic material containing graphite flake. Characteristics of green and of fired materials show a decrease in the rate of change of density as the pressure increases. A plot of the density distribution field for an isostatically pressed specimen is compared with that for a similar size of die pressed piece. The effect of bag movement on the structure of the pressed body is illustrated. Techniques for preparing material for isostatic pressing and the selection of binders are considered.*

### Utilisation du pressage isostatique pour la fabrication de produits céramiques denses

*Une presse isostatique de 250 mm et de 750 mm de diamètre, de même que l'outillage pour la fabrication d'ébauches simples et de pièces finies compliquées sont décrits. Les avantages des systèmes à "moule humide" et à "moule sec" sont indiqués. L'élaboration des enveloppes au laboratoire à l'aide de méthodes par immersion et par coulage est décrite et les avantages du caoutchouc naturel et des matériaux synthétiques sont discutés. Des courbes illustrent la variation de la densité en fonction de la pression appliquée pour des compositions à base d'alumine et pour une matière fortement anisotrope contenant des paillettes de graphite. On constate aussi bien pour les matériaux crus que pour ceux ayant été cuits une diminution de la vitesse de variation de la densité en fonction de l'augmentation de la pression. Un diagramme de répartition de la densité, fourni par un échantillon pressé isostatiquement, est comparé à celui obtenu avec une pièce de taille analogue, produite par pressage traditionnel. L'effet du déplacement de l'enveloppe sur la structure de la composition pressée est illustrée. Les techniques de préparation de la matière destinée au pressage isostatique et le choix des liants sont étudiés.*

**Anwendung der Technik des isostatischen Pressens zur Herstellung dichter keramischer Körper**

*Zwei isostatische Pressen mit zugehörigem Gerät, zur Herstellung einfacher Formkörper und komplizierterer Stücke, werden beschrieben. Die Vorteile der verschiedenen Hüllmaterialien werden erörtert. Die Herstellung der Hüllen im Labor durch Tauchen oder Gießen und die Verwendung von natürlichem Gummi und synthetischen Materialien werden erörtert. Einige Kurven zeigen die Änderung der Dichte mit dem Formgebungsdruck für Al₂O₃-Materialien und für einen anisotropen Stoff mit Graphitflocken. Die Messungen von Grünlingen und gebranntem Material zeigen eine Abnahme in der Dichteänderung mit zunehmendem Formgebungsdruck. Das Feld der Dichteverteilung für eine isostatisch und eine normal gepreßte Probe gleicher Abmessung wird verglichen. Die Technik der Herstellung von geeigneten Pulvern wie die Auswahl des Bindemittels wird erörtert.*

## 1. INTRODUCTION

Morganite Research and Development Limited (M.R.D.) has been interested in the fabrication of a wide range of ceramic products and materials for many years. Recently it has devoted attention to evaluating the possibilities of isostatic pressing techniques as a means of producing dense and uniform ceramic bodies. This paper describes some of this work, which has been associated mainly with a study of the mechanism of compaction by isostatic pressing.

## 2. EQUIPMENT FOR ISOSTATIC PRESSING

The main features of an isostatic press are the pressure vessel and the pumping system. The presses used by M.R.D. are:

(1) A proprietary piece of equipment having a pressure chamber with a working space 10 in. diameter × 18 in. deep for use of up to 20,000 p.s.i.

(2) Another larger piece of proprietary equipment with a pressure vessel having a working space 30 in. diameter × 36 in. deep and capable of reaching a maximum working pressure of 10,000 p.s.i. An additional feature is the provision of external heaters capable of raising the working chamber to a temperature of 100°C.

Figure 1 shows a top view of the 30-in. unit with the plug in the open position. It illustrates clearly the massive nature of the plug and the breach block type, interrupted thread, which facilitates opening and closing. This vessel, which weighs some 6 tons, is heated by electrical strip heaters of approximately 50 kW electrical loading to provide working temperatures up to 100°C. Maximum working pressure is

FIGURE 1.—Top view of 30-in. isostatic pressing unit.

10,000 p.s.i., thus any pressure up to this value can be maintained by a sensitive control valve. Oil is used as the hydraulic fluid.

## 3. TOOLING REQUIREMENTS

Figure 2 shows the basic tooling required for pressing a rod. It consists of a bag containing the material to be compacted and a perforated metal tube which surrounds the bag and supports it during the filling operation. The whole assembly is immersed in the oil and compaction takes place evenly over the whole area of the bag.[1]

Figure 3 shows two simple arrangements for making a hollow article. On the left is "wet-bag" tooling in which the material to be compacted is contained inside a flexible bag or sack (usually made of rubber), and a former or mould shape is inserted into the powder,

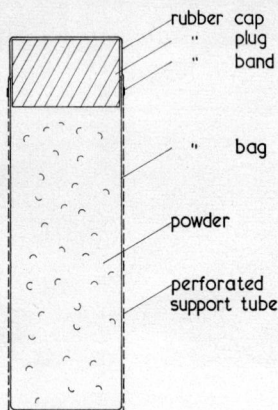

FIGURE 2.—Basic tooling required for the isostatic pressing of a rod or cylinder.

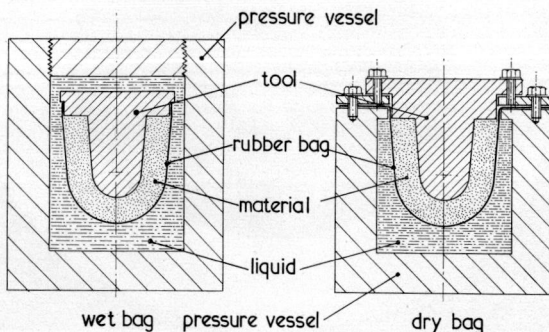

FIGURE 3.—Arrangement of wet- and dry-bag tooling for isostatic pressing.

the whole being sealed at the top.[2-4] The assembly is then inserted into the pressure vessel, and the hydraulic pressure of the fluid compresses the bag plus powder in an even manner. In the "dry-bag" method, shown on the right, the inner surface of the bag in contact with the powder is always dry and loading/unloading of the tool can be accomplished without contact with the fluid. Dry-bag methods allow greater ease of handling.

Figure 4 shows an alternative wet-bag tooling arrangement in which the bag is expanded by the hydraulic fluid in order to produce a uniform-walled article, such as a tube or sheath.

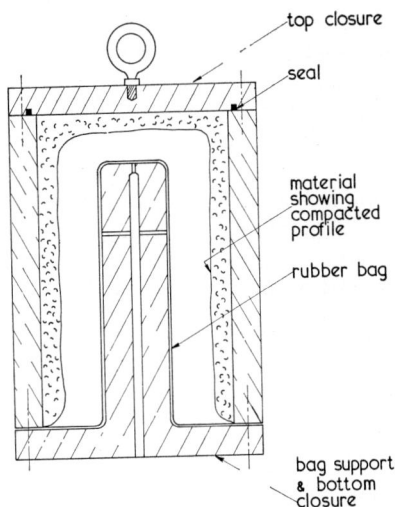

FIGURE 4.—Wet-bag tooling, in which the bag is expanded by the hydraulic fluid.

## 4. COMPACTION OF POWDERS BY ISOSTATIC PRESSING

It is necessary to establish certain fundamental properties of the materials being pressed. The compression ratio required is of considerable importance; this is the ratio of a typical dimension of the tool in the loosely filled condition to the same dimension after compaction. The importance lies not only in the nature of the product itself, but in the way in which the flexible member of the tool (flexible bag) is called upon to perform. Compression ratios are typically between 2:1 and $2\frac{1}{2}$:1, and when values become very much greater than $2\frac{1}{2}$:1, precautions have to be taken in the design of the tool to avoid excessive strain or movement on the bag. The strain is related to the wall-thickness/piece-diameter ratio.

The uniformity of filling of the tool can be very important since it affects the final dimensions and density distribution of the product. Isostatic pressing cannot rectify a badly filled tool and considerable care is required to achieve the best results. Clearly, the higher the packing density of the initial charge, the lower the compression ratio. Vibration filling can assist in achieving a high initial density, but there is a danger of segregation of the powder. Often a pre-form is made by some other technique and then isostatically pressed to give increased density. Again one can densify the raw material by making a slug by isostatic pressing followed by grinding to powder and repressing.

The material of the bag must be compatible with the fluid used in the pressure vessel and with the material being compacted. For most powders, however, no problem arises and selection depends on the nature of the hydraulic fluid. For use with water, or soluble oils in water, simple rubber compounds are adequte. Rubber is very satis-factory, since it has excellent elastic properties which offer the tool designer maximum freedom. Where oil is used as the pressure fluid, a neoprene bag gives good results, but it is not as flexible as ordinary rubber latex. In some cases a composite bag consisting of rubber coated with neoprene may give better results. Bag manufacture may be achieved by:

*1. Dipping*

Useful for making relatively thin-walled bags for experimental purposes.

*2. Casting*

Using a metal mould for thicker sections.

*3. Fabricating*

A sheet of rubber is cut and welded to shape. Can be used for thick sections; alternatively, a solid block of rubber may be machined.

## 5. EXPERIMENTAL RESULTS FOR ISOSTATICALLY PRESSED CERAMIC MATERIALS

### 5.1 Effect of Pressure on Compacted Density

Using a relatively simple tool of the type shown in Figure 3 we have examined the behaviour of a number of materials of interest to our Group. The materials can be classified broadly into:

(1) Single-component bodies such as alumina and other oxides containing a small proportion of an organic binder to impart strength to the pressed compact.

(2) A multi-component system in which one or more types of hard particle are distributed through a softer matrix. Typically, this might be a system having silicon carbide and other hard materials distributed through a tar or pitch matrix.

For materials of type (1) and (2), the pressure/density relationship of the type classically derived for simple pressing was obtained.

Figure 5 illustrates a pressure/density curve obtained for a commercial alumina body in the green state. This alumina is 97·5% pure and 4% paraffin wax binder was added. The compacted density increases continuously up to the limit of the pressure used (20,000 p.s.i.), but the rate of density increase falls off fairly rapidly at pressures over 5,000–6,000 p.s.i. For comparison, the pressure/density relationship for conventional pressing is included; the curve is basically similar.

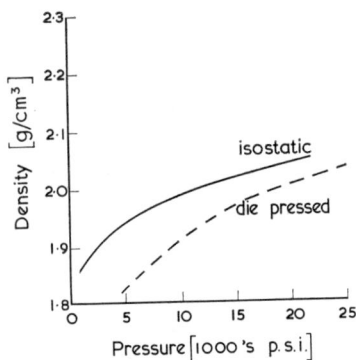

FIGURE 5.—Pressure/density curve for a commercial alumina body for die pressing and isostatic pressing.

Figure 6 indicates the above relationships for alumina which was fired to 1,650°C after isostatic pressing: the fired density increases with forming pressure, but the curve flattens out more markedly than for the green state. Thus an optimum fired density of 3·85 g/cm$^3$ is reached at a green forming pressure in the region of 6,000–7,000 p.s.i. Clearly, therefore, this pressure is one which is of interest to the ceramic manufacturer, since maximum fired product density is obtained; little is gained by using higher pressures. A curve for purer $Al_2O_3$ (99·3%) indicates the same pattern but lower densities were obtained, due to reduced rates of sintering.

Figure 7 shows the density/pressure curve for a multi-component system having hard particles embedded in a soft matrix. The curve is basically similar to that in Figure 5, but the fall off in density with increased pressure is more marked than with the alumina and the fall-off occurs at very low pressures (1,200 p.s.i.), which is due to the improved flow characteristics of the soft matrix. The curve for a normally pressed specimen indicates a similar behaviour. The equivalent curve examined after the specimen has been fired shows that the

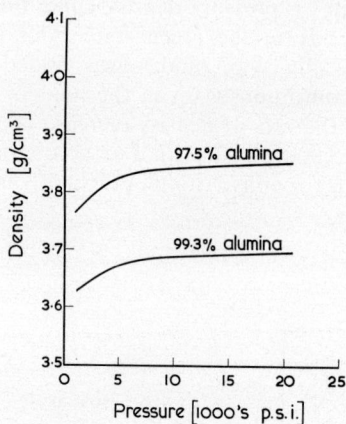

FIGURE 6.—Pressure/density curves for alumina of two purity levels formed by isostatic pressing, after firing to 1650°C.

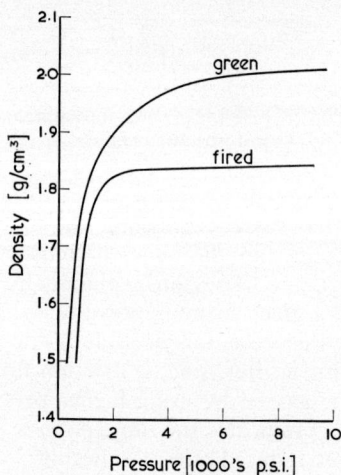

FIGURE 7.—Pressure/density curve for a multi-component system after isostatic pressing and subsequent firing.

critical pressure is displaced towards the left of the curve, and that pressing-pressures over 2,000 p.s.i. do not result in increase of fired density.

To summarize Figures 5, 6 and 7, the density/pressure relationships are basically similar to those obtained for normal pressing dies. Therefore why use isostatic pressing?

The interest in isostatic pressing lies not in the similarities to simple pressing, but in the differences, which are most manifest in:

(1) Density distribution in a compacted piece.
(2) Manner in which the material moves during compaction.

## 5.2  Density Distribution of Isostatically Pressed Bodies

Figure 8 shows the density distribution in an isostatically pressed cylinder together with pressure contours in a metal powder cylindrical specimen pressed from one end in a die.[5] The figures on the diagram indicate the density distribution levels. The isostatically pressed specimen is much better around the edges, and across the top and bottom where wall and die friction are most significant in the simple pressed specimen. Overall, the isostatically pressed piece is much more uniform. This difference, or improvement with isostatic pressing, is even more marked if we consider the forming of a more

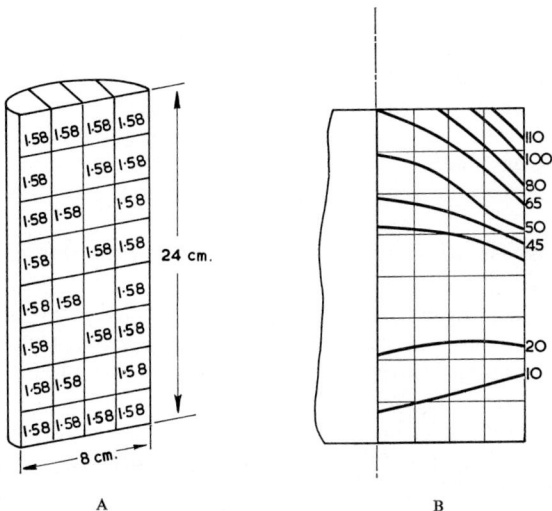

FIGURE 8

A. Density distribution in an isostatically pressed cylinder.
B. Pressure contours in metal powder specimen after die-pressing.

complicated hollow shape. Figure 9 indicates the density distribution in a closed-ended piece. The material in this case is of a type which has almost no inherent plasticity and therefore does not flow easily in the die. The distribution obtained in the simply pressed specimen reflects this; the compaction obtained in the walls is, of course, very poor indeed, whereas the material at the bottom of the die (normal to die movement) is compacted to a high degree. We get a dense bottom and weak porous sides. However, for the isostatically pressed piece the density distribution is very nearly uniform throughout the wall— all the way down the sides and across the bottom. Incidentally, the specimen shown here was formed by collapsing a rubber bag on to a centre former.

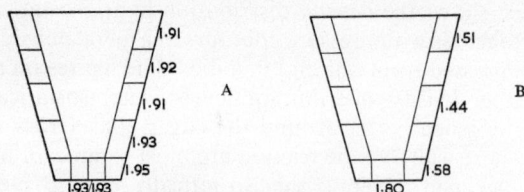

FIGURE 9.—Density distribution (g/cm³) of a closed-ended component: A. After isostatic pressing. B. After die pressing.

For ceramic fabrication an important advantage of isostatic pressing is that a high degree of density uniformity is achieved which is reflected in the minimum of distortion during firing. This is of great value for fabricating articles in high-shrinkage bodies such as alumina. Figure 10 shows a thin-walled alumina tube made by isostatically pressing, before and after firing; the distortion has been reduced to negligible proportions and no cracking was experienced.

### 5.3 Mechanism of Compaction and Powder Flow during Isostatic Pressing

With a dilating-bag tool, the surface formed next to the bag is expanding, and in certain materials may lead to cracking. In general the material distorts by moving along slip planes when the yield shear stress is exceeded and then reconstitutes to give a continuous matrix when the stress falls below the critical value. If reconstitution does not take place, discontinuities remain and may appear as cracks.

The movement required of a material at a gross change of section is considerable even when we are considering a fairly non-plastic mix. Figure 11 shows a closed-ended cylinder in which traces of white

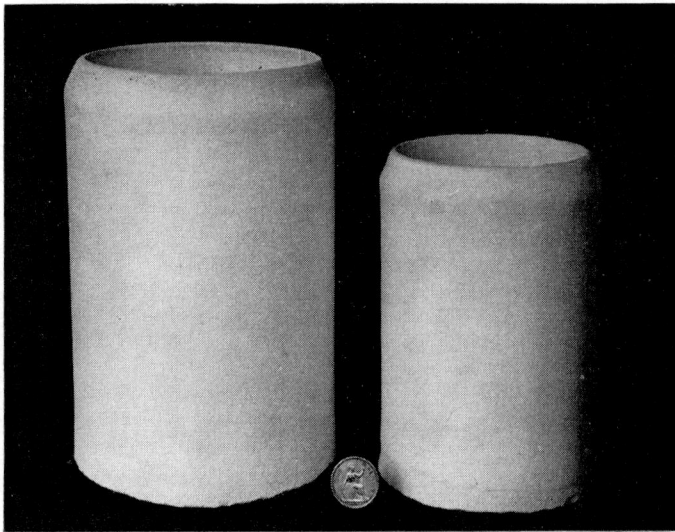

FIGURE 10.—Thin-walled alumina tube made by isostatically pressing: before and after firing.

material have been inserted into black. The left-hand side indicates the trace on the uncompressed piece while the right-hand side shows the situation after compression. The material is required to flow first sideways and then partially upwards into the wall where the distortion is low. There is still some movement of material normal to the line of action of the compressive force.

## 5.4 Filling of Tool Prior to Pressing

Generally, of course, it is desirable to fill the tool homogeneously; all parts should be filled with material representative of the thoroughly mixed batch, and the packing density should be as uniform as possible throughout the tool.[6] In practice this is achieved by pre-mixing, by preparing the material in such a way that a free-flowing powder is obtained, and by assisting filling by means of jolting or vibration of the tool. Spray-dried powder is ideal for isostatic pressing, particularly since a temporary binder can be incorporated.

Alternatively, for fairly small-scale activity, as in laboratory work, the material may be granulated by starting with a paste-like consistency and mixing in a heated sigma-blade mixer while moisture is slowly driven off. Where the nature of the material is such that it is difficult to get the required weight of charge into the tool even with vibration-assisted free-flowing powder, it may be necessary to

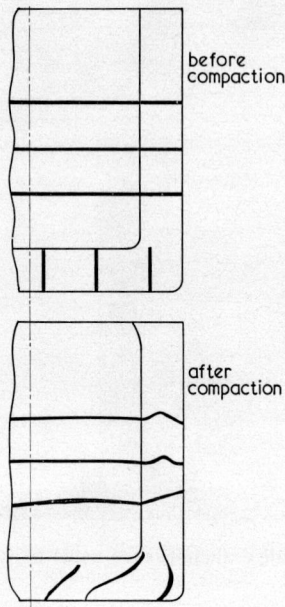

FIGURE 11.—Movement of material during isostatic pressing.

pre-press a block of material isostatically and then to crush and granulate the block before loading into the tool.

## 5.5 Binders

Binders can play an important part in any pressing operation and isostatic pressing is no exception. The function of a binder is to provide good green strength for the compacted piece so that handling, subsequate machining, etc., can be carried out. A binder should also provide some lubrication of the particles to facilitate flow and movement during compaction, and this can be achieved by additions to the main binder. A binder should be easy to burn off during firing and should leave a negligible residue. Examples of binders having desirable properties are:

Polyethylene glycol, di-ethylene glycol monostearate, paraffin wax, gum arabic, polyvinyl alcohol.

Stearate compounds added to spray-dried or granulated powder which already contains binder in the granules provides an effective way of lubricating the particles to give better flow.

## 6. CONCLUSIONS

Isostatic pressing is a relatively new technique for ceramics, but it offers a number of advantages over conventional pressing equipment, namely:

(1) Greater uniformity of density of the compacted material.
(2) Less likelihood of lamination and cracking.
(3) Certain shapes can be produced easily which would be difficult with simple pressing techniques.
(4) Size is limited only by the equipment available, rather than by the nature of the material being pressed.

Isostatic pressing requires good tool design in order to ensure uniformity of the product and the material characteristics have to be taken into account in arriving at the optimum design. It is important to ensure uniformity of filling of the tool, otherwise a non-uniform wall thickness will occur. Pressing-pressures required depend to a considerable extent on the type of powder being compacted; the easier the material flows and moves under pressure, the lower the actual forming pressure required. In practice, values between 2,000 and 20,000 p.s.i. are used.

Today isostatic pressing has become accepted as a viable production method for fabricating a wide range of ceramic articles, largely owing to developments in rapid cycling and automatic control making fast rates of production possible.[7] As a result many ceramic manufacturers now use isostatic pressing for routine production, and the technique has been extended to incluce products of low value per unit volume.

### REFERENCES

There is a good deal of literature concerned directly with isostatic pressing or dealing with particular processes related to some aspect of isostatic pressing. A comprehensive list of such references is to be found in:

*Berichte der Deutschen Keramischen Gesellschaft*
"Das Isostatische Pressverfahren in der Keramik"
Volume 44, 1967, No. 3, page 97.

Publications relating directly to the topics covered in this paper are:

1. KINGERY, W. D., "Ceramic Fabrication Methods" (J. Wiley & Sons, 1958).
2. JACKSON, H. C., "Hydrostatic Isostatic Pressing", *Precision Metal Molding*, 1963.
3. LOOMIS, D. G., "Isostatic Pressing for Ceramics", *Ceram. Age*, July 1962.

GILL AND BYRNE

4. HARMON, C. G., "Hydrostatic Pressing as a Fabrication Technique", *Bull. Amer. Ceram. Soc.*, **30**, 341, 1951.
5. DUWEZ, P., and ZWELL, L., *A.I.M.E. Techn. Publ. 2515, Metals Trans.*, No. 1, 137, 1949.
6. CATCHPOLE, C., "Isostatic Pressing", *J. Brit. Ceram. Soc.*, **1**, 385, 1964.
7. MULLER, C., "Isostatisches Pressen", *Sprechsaal*, **99**, 467, 1966.

**ACKNOWLEDGMENTS**

The authors wish to thank the directors of Morganite Research and Development Ltd for their permission to publish this paper.

# 8.—Stabilized Zirconia Particles by Sol–Gel Processes

## By J. L. WOODHEAD

*Chemistry Division, Atomic Energy Research Establishment, Harwell*

*ABSTRACT*                                                      E181c

*The general principles of a sol–gel process for the preparation of particles are: (1) the conversion of a metal salt or oxide to a concentrated sol, (2) dehydration and/or de-ionization of the sol to a gel, and (3) calcination and sintering of the gel to oxide. Recently the versatility of this process has been exploited chiefly for the production of experimental quantities of $ThO_2$, $UO_2$, $PuO_2$, $ThC_2$, $UC_2$, either alone or as mixtures. This sol–gel process has now been examined for the preparation of particles for non-nuclear ceramics, and a typical example of the preparation of particles of zirconia is described. Zirconia sols were prepared and converted to microspheres (20–100 μm) by gelation in an organic phase. The compatibility of zirconia sols with calcium and yttrium salts (calcia and yttria as stabilizers) was established over a range of concentrations and such sols were converted to gels by evaporation. When calcined, the products densified rapidly, and stabilized zirconia of 95% theoretical density was prepared at 1150°C.*

### *Préparation de particules céramiques suivant le procédé "sol-gel"*

*Les principes généraux d'un procédé sol-gel destiné à la préparation de particules sont les suivants: (1) transformation d'un sel ou d'un oxyde métallique en un sol concentré; (2) déshydratation et/ou désionisation du sol en gel et (3) calcination et frittage du gel en oxyde. L'universalité de ce procédé a été utilisée récemment surtout pour la production de quantités expérimentales de $ThO_2$, $UO_2$, $ThC_2$, $UC_2$, ces composés étant préparés soit seuls, soit sous forme de mélanges. Ce procédé "sol-gel" vient d'être étudié en vue de la préparation de particules pour céramiques non nucléaires et un exemple de la préparation de particules de zircone est décrit. Des sols de zircone ont été préparés et transformés en microsphères (20–100 μm) par gélatinisation dans une phase organique. La compatibilité des sols de zircone avec des sels de calcium et d'yttrium (l'oxyde de calcium et l'oxyde d'yttrium jouant le rôle de stabilisateurs) a été établie pour un domaine étendu de concentrations et les sols ainsi obtenus ont été transformés en* *gels par évaporation. Soumis à la calcination, les produits se densifient rapidement et à 1150°C on prépare une zircone stabilisée d'une densité égale à 95% de la densité théorique.*

### Keramische Partikel durch Sol-Gel-Prozesse

*Die allgemeinen Prinzipien eines Sol-Gel-Prozesses für die Anfertigung von Teilchen sind: 1. die Umwandlung von Metallsalz oder Oxiden in ein konzentriertes Sol, 2. Dehydratation und/oder Deionisation des Sols zu einem Gel und 3. Kalzinierung und Sinterung des Gels zum Oxid. Kürzlich wurde die Verwendbarkeit dieses Prozesses erforscht, hauptsächlich für die Herstellung experimenteller Mengen von $ThO_2$, $UO_2$, $PuO_2$, $ThC_2$, $UC_2$, entweder allein oder als Mischungen. Dieser Sol-Gel-Prozess wurde nun für die Herstellung "nichtnuklearer" Keramik untersucht. Es wird als typisches Beispiel die Herstellung von Zirkonoxidteilchen beschrieben. Zirkonoxidsole wurden angefertigt und durch gelieren in einer organischen Phase in Mikrokugeln (20–100 μm) umgewandelt. Die Verträglichkeit der Zirkonoxidsole mit Kalzium- und Yttriumsalzen (Kalziumoxid und Yttriumoxid als Stabilisatoren) wurde in einem Bereich verschiedener Konzentrationen bewiesen, und solche Sole wurden durch Vakuumtrocknung in Gele umgewandelt. Die kalzinierten Produkte verdichteten rasch, und stabilisiertes Zirkonoxid von 95% der theoretischen Dichte wurde bei 1150°C hergestellt.*

## 1. INTRODUCTION

The fabrication of nuclear ceramic particles by sol–gel processes has received considerable attention recently. In the U.S.A., processes have been developed for preparing particles of $ThO_2$, $ThO_2$–$UO_2$, $UO_2$–$ZrO_2$, $UO_2$, $PuO_2$ and $ThC_2$ and $ThC_2UC_2$.[1,2,3] Other processes involving different sol-forming techniques have been reported from Italy[4,5] and from Holland.[6] A recent British report[7] describes the preparation of $UO_2$ particles.

The basic concept[8] which is common to these contemporary sol–gel processes is that the precursor to the oxide or carbide product is a concentrated hydrous sol which on dehydration forms a glassy gel. Calcination of the gel under the appropriate conditions yields the oxide or carbide. Significant advantages of such processes include:

(1) Binary or ternary systems can be produced with excellent homogeneity by the mixing of sols.

(2) The gels can be fired to yield oxides with densities close to the theoretical crystal value, sometimes at abnormally low temperatures, e.g. for $ThO_2$, $>99\%$ theoretical density can be achieved after firing at 1,150°C.

(3) The hydrous sols can be made into interesting forms, e.g. dehydration via organic solvents gives spherical particles with diameters in the range 25–1000 $\mu$m.

(4) Controlled porosity can be introduced into the final product by use of additives at the sol stage.

These advantages have been exploited for the production of nuclear ceramics: the control of such factors as particle shape, size and quality is no less important in non-nuclear ceramics.

This paper deals with some aspects of our work on a sol–gel process for the production of stabilized zirconia, a ceramic oxide of relatively high cost.

## 2. RESULTS AND DISCUSSION

### 2.1 Densification of Sol–Gel Zirconia

Zirconia sol (3·5 M) was prepared by dispersing zirconium hydroxide in dilute acid. Aliquots of this sol were blended with calcium and with yttrium salt solutions in order to give a predetermined range of $Y_2O_3/ZrO_2$ and $CaO/ZrO_2$ concentrations. The quantities of calcia or yttria added covered the "sensitive" region for obtaining tetragonal or cubic phase zirconia in the calcined products. No coagulation occurred in the mixtures. These solutions were evaporated at 80° to give glassy gels which were then calcined in air up to 1,150°C. Samples were removed after successive intervals of time and examined for crystal form by X-ray diffraction, mercury density and surface area. The results are given in Table 1, and Figure 1 illustrates the glassy form of the calcined product compared with calcined zirconium hydroxide.

The pattern of densification and decreasing surface area show a marked similarity to those obtained for thoria[3] and for urania[2] made by sol–gel techniques. Preliminary experiments have been made on the compacting of the sol–gel zirconia and cylindrical test-pieces have been subjected to repeated firing and cooling without breakage.

### 2.2 Production of Spherical Particles of Stabilized Zirconia

The techniques developed recently in sol–gel processes for the formation of spherical ceramic particles for use as nuclear fuels from sols or from salt solutions include incorporation of internal gelling agents,[6] dehydration with organic solvents,[2] spraying[3] and gelation with organic amines.[3] In this work we have prepared spherical zirconia particles on a kilogram scale with a yield without recycle of $>95\%$ in the size range 37–75 $\mu$m in which gelation was effected by an anion-extracting organic amine. Figure 2 illustrates typical (ungraded) specimens of the spherical gel and calcined spheres.

**Table 1**

**Densification, Crystallinity and Surface Area for $ZrO_2$, $Y_2O_3/ZrO_2$ and $CaO/ZrO_2$ calcined to Different Temperatures**

| Sample No. | $Y_2O_3$ (w/o) | CaO (w/o) | Calcination temperature | | | | | | | | |
|---|---|---|---|---|---|---|---|---|---|---|---|
| | | | 600°C | | | 870°C | | | 1,150°C | | |
| | | | Density (Hg) | Crystal habit | S.A. ($m^2/g$) | Density (Hg) | Crystal habit | S.A. ($m^2/g$) | Density (Hg) | Crystal habit | S.A. ($m^2/g$) |
| 1. | — | — | 4·49 | M+T | 36 | 5·27 | M+T | 1·93 | — | — | — |
| 2. | 3·5 | — | 4·45 | — | 25·0 | 4·95 | — | 1·60 | 5·29 | M+T | 0·26 |
| 3. | 7·0 | — | 4·39 | — | 22·2 | 4·85 | — | 1·51 | 5·33 | T | 0·20 |
| 4. | 14·0 | — | 4·21 | — | 15·5 | 4·63 | — | 1·38 | 4·84 | C | 0·11 |
| 5. | — | 3·0 | 4·16 | T | 33·1 | 4·51 | T | 1·89 | — | — | — |
| 6. | — | 6·0 | 4·33 | T+C | 22·2 | 4·66 | T+C | 1·74 | — | — | — |
| 7. | — | 9·0 | 4·10 | T+C | 13·7 | 4·47 | T+C | 1·61 | — | — | — |

M = Monoclinic; T = Tetragonal; C = Cubic

A

B

FIGURE 1

A. $ZrO_2$ from $Zr(OH)_4$ calcined at 1150°C (×10)

B. $ZrO_2$ by sol–gel process (×10)

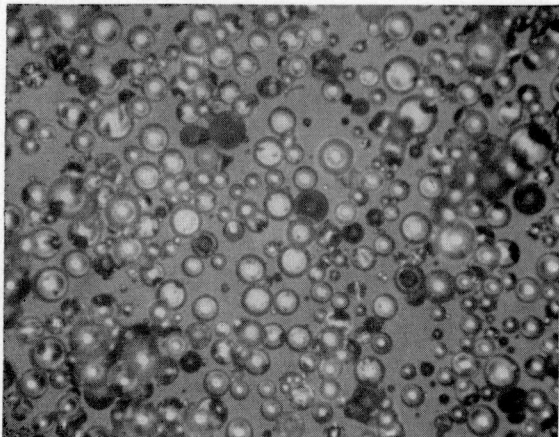

A

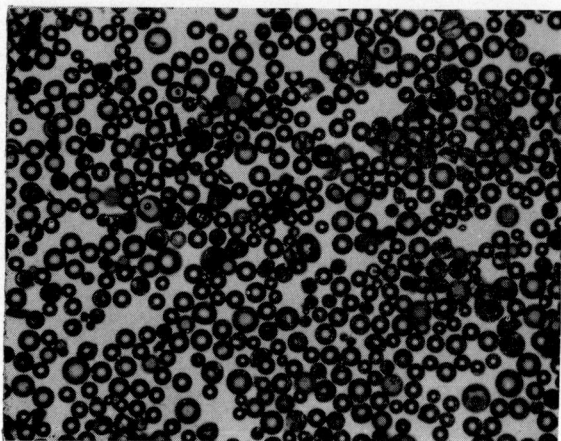

B
FIGURE 2
A. $5^w/_o$ CaO–ZrO$_2$ gel microspheres ($\times$ 100)
B. $5^w/_o$ CaO–ZrO$_2$ microspheres calcined at 600°C ($\times$ 100)

## 3. CONCLUSIONS

The investigations carried out to date indicate that the main advantages inherent in the contemporary sol–gel processes developed for nuclear ceramics apply to zirconia stabilized with calcia or with yttria. Such material could find useful applications, e.g. flame spraying and vibrocompaction of artefacts.

## ACKNOWLEDGMENTS

The author wishes to record his thanks to Mr A. C. Fox for X-ray diffraction measurements and Miss V. C. Baugh for the density determinations and other experimental assistance.

## REFERENCES

1. FERGUSON, D. E., DEAN, C. C., and DOUGLAS, O. A., Geneva Conference Paper A/CONF.28/P. 237 (1964).
2. WYMER, R. G., and COOBS, J. H., Proc. Brit. Ceram. Soc., No. 7, 61, 1967.
3. HAAS, P. A., CLINTON, S. D., and KLEINSTUBER, A. T., Can. J. Chem. Eng., p. 348 (1966).
4. COGLIATI, G., DE LEONE, R., GINDOTTI, G. R., LANZ, R., LORENZINI, L., MEZI, E., and SCIBONA, G. Geneva Conference Paper A/CONF.28/P.555 (1964).
5. COGLIATI, G., LANZ, R., and MEZI, E., CNEN, RT/CHI(65) 30 (1965).
6. VAN DER BRUGGHEN, F. W., HERMANS, M. E. A., KANIJ, J. B. W., NOOTHOUT, A. J., VAN DER PLAS, TH., and SLOOTEN, H. S. G., Nr Ch–E/PV–6/66, N. V. Kema, Arnhem, Nederlands.
7. LANE E. S., FLETCHER, J. M., HOLDOWAY, M. J., HYDE, K. R., LYON, C. E., and WOODHEAD, J. L., A.E.R.E. Document R.5241, 1966.
8. SZILARD, B., J. Chem. Phys., 5, 488, 1907; Compt. Rend., 143, 1145, 1906.

# 9.—The Capillary Action of Plaster Moulds

By P. H. DAL and W. J. H. BERDEN

*Laboratory N. V. Koninklijke Sphinx-Céramique,*
*Maastricht, Netherlands*

*ABSTRACT*                                              H174/D43

*The formation of a cast from casting-slip on a plaster surface
takes place under the influence of the capillary action of the pores
in the plaster mould. By using casting-slips from which, as is usual
in practice, a cast with a permeability of about $20 \times 10^{-14}$ cm² will
be formed, the capillary force of the pores is almost completely used for
the formation of the cast. Therefore the pressure at the plaster/cast
boundary is nearly equal to the total capillary force of the pores
per unit surface of the plaster/cast boundary. This pressure ($P_s$)
can be calculated from: $P_s = S.\sigma.\cos \varphi$ in which S is the specific
surface per unit bulk volume of the plaster, $\sigma$ is the surface tension
of water and $\varphi$ the contact angle (cos $\varphi = 1$ for complete wetting).
The specific surface and the pressure at the plaster/cast boundary
of some plaster samples were calculated from pore distribution and
from permeability data. The impression was established that the
pressures at the plaster/cast boundary (calculated from the pore-
distribution curves) agree better with the filtration pressure of about
1 atm. (required for the cast formation) than the lower values calculated
from permeability. Although some experimental results are not
completely satisfactory, it is nevertheless concluded that these results
confirm the filtration theory. It should be noted that pores $<0.1$ μm
are rare. Consequently our hypothesis (published several years ago)
that the formation of cast happens under the influence of very fine pores
with a hydraulic radius of 30 to 50 Å was incorrect.*

## L'effet de capillarité exercé par les moules de plâtre

*La formation du tesson à partir d'une barbotine de coulage au
contact d'une surface de plâtre est due à l'effet de capillarité des
pores du moule de plâtre. En utilisant des barbotines de coulage
fournissant un tesson avec une perméabilité de $20 \times 10^{14}$ cm², la
force capillaire totale exercée par les pores est presque entièrement
utilisée pour la formation du tesson. La pression à l'interface plâtre/
tesson est par conséquent presque égale à la force capillaire des pores*

*par unité de surface de l'interface plâtre/tesson et peut être calculée grâce à l'équation suivante: $P_s = S. \sigma. \cos \varphi$, où $S = $ surface spécifique par unité de volume du plâtre, $\sigma = $ tension superficielle de l'eau et $\varphi = $ angle de mouillage ($\cos \varphi = 1$ pour un mouillage complet.) La surface spécifique et la pression à l'interface plâtre/tesson ont été calculées pour quelques échantillons de plâtre à l'aide de données relatives à la distribution des pores et à la perméabilité. Les valeurs pour la pression, calculées à partir des courbes de répartition des pores concordent mieux avec la pression de filtration (env. 1 atm.) nécessaire à la formation du tesson, que les valeurs plus faibles calculées en tenant compte de la perméabilité. Bien que certains résultats ne soient pas tout à fait satisfaisants, les auteurs concluent de ces expériences que la validité de la théorie de la filtration est démonstrée expérimentalement. Il n'existe pratiquement pas de pores $< 0,1 \mu m$. La thèse avancée il y a quelques années par les auteurs, selon laquelle la formation du tesson serait due à l'influence de pores très fins avec un rayon hydraulique de 30–50 Å était par conséquent erronée.*

## Die Kapillaritätswirkung von Gipsformen

*Die Bildung eines Rohscherbens auf einer Gipsoberfläche beim Schlickerguß rührt von der Kapillaritätswirkung der Poren in der Gipsform her. Bei der Verwendung von Schlicker, aus dem ein Scherben mit einer Durchlässigkeit von $20.10^{-14} cm^2$ gebildet wird, wird fast die gesamte von den Poren erzeugte Kapillaritätskraft zur Bildung des Scherbens benötigt. Daher ist der Druck an der Gips-Schlicker-Grenze nahezu gleich der gesamten Kapillaritätskraft der Poren pro Grenzflächeneinheit und kann aus $P_s = S. \sigma. \cos \sigma$ berechnet werden, worin S die spezifische Oberfläche pro Einheitsvolumen des Gipses, $\sigma$ die Oberflächenspannung von $H_2O$ und $\varphi$ den Benetzungswinkel ($\cos \varphi = 1$ für vollständige Benetzung) bedeutet. Die spezifische Oberfläche und der Druck an der Grenzfläche Gips-Schlicker einiger Gipsproben wurden aus Porenverteilungen und aus Durchlässigkeits-koeffizienten berechnet. Es wurde der Eindruck gewonnen, dass der vorhandene Filtrationsdruck (etwa 1 atm) sich besser aus den Porengrössenverteilungskurven wie aus den Durchlässigkeitswerte berechnen liess. Obwohl einige Ergebnisse noch nicht ganz befriedigen, wird die Filtrationstheorie als experimentell bewiesen angesehen. Poren unter 0, 1 $\mu m$ sind kaum vorhanden. Folglich ist die einige Jahre vorher von den Autoren aufgestellte Theorie, daß Scherbenbildung durch den Einfluß sehr feiner Poren mit einem hydraulischen Radius von 30-50Å hervorgerufen wird, falsch.*

## 1. FILTRATION THEORY

In the process of slip casting, plaster moulds withdraw water from the liquid slip by suction and simultaneously a cast is formed on the plaster surface. Though this procedure has been known for a long time, not until 1950 were serious investigations made into the mechanism of cast formation.[1,2] After an intricate mathematical theory had been worked out by DEEG,[1] in which cast formation was treated as a matter of diffusion, in 1958 DEEN and one of the present investigators[3,4] worked out another theory, which treats the formation of a cast on a plaster surface as being caused by filtration of the casting slip. They started from the idea that the capillary action of the pores in the plaster results in a constant pressure, which constitutes the driving-force in cast formation and in the capillary suction of the filtrant into the plaster mould.

Now one can describe cast formation on a (infinitely large) plane plaster-wall by Darcy's Law, analogous to a current through two resistances connected in series.

$$\frac{1}{O}\frac{dQ}{dt} = \frac{P}{\eta\left(\frac{L_s}{K_s} + \frac{L_g}{K_g}\right)} \qquad . \qquad . \qquad . \qquad . \quad (1)$$

in which
$O$ = Surface of the cast on the plaster wall (cm$^2$)
$Q$ = Quantity of water absorbed (cm$^3$)
$t$ = Time (sec.)
$P$ = Pressure of the plaster mould (dyne/cm$^2$)
$K_s$ = Permeability of the cast (cm$^2$)
$K_g$ = Permeability of the plaster mould (cm$^2$)
$L_s$ = Thickness of the cast formed from the casting slip (cm)
$L_g$ = Limit of penetration of the water front in the plaster mould (cm)
$\eta$ = Viscosity of the water (dyne. sec/cm$^2$)

Since the water flowing in the cast and in the plaster mould encounters resistance, only part of the pressure exerted by the pores of the mould will contribute to cast formation, i.e. the pressure exerted by plaster mould can be divided into a pressure drop in the cast, $P_s$, and a pressure drop in the plaster mould, $P_g$.

From this it follows that

$$P = P_s + P_g \qquad . \qquad . \qquad . \qquad . \quad (2)$$

$$\frac{1}{O}\frac{dQ}{dt} = \frac{P_s K_s}{\eta L_s} = \frac{P_g K}{\eta L_g} \qquad . \qquad . \qquad . \quad (3)$$

The quantity of water withdrawn from the slip and absorbed in the mould:

$$Q = \frac{1 - c - n_s}{c} \, OL_s = n_g OL_g \quad . \quad . \quad (4)$$

in which    $c$ = Volume concentration of the solid substance in the casting-slip.

$n_s$ = Volume fraction of water in the cast.

$n_g$ = Volume fraction of water in the wetted part of the plaster mould.

Henceforth $(1 - c - n_s)/c$ will be replaced by the factor $\alpha$.

From Equations (2), (3) and (4) it follows that

$$\frac{P_s}{P_g} = \frac{n_g K_g}{\alpha K_s} \quad . \quad . \quad . \quad . \quad . \quad (5)$$

and

$$P_s = P \left( \frac{n_g K_g}{\alpha K_s + n_g K_g} \right) \quad . \quad . \quad . \quad (6)$$

which means that the total pressure drop and the ratio of the two fractional pressure drops are constant, and therefore the pressure drop $P_s$ in the cast (i.e. the fractional pressure at the plaster/cast boundary, or the filtration pressure) is also constant.

From the above equations it follows that the thickness of the cast, $L_s$, is given by:

$$L_s = \frac{1}{\alpha} \sqrt{\frac{2Pt}{\eta \left( \frac{1}{\alpha K_s} + \frac{1}{n_g K_g} \right)}} = \sqrt{\frac{2 P_s K_s t}{\eta \alpha}} \quad . \quad . \quad (7)$$

From this theory it follows that cast formation is proportional to the square root of time. This can be easily proved experimentally, especially with an optimum deflocculated casting slip. The relationship with the pressure exerted by the plaster mould is more complicated, as its magnitude depends on the pore size distribution, and so on the values of $n_g$ and $K_g$ of the plaster mould. This dependence becomes clearer when in equation (7) we replace the pressure of the plaster mould ($P$) by the partial pressure at the plaster/cast boundary ($P_s$). The equation for cast formation then becomes the usual filtration equation, in which the thickness of the cast (or filtered layer) is expressed not only as proportional to the square root of time, but also as proportional to the square root of the partial pressure.

## 2. CRITICAL ANALYSIS OF THE THEORY

The amount of the filtration pressure ($P_s$) can be deduced from separate tests on filter-paper with the same casting-slip and over the same period of time, applying pressures ranging from 0·25 to 4 atm. The pressure at the plaster/cast boundary must now be equal to the applied pressure, resulting in an equally thick filter-layer being formed on the filter-paper. From these filtration tests the permeability of the growing cast can also be calculated, and from such measurements the filtration pressure of casting moulds can be derived as approximately 1 atm. To prove the filtration theory, this filtration-pressure of 1 atm must be deducible from the properties of the porous mould. According to the above-mentioned formulas, this should be possible if the pressure of the porous mould ($P$), the volume of water removed by capillary suction (expressed as the volume fraction $n_g$) and the permeability ($K_g$) are known.

Apparently this did not look to be such a problem, because from the free water-suction (e.g. from a saturated solution of gypsum) hardly any drop in water content in the plaster-layers could be detected. This can be clearly seen by the sharp "wet-dry" dividing line of the rising water front, which disappears only at greater heights. A constant value can therefore be taken for the water content ($n_g$), irrespective of the rise in height of the water. The permeability ($K_g$) of a plaster sample saturated by means of water absorption can be measured directly. However, the "pressure of the plaster mould" ($P$) should be reflected in the capillary suction, which can be expressed by the square of the height the water rises and the time ($L^2/t$ value).

$$\frac{1}{O}\frac{dQ}{dt} = \frac{K_g P}{\eta\, L} \text{ with } dQ = n_g O L_g \text{ and thus } P = \frac{L^2}{t}\cdot\frac{\eta n_g}{2K_g} \qquad . \quad (8)$$

When these equations are used to calculate the "pressure of the plaster mould", the resulting values lie between 0·15 and 0·25 atm., depending on the kind of plaster. This is inadequate to explain a filtration pressure of 1 atm.

Another way to tackle this problem is to apply Poiseuille's Law instead of Darcy's Law, which in essence gives the same result but can only be applied when the shape and size of the pores are defined. If for the sake of simplicity we assume that plaster moulds consist of cylindrical capillaries with a capillary pressure $P_k$, then it follows that

$$\frac{dL}{dt} = \frac{r^2 P_k}{8\eta L} \text{ with } P_k = \frac{2\sigma}{r} \text{ and thus } P_k = \frac{1}{L^2/t}\cdot\frac{\sigma}{\eta} \qquad . \quad (9)$$

From the $L^2/t$ values of the water absorption, capillary pressures can be calculated to be about 10 atm. This contradiction can only be explained if the capillary rise takes place in pores the diameter of which varies. Indeed it can be proved that the $L^2/t$ relation applies not only to capillaries of constant diameter but also to capillaries with a changing diameter—or in general to porous materials with a regular pore system. This, however, does not exclude the fact that for the same $L^2/t$ value the diameter of a cylindrical capillary may be narrower than the entrance diameter of capillaries of variable diameter. So it is impossible to calculate a mean capillary-pressure from the $L^2/t$ value, because this value corresponds to a capillary diameter that is too small and not to the average pore diameter (also see Appendix 1).

Another problem is that the water content in the plaster mould gradually diminishes as the water rises, especially where a cast is forming at the same time (Figure 1). Apart from the much lower rate at which the water is withdrawn, there immediately appears a distinct decrease in the water content from the plaster/cast boundary to the deeper layers of plaster, because there is ample opportunity for the water to flow from the wider pores to the narrower ones. The progression of the gradually blurring waterfront is again proportional to the square root of time. As cast formation proceeds, the water con-

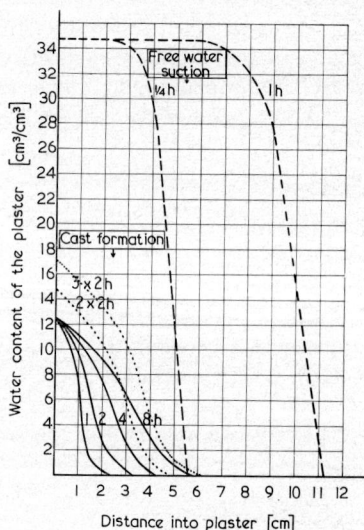

FIGURE 1.—Water distribution in plaster cylinder (porosity 47·7 vol%) during suction.

tent in the first plaster-layer remains constant, but also remains far
below the value for free water absorption. Only if the plaster had been
moistened beforehand, e.g. by the formation of a previous cast,
would the water concentration in the first plaster-layer have been
higher (Figure 1).

Incorrectly we were of the opinion a few years ago that the
capillary pressure of the pores could be calculated from the $L^2/t$
value measured in the plaster-mould during cast formation.[5]
We had unwittingly assumed that the pores through which the water
flowed were of equal size, consequently we found that the cast
formation depended on the influence of very fine pores with a
hydraulic radius of 30 to 40 Å [5] (also see Appendix 2).

Later on, however, tests on capillary condensation and on pore
size distribution of a number of plaster-moulds (Figure 2) established

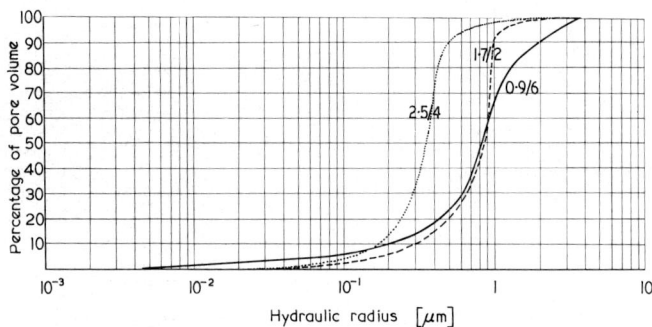

FIGURE 2.—Pore-size distribution curves of 3 samples (measured by Institut für
Gesteinshüttenkunde RWTH, Aachen).

that such fine pores in plaster must be rare. Also in the previous
theory, the capillary pressure of the pores and what could be called
the capillary action of the complete plaster mould were not clearly
distinguished. As we see it now, however, the latter concept could
have no physical significance because only the pores exert capillary
pressure, the value of which depends on their size.

## 3. EXTENSION OF THE THEORY

As the concept "pressure of the plaster mould" appeared to be
unrealistic, the theory had to be revised somewhat, so that only
the capillary pressure of the pores was significant. The pore-space
in a plaster-mould consists of a collection of open channels connecting large and small holes between the plaster crystals. The capillary

pressure with which water is drawn into those pores changes as the water level rises in a wider or a narrower part of the pore.

In order to tackle the problem mathematically, we assumed a hypothetical texture of isotropic connected pores of equal size.

Capillary suction is exerted as a result of the surface tension of the water ($\sigma$) and the degree of wetting of the capillary walls. With complete wetting (cos $\varphi = 1$), the capillary pressure of the pores amounts to $\sigma/m$, ($m =$ the hydraulic radius of the pores defined as the quotient of the volume and the internal surface of the pores (for cylindrical capillaries: $m = \frac{1}{2}r = \frac{1}{4}d$).

In cast formation, part of the capillary pressure is required to overcome the resistance to flow in the plaster pores. If we call this part of the capillary pressure $P_c$, the pressure $[(\sigma/m) - P_c]$ remains for the cast formation, and pressure is exerted over the entire plaster surface. As in an isotropic porous material the surface porosity is equal to the volume porosity ($n_g$), the pressure at the plaster/cast boundary amounts to:

$$P_s = n_g \left( \frac{\sigma}{m} - P_c \right) \quad . \quad . \quad . \quad . \quad (10)$$

According to Darcy it holds unchanged for water movement through the cast:

$$\frac{1}{O} \frac{dQ}{dt} = \frac{K_s P_s}{\eta L_s} \quad . \quad . \quad . \quad . \quad (11)$$

To describe the flow of water in the mould we adapt Poiseuille's Law. For a piece of plaster, thought of as consisting of a bundle of cylindrical capillaries perpendicular to the contact plane,

$$\frac{dL_c}{dt} = \frac{m^2}{2\eta L_c} \cdot P_c \cdot \quad . \quad . \quad . \quad (12)$$

For flattened capillaries the factor 2 changes to values between 2 and 3, depending on the degree of flattening. Following the work of KOZENY and CARMAN,[6] the factor 5/2, more workable in practice, is introduced. When the parallel pores are replaced by an isotropic system of connected pores of equal size, we must distinguish between the rate of flow in the pores ($dL_c/dt$) and the speed of advance of the water level ($dL_g/dt$).

Statistically the direction of the pores is at an angle of 45° to the direction of advance. Therefore:

$$\frac{L_c}{L_g} = \sqrt{2} \text{ and } \frac{dL_c}{dL_g} = \sqrt{2} \quad . \quad . \quad . \quad (13)$$

Hence it also holds:

$$\frac{dL_c}{dt} = \frac{2m^2}{5\eta L_c} \cdot P_c \text{ and } \frac{dL_g}{dt} = \frac{m^2}{5\eta L_g} \cdot P_c = \frac{m^2}{5\eta L_g}\left(\frac{\sigma}{m} - \frac{P_s}{n_g}\right) \quad . \text{ (14)}$$

From Equation (14) and the results of the equilibrium ($Q = \alpha\, OL_s = n_g OL_g$) the thickness of the cast ($L_s$):

$$L_s = \sqrt{\frac{2\sigma n_g t}{\eta\left(\dfrac{\alpha m}{K_s} + \dfrac{5\alpha^2}{mn_g}\right)}} = \sqrt{\frac{2 P_s K_s t}{\eta\alpha}} \qquad . \qquad . \text{ (15)}$$

in which

$$P_s = \frac{\sigma n_g}{m}\left(\frac{1}{1 + \dfrac{5\alpha K_s}{m^2 n_g}}\right) \qquad . \qquad . \qquad . \qquad . \qquad . \text{ (16)}$$

Furthermore it follows that

$$\text{if } m \to 0 \qquad : \frac{\sigma}{m} \to \infty; \text{ but } P_s \to 0$$

$$\text{is } 5\alpha\, K_s \ll m^2 n_g : P_s \to \frac{\sigma n_g}{m} \text{ and thus } P_c \to 0$$

This means that very fine pores exert a very high capillary pressure but also a very high resistance to flow, resulting in very little water being withdrawn from the slip. Wider pores exert a lower capillary pressure but also less resistance to flow, so their capillary pressure contributes significantly to cast formation. It also holds that, for each porosity, a maximum cast-formation exists. The hydraulic radius of the pores in this maximum cast-formation is derived from $dL_s/dm = 0$. The solution then is:

$$L_{s(max)} = \sqrt[4]{\frac{\sigma^2 K_s n_g^3 t^2}{5\alpha^3 \eta^2}} \text{ for } m = \sqrt{\frac{5\alpha\, K_s}{n_g}} \qquad . \qquad . \text{ (17)}$$

The plot of cast thickness as a function of the hydraulic radius of the pores for various porosities is shown in Figure 3. Each slip and each cast-formation time will only shift the curve, but not alter its shape.

If the total capillary pressure exerted by the pores could be utilized completely for cast formation (100% pressure output) the pressure at the plaster/cast boundary comes to:

$$P_s = \frac{\sigma n_g}{m} \qquad . \qquad . \qquad . \qquad . \qquad . \qquad . \qquad . \text{ (18)}$$

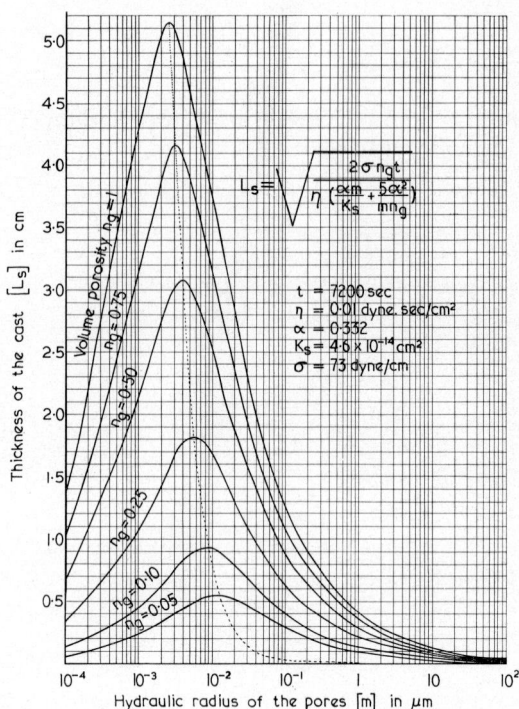

FIGURE 3.—Formation of cast on a flat surface of a porous material with an isotropic system of pores of equal hydraulic radius.

The extent to which this situation can be approximated can be calculated from data previously obtained. The permeability of the growing cast is usually not higher than $20 \times 10^{-14}$ cm², for the factor $\alpha$ a good average value of $0.3$ can be taken, whereas the porosity of a plaster mould in general comes to approximately 50 vol. %. Starting with these data and depending on the hydraulic radius of the pores, one arrives at the following values:                    . .

| Hydraulic radius | $0.01$ $\mu$m | $0.03$ $\mu$m | $0.1$ $\mu$m | $0.3$ $\mu$m | $1$ $\mu$m |
|---|---|---|---|---|---|
| $P_s = (\sigma n_g / m)$. factor | $0.625$ | $0.937$ | $0.994$ | $0.9993$ | $0.999$ |

From the pore-size-distribution curves—determined on a number of plaster samples—it appears that pores with a hydraulic radius of less than $0.03$ $\mu$m seldom exist, which means that the pressure at the plaster/cast boundary is approximately equal to the total tension of the pores per unit surface in the plaster/cast boundary. If the pore-distribution curve, in which the volume-fraction $n_g$ is for pores with a hydraulic radius $< m$, is given by the equation:

$$n_g = f(<m)$$

then the specific surfaces as well as the boundary pressure (by a 100% output) can be deduced:

$$S = \int_{m_1}^{m_2} \frac{dn_g}{m} \qquad . \qquad . \qquad . \qquad . \qquad . \qquad . \qquad (19)$$

and

$$P_s = \int_{m_1}^{m_2} \frac{\sigma}{m} dn_g = S \cdot \sigma \qquad . \qquad . \qquad . \qquad . \qquad . \qquad (20)$$

in which  $S$ = Specific surface (cm²/cm³ bulk material)

$n_g$ = Pore volume (cm³/cm³ bulk material)

$m_1$ and $m_2$ = Pore limits for the widest and the narrowest capillary.

As long as the pores are not too narrow, the boundary pressure is proportional to the specific surface of the pore volume. Hence it holds in general, and also in the case of incomplete wetting of the capillary walls, that

$$P_s = S . \sigma \cos \varphi \qquad . \qquad . \qquad . \qquad . \qquad (21)$$

## 4. EXPERIMENTS

In checking the theory against experimental data, three different porous plaster materials were used, obtained from a plaster with a predominant proportion of β-hemi-hydrate (in practice also used for the manufacture of moulds) by mixing it in plaster: water ratios of 0·9 and 1·7, stirred for 6 and 2 minutes respectively. Furthermore an autoclave plaster with a predominant proportion of α-hemi-hydrate was mixed in with water ratio of 2·5 and stirred for 4 minutes.[5] The test-specimens from the set plasters were dried at 40°C. To compare them with a very coarse porous texture, a refractory material was used (Table 1).

The following should be noticed:

All the measurements have been made as far as possible on the same test-pieces (cylinders 4 cm diam. and 3 cm high). On various test-pieces the pore-size distribution was measured by means of a mercury porosimeter at the Institut für Gesteinshüttenkunde RWTH, Aachen (Figure 2). For the porous refractory material it

9

**Table 1**

**Texture of the Porous Materials and the Formation of Cast**

| | Plaster | | | Coarse porous material |
|---|---|---|---|---|
| | 0·9/6 | 1·7/2 | 2·5/4 | |
| Volume porosity (cm³/cm³ bulk material) | 0·613 | 0·477 | 0·358 | 0·320 |
| Free water suction (cm³/cm³ bulk material) | 0·450 | 0·347 | 0·255 | 0·308 |
| $L^2/t$ ratio (cm²/sec of the free water suction at 20°C) | 0·062 | 0·035 | 0·008 | 0·24 |
| Capillary pressure of the widest pore-channels (atm) | 0·6 | 0·7 | 1·1 | 0·05 |
| Water permeability ($10^{-10}$cm²) | 8·7 | 4·4 | 1·04 | 1140 |
| Air permeability ($10^{-10}$ cm²) | 11·4 | 5·8 | 1·2 | 530 |
| Specific surface (cm²/cm³ bulk material) | | | | |
| Calculated from: Pore-size distribution curve | } 22900 (12600) | 13400 | 14400 | — |
| Water permeability | 7280 | 7030 | 9400 | 240 |
| Air permeability | 6980 | 6870 | 10880 | 350 |
| Pressure at the plaster/cast boundary (atm) | | | | |
| Calculated from: Pore-size-distribution curve | } 1·63 (0·90) | 0·82 | 1·04 | — |
| Water permeability | 0·53 | 0·51 | 0·68 | 0·02 |
| Air permeability | 0·51 | 0·50 | 0·79 | 0·03 |
| Formation of cast in 2 h (cm) | 0·47 | 0·45 | 0·39 | 0·07 |
| Water content of the cast (% of the wet state) | 15·5 | 15·5 | 15·6 | 17·0 |
| Permeability of the cast ($10^{-14}$cm²) | 5·8 | 5·9 | 6·2 | 7 to 8 |
| Filtration pressure (atm.) | 1·0 | 0·95 | 0·65 | 0·02 |

could only be established that 98·6% of the pores are wider than 7·5 $\mu$m.

The capillary pressure of the widest pores was measured by BESKOW's method,[10] based on the measurement of the minimal pressure required for the first air-bubble to burst through the water-saturated and submerged test-object (a saturated solution of gypsum was used instead of water).

The specific pore-surface was calculated from the water-permeability (measured in a saturated solution of gypsum), by means of the KOZENY-CARMAN equation:[6]

$$K = \frac{n^3}{5S^2} \qquad \qquad (22)$$

For the calculation of the specific surface from air permeability, a correction was still needed for the molecular diffusion of air by flow in the finest pores[1-9]:

$$K = \frac{n^3}{5S^2} + \frac{ZFn^2}{5S} \qquad \qquad (23)$$

in which  $K$ = Permeability (cm²)
$n$ = Volume porosity (cm³/cm³ bulk material)
$S$ = Specific surface (cm²/cm³ bulk material)
$Z$ = Slip factor = 3
$F$ = Mean free path of the air molecules ($6 \times 10^{-6}$ cm).

The specific surfaces were also calculated from the pore-size-distribution curves. These curves cannot be represented in the form of simple functions. In order to calculate the specific surface, the curves were replaced by fractional lines which approximated the curves as closely as possible. For the straight pieces, the relationship between $n_g$ and $m$ could be established and so an equation of the following kind developed:

$$n_g = (am+p) \Big|_{m_1}^{m_2} + (bm+q) \Big|_{m_2}^{m_3} + \quad . \quad . \quad . \quad (24)$$

Hence it followed for the specific surface that:

$$S = a . \log_e \frac{m_2}{m_1} + b . \log_e \frac{m_3}{m_2} + \quad . \quad . \quad . \quad (25)$$

Furthermore Table 1 presents data on the casts being formed on moulds from the previously mentioned porous materials. Comparative measurements have been made on an optimum deflocculated casting-slip (179 g/l, with a water concentration of 51·5 vol.%).

The filtration pressure and permeability of the developing cast have been deduced from comparative filtration tests.

## 5. REVIEW OF THE RESULTS

The high value of $1 \cdot 63$ atm., which can be calculated from the pore-size-distribution curve of the plaster sample $0 \cdot 9/6$, is due to the relatively high proportion of fine pores in this material. However, we question this result because there is no obvious reason why this sample should contain more fine pores than the other two samples, for which—because of the mixing-procedure—finer pores could be predicted. If we assume that sample $0 \cdot 9/6$ does not contain more fine pores than sample $1 \cdot 7/2$, the more acceptable value (in brackets) of $0 \cdot 90$ atm. for the boundary pressure results. Furthermore the boundary pressure of sample $2 \cdot 5/4$—calculated from the air- and water-permeability—agrees better with the required filtration-pressure than the pressure calculated from the pore-size distribution curve. On the other hand, the boundary-pressures of the porous samples $0 \cdot 9/6$ and $1 \cdot 7/2$—calculated from the two permeabilities— are too low. This means that the permeability value given does not always have to be correct to calculate the specific surface of the pore-space. In particular, this arises when the porous object contains some wider passages, through which extra air or liquid leaks in a permeability measurement.

Recapitulating, it can be concluded that there exists reasonably good agreement between the pressure at the plaster/cast boundary— calculated from the pore texture—and that derived from cast formation. In principle it is proved that the formation of a ceramic cast can be regarded as by a filtration process. Probably the problem will never be solved to complete satisfaction so long as the specific surface of the pore-space cannot be calculated more accurately. This is not possible now if we base it on the permeability and the pore-distribution curve.

### REFERENCES

1. DEEG, E., *Ber. dtsch. keram. Ges.*, **30**, 129, 1953.
2. DIETZEL, A., and MOSTETZKY, H., *Ber. dtsch. keram. Ges.*, **33**, 7, 47, 73, 115, 1956.
3. DAL, P. H., and DEEN, W., *Klei*, **9**, 59, 1959.
4. DAL, P. H., and DEEN, W., 6th International Ceramic Congress, Wiesbaden 1958. "Die Scherbenbildung beim keramischen Gieszverfahren".
5. BERDEN, W. J. H., and DAL, P. H., "Science of Ceramics", Vol. 2 (Academic Press for British Ceramic Society), pp. 171–188.
6. CARMAN, P. C., *J. Agr. Sci.* **29**, 262, 1939.
7. MICHAELS, A. S., and LIN, C. S., *Ind. Engng Chem.*, **46**, 1239, 1954.
8. DEEN, W., *Ber. dtsch. keram. Ges.*, **37**, 489, 1960.
9. DEEN W., *Ber. dtsch. keram. Ges.*, **38**, 107, 1961.
10. BESKOW, G., "Om Jordarternas Kapillaritet" (Stockholm, 1930).

## Appendix I

### WATER SUCTION IN A CAPILLARY OF VARYING DIAMETER

For a mathematical discussion of the water suction in a capillary of varying diameter we restrict ourselves to the simple case of a cylindrical capillary built up of alternate units of length $h$ and radius $r$ + length $H$ and radius $R$.

Furthermore we assume that the flow in those capillaries is laminar and that no resistance occurs at the point of widening or narrowing. Let us first trace the flow of water in the first capillary-unit, where we can start the flow in a narrow as well as in a wide part. We choose arbitrarily the narrow part ($h$ with $r$), in which—after Poiseuille—the flow of water takes place as follows:

$$\frac{dQ_r}{dt} = \frac{\pi r^4 P_r}{8\eta h} \text{ with } dQ_r = \pi r^2 dh \text{ and } P_r = \frac{2\sigma}{r}$$

It follows that the time the narrow part (1 h) takes to fill is

$$t_{h_1} = \frac{2\eta h^2}{\sigma r}$$

The further rise of the water in the capillary to height $h + H$ takes place as follows:

$$\frac{dQ_R}{dt} = \frac{P_R}{8\eta\left(\dfrac{H}{\pi R^4} + \dfrac{h}{\pi r^4}\right)} \text{ with } d Q_R = \pi R^2 dH \text{ and } P_R = \frac{2\sigma}{R}$$

from which it follows that the time for the wide part to fill is:

$$t_{H_1} = \int_0^H \frac{4\eta}{\sigma}\left(\frac{H}{R} + \frac{hR^3}{\pi r^2}\right)dH = \frac{4\eta}{\sigma}\left(\frac{H^2}{2R} + \frac{hHR}{r}\right)$$

The total time required for the filling of the first capillary-unit amounts to:

$$t_1 = \frac{4\eta}{\sigma}\left(\frac{h^2}{2r} + \frac{H^2}{2R} + \frac{hHR^3}{r^4}\right)$$

From now on the resistance to flow in each filled narrow and wide capillary section remains constant and is equal to

$$W = 8\eta\left(\frac{H}{\pi R^4} + \frac{h}{\pi r^4}\right)$$

When a total of $(n-1)$ capillary units have been filled, for the flow in the next $(n^{th})$ narrow and wide capillary-piece:

$$\pi r^2 \frac{dh}{dt} = \frac{P}{(n-1)\,W+(8\eta h/\pi r^4)} \text{ with } P_r = \frac{2\sigma}{r}$$

$$t_{h_n} = \int_0^h \frac{\pi r^3(n-1)W+(8\eta h/r)}{2\sigma}\, dh = \frac{4\eta}{\sigma}\left\{\frac{(n-1)\,Hhr^3}{R^4} + \frac{h^2(n-\frac{1}{2})}{r}\right\}$$

and furthermore

$$\pi R^2 \frac{dH}{dt} = \frac{P_R}{(n-1)W+(8\eta h/\pi r^4)+(8\eta H/\pi R^4)} \text{ with } P_R = \frac{2\sigma}{R}$$

$$t_{H_n} = \int_0^H \frac{\pi R^3(n-1)\,W+(8\eta hR^3/r^4)+(8\eta H/R)}{2\sigma}\, dh$$

$$= \frac{4\eta}{\sigma}\left\{\frac{H^2(n-\frac{1}{2})}{R} + \frac{HhR^3 n}{r^4}\right\}$$

The time required to fill the $n^{th}$ capillary unit amounts to:

$$t_n = \frac{4\eta}{\sigma}\left\{\frac{h^2(n-\frac{1}{2})}{r} + \frac{H^2(n-\frac{1}{2})}{R} + \frac{HhR^3 n}{r^4} + \frac{(n-1)\,Hhr}{R^4}\right\}$$

The total time required to fill the complete capillary of a length $n(h+H)$ amounts to:

$$t_{\frac{1}{2}n(n+1)} = \frac{2\eta}{\sigma}\left\{\frac{n^2 h^2}{r} + \frac{n^2 H^2}{R} + \frac{n(n+1)\,HhR^3}{r^4} + \frac{n(n-1)\,Hhr^3}{R^4}\right\}$$

For higher values of $n$:

$$t_{\frac{1}{2}n^2} = \frac{2\eta n^2}{\sigma}\left(\frac{h^2}{r} + \frac{H^2}{R} + \frac{HhR^3}{r^4} + \frac{HhR^3}{R^4}\right)$$

The time in which the height of rise $n(h+H)$ is reached therefore is proportional to $n^2$, which means that—macroscopically—the height the water rises is proportional to the square root of time.

Furthermore the radius can be calculated of the capillary with a constant diameter in which the same water-rise would take place (rise $n(h+H)$ in time $t$). Let the radius of this cylindrical capillary be $x$.

$$t = \frac{2\eta}{\sigma}\cdot\frac{n^2(H+h)^2}{x} = \frac{2\eta n^2}{\sigma}\left(\frac{H^2}{R} + \frac{h^2}{r} + \frac{HhR^3}{r^4} + \frac{Hhr^3}{R^4}\right)$$

$$x = r \left\{ \frac{H^2 + 2Hh + h^2}{\dfrac{H^2 r}{R} + Hh \left( \dfrac{R^3}{r^3} + \dfrac{r^4}{R^4} \right) + h^2} \right\}$$

Hence it follows, that:

$$x \leqq r \quad \text{if } \frac{H}{h} \leqq \frac{(R^3/r^3) + (r^4/R^4) - 2}{1 - (r/R)}$$

$$r < x < R \text{ if } \frac{H}{h} > \frac{(R^3/r^3) + (r^4/R^4) - 2}{1 - (r/R)}$$

This means that a capillary with a constant diameter can sometimes be even narrower than the narrow mouths of wider pores to get the same speed of advance of the water level.

In any other capillary with repetitive pore-structure the height the water rises must obey the $\sqrt{t}$ relation. This can be reasoned out very simply as follows:

By periodic recurrent increase and decrease of the capillary pressure on average this pressure remains constant. When we call this "average" pressure $\bar{P}$ and the flow resistance in each capillary-unit $W$, then the time that the $n^{th}$ capillary unit takes to fill amounts to

$$t_n = \frac{\bar{P}}{nW}$$

For a high value of $n$ the total time required to fill the complete capillary amounts to:

$$t_{\frac{1}{2}n(n+1)} = \frac{\bar{P}}{\frac{1}{2}n(n+1)\,W} = \frac{\bar{P}}{n^2 W}$$

## Appendix 2

### CRITICAL APPRAISAL OF OUR FORMER "FINE PORES" HYPOTHESIS.

In our earlier paper we believed that cast formation was due to the influence of very fine pores. This opinion has since proved to be incorrect.

We started with the data on the rate at which water rises in the plaster when a cast is being formed from casting-slip. For all samples this rise could be characterized by a measurable accuracy equal to

$L_g{}^2/t$ value of $0.0009 \pm 0.0002$ cm²/sec. The reasoning at the time was as follows:

The capillary pressure $(P)$ and the hydraulic pore radius $(m)$ in which the water flows can be calculated from the $L_g{}^2/t$ value of $0.0009$ cm²/sec.

$$\frac{L_g{}^2}{t} = \frac{2m^2 P}{5\eta} \quad \text{with } P = \frac{\sigma}{m}$$

$$\eta = 0.01 \text{ poise}$$
$$\sigma = 73 \text{ dyne/cm}$$

from which it follows that $P = 230$ atm. in pores with $m = 30$ Å.

By equating the capillary pressure of those pores to the "pressure of the plaster mould", the cast thickness $L_s$ was calculated by means of the following data:

Cast-formation time                                  $t\ \ = 7200$ sec.
Permeability of the cast                             $K_s = 4.6 \times 10^{-14}$ cm²
Volume fraction of water in the cast          $n_s = 0.354$
Volume fraction of solids in the casting-
   slip                                              $c\ \ = 0.485$     $\left.\right\} \alpha = 0.332$
Volume fraction of water in the wetted
   part of the plaster mould:                $n_g = 0.0653$

$$L_s = \frac{1}{\alpha} \sqrt{\frac{2Pt}{\eta\left(\dfrac{1}{\alpha K_s} + \dfrac{5\alpha^2}{n_g{}^2 m^2}\right)}}$$

This gave the value $0.50$ cm, in accordance with the value found by experiment.

Apart from the incorrect hypothesis that the water flow through pores of equal size and that the capillary pressure of these pores could be equated to "the pressure of the plaster mould", there is another mistake in the reasoning. The $L_g{}^2/t$ value—according to the original theory—should be related to $P_g$ (pressure decrease in plaster) and not to $P$. The solution then is:

$$\frac{L_g{}^2}{t} = \frac{2m^2 P_g}{5\eta} = \frac{2m^2(P - P_s)}{5\eta} = \frac{2\sigma^2}{5\eta P}\left(1 - \frac{P_s}{P}\right)$$

The only way to solve $P$ now is to know the value of $P_s$. The value of $P_s$, however, can only be derived from the cast thickness, which we are trying to calculate from the relation:

$$P_s = \frac{L_s{}^2 \eta \alpha}{2K_s t}$$

From the cast thickness $L_s=0.50$ cm (measured experimentally) it follows that $P_s=1.235$ atm., whereas for $P$, $P_g$ and $m$ two solutions are found:

$$P=235.6 \text{ atm} \qquad P_g=234.3 \text{ atm} \qquad m=\ \ \ 31 \text{ Å}$$
$$\text{or } P=\ \ \ 1.256 \text{ atm} \quad P_s=\ \ \ 0.004 \text{ atm} \quad m=5800 \text{ Å}$$

Apart from the fact that there prove to be two solutions, which in principle is not impossible (see Figure 3), it is now clear that a circular argument was used. The cast thickness determined experimentally had to be used in order to calculate whether this thickness could also be derived by the theory. In our original calculation, in which $P$ was used instead of $P_g$, this was not obvious. As the very high value of $P_g$ could be almost equated to $P$, here too the experimental value of $0.50$ cm was found for $L_s$. Apparently no use was made of the value of $L_s$ found experimentally; but the value of $L_s$ was already fixed, as should have followed immediately from the equilibrium equation

$$\alpha.L_s=n_g.L_g$$

when the data of $L_g^2/t$, $t$, $n_g$ and $\alpha$ are known, independent of $P$ and $m$.

# 10.—The Effect of Very High Pressures on Ceramic Systems

By C. J. M. ROOYMANS

*Philips Research Laboratories,*
*N.V. Philips Gloeilampenfabrieken, Eindhoven, Netherlands*

***ABSTRACT*** **C32d**

*The influence of very high pressures on the crystal structure of ceramic materials, on crystal defects, and on reactivity is dealt with. Attention is directed to the great difference in actual compression of a material with or without a phase transition taking place in the structure. Mention is made of the change in co-ordination often accompanying the transition, and the change sometimes occurring in the occupation of the electron shells. The concentration of defects in the solid state varies with pressure, the sign of the variation depending on whether the defects under consideration are characterized by a volume decrease (interstitials) or a volume expansion (vacancies). The variation in the concentration of defects can have a marked influence on the reactivity; other important factors changing the reactivity of the material are the closer contact between the reacting grains and the occurrence of phase transitions. Finally, some applications of very-high-pressure techniques to ceramic materials are reviewed, and the importance of these techniques in science and technology is discussed.*

## L'influence des très hautes pressions sur les systèmes céramiques

*Nous nous bornerons à traiter: l'influence des très hautes pressions sur la structure cristalline des matières céramiques, sur les imperfections cristallines, et sur la réactivité. L'attention est attirée à la grande différence qui existe dans la compression réelle de la substance, qu'il ait ou non une transition de phase au sein de la structure. Le changement de co-ordination, accompagnant souvent la transition, est mentionné également ainsi que les changements qui se produisent parfois dans l'occupation des enveloppes d'électrons. La concentration des imperfections dans l'état solide varie avec la pression; le signe mathématique de cette variation sera différent selon que les imperfections considérées se caractérisent par une diminution de volume (interstices) ou une expansion (lacunes). La différence de concentration des imperfections peut produire une influence considérable sur la réactivité; parmi les autres facteurs importants, nous citons le contact plus étroit entre les grains en réaction et la présence de transitions de phase rehaussant la réactivité de la substance. Enfin, certaines applications des techniques à très haute pression pour les matériaux céramiques sont passées en revue et leur importance aux points de vue scientifique et technologique est exposée.*

## Der Einfluss sehr hoher Drucke auf die Keramiksysteme

*Nur drei Themen werden behandelt, nämlich der Einfluss sehr hoher Drucke auf die Kristallstruktur keramischer Werkstoffe, auf die Gitterfehlstellen, und auf die Reaktivität. Es wird darauf hingewiesen, dass die tatsächliche Kompression eines Materials einen grossen Unterschied aufweist, je nachdem ob eine Phasenumwandlung stattfindet oder nicht. Die damit verbundene Änderung der Koordination, und die manchmal auftretende Änderung der Besetzung der Elektronenschalen wird erwähnt. Es wird gezeigt, dass die Fehlstellenkonzentration im Festkörper mit dem Druck schwankt, wobei das Vorzeichen der Änderung davon abhängt, ob die betreffenden Defekte durch eine Verringerung des Volumens (Zwischengitterplätze) oder eine Ausdehnung des Volumens (Leerstellen) gekennzeichnet sind. Die Änderung der Fehlstellenkonzentration kann einen merkbaren Einfluss auf die Reaktivität haben; weitere wichtige Faktoren sind der engere Kontakt zwischen den reagierenden Körnern und das Auftreten von Phasenübergängen, die zu einer verstärkten Reaktivität des Materials führen. Abschliessend werden einige Anwendungen der Höchstdruck-Verfahren auf keramische Werkstoffe behandelt, wobei die Bedeutung dieser Verfahren für Wissenschaft und Technik herausgestellt wird.*

## 1. INTRODUCTION

Advances during the last ten years in the construction of apparatus suited for high-pressure high-temperature research have led to a greatly increased interest in the consequences of applying very high pressures to different materials. This paper is concerned with the effects on ceramic systems. It seems appropriate to start by defining the terms "ceramic systems" and "very high pressures" as used in this paper.

The term "ceramic systems" covers the inorganic non-metallic materials, generally with high melting-points. The lower limit of the "very-high-pressure" area is mostly put near 20 kb.* There is no upper limit apart from experimental possibilities. In static experiments it is now possible to reach 500 kb in room-temperature

*1 kb (kilobar)=1000 bars; one bar=0·98 atm.

conditions, and pressures near 300 kb at temperatures up to 2500°C. In dynamic experiments, using shock waves, much greater values of pressure and temperature can be reached. Pressures of over 3·6 Mb, which is the pressure at the centre of the earth, have been obtained. The temperatures in these experiments, however, are mainly determined by the experimental set-up and by the density of the material through which the shock-wave propagates, and cannot be chosen independently. It is easy to speak of 10 or 100 kb, but a reader who is not familiar with high pressures might recall that the tensile strength of the best steels lies at about 25 kb. Compressed gas in a cylinder is contained at about 0·1 kb; the deepest parts of the ocean reach about 1 kb; a uniform column of rock about 35 km high would exert a pressure on its base of about 10 kb.

The definitions used here give the freedom to mention an overwhelming abundance of data obtained with different disciplines. The very-high-pressure techniques have been used as a tool for investigating almost all aspects of the behaviour of solids and liquids, e.g. crystal structure; polymorphism; electrical, magnetic and optical properties; diffusion; and others. In this paper, however, only three topics will be dealt with:

(1) The influence of very high pressures on crystal structures of ceramic materials (mainly oxides).

(2) The influence of very high pressures on crystal defects.

(3) The influence of very high pressures on reactivity.

Finally some practical applications of modern high-pressure research will be reviewed.

## 2. THE INFLUENCE OF VERY HIGH PRESSURES ON CRYSTAL STRUCTURES

The direct effect of pressure on a material is a compression, a shortening of the inter-atomic distances by a force directed opposite to the repulsion forces normally present in the material. The compressibilities of the materials under consideration here, however, are rather small, as can be seen from the data in Table 1. Very high pressures are required to obtain a volume compression of, for instance, 10%. Much greater effects, however, do occur if the material undergoes a phase transformation induced by pressure. Two different types of such transformations can be distinguished, viz. a modified arrangement of the atoms or ions involved, leading to a different, more closely packed, crystal structure, or a discontinuous change in the radii of the atoms or ions as a consequence of a pressure-induced shift in the occupation of the various electron shells. Of course, a combination of these two possibilities is also imaginable. Let us first consider in more detail the phase transition involving another arrangement of the ions. In Table 2 some typical examples have been compiled.

**Table 1**
**Volume Compression of Some Oxides**[1]

| Compound | Pressure at which $V/V_0 = 0\cdot9$ (kb) |
|:---:|:---:|
| MgO | 230 |
| CaO | 140 |
| CdO | 190 |
| $Al_2O_3$ | > 320 |
| $Fe_2O_3$ | 250 |

This table, in which the volume differences have been calculated from the lattice constants of the two phases at one atmosphere and at room temperature, shows that the density increase obtained in this way clearly surpasses the increase caused by normal compressibility.

Inspection of the changes occurring in the crystal structure reveals that in nearly all cases the volume decrease is accompanied by an increase of the cation co-ordination number. This is, for ionic compounds at least, quite a general rule.

Secondly, it is striking that, apart from the medium-pressure phase of $SiO_2$, coesite, no novel high-pressure-structure types have been observed. This, too, is quite a general phenomenon. The high-pressure phases nearly always show crystal structures known for related compositions. Figure 1 shows the pressure/temperature diagram for $SiO_2$, one of the most thoroughly investigated substances.

Starting with a simple model of hard spheres of equal size it seems reasonable to assume that an increase in density goes parallel with an increase in co-ordination. It is known that, with spheres of equal radius, a space filling of 74% can be reached in an f.c.c. arrangement, while in a b.c.c. array this is only 68%. In four-co-ordinated structures the filling is even less than that. However, the radius of the spheres is in fact dependent on the co-ordination number, $R_{f.c.c.} > R_{b.c.c.}$ which means that, in practice, elements that undergo temperature-induced phase transitions from f.c.c. to b.c.c., or vice versa, show only minor volume changes in the order of 1%.[3]

Table 2

| Compound | Structure in ambient conditions | Cation co-ordination number | Necessary (P/T) conditions | High-pressure structure | Cation co-ordination number | $-\Delta V/V(\%)$ |
|---|---|---|---|---|---|---|
| ZnO | Wurtzite | 4 | 105kb/200°C | Rock salt | 6 | 18 |
| $Fe_2SiO_4$ | Olivine | 4.6.6 | 38kb/600°C | Spinel | 4.6.6 | 10·5 |
| $LiAlGeO_4$ | Phenacite | 4.4.4 | 25kb/450°C | Spinel | 4.6.6 | 19 |
| $MgGeO_3$ | Pyroxene | 4.6 | 40kb/700°C | Ilmenite | 6.6 | 15·4 |
| $CdTiO_3$ | Ilmenite | 6.6 | 25kb/500°C | Perovskite | 12.6 | 8·5 |
| $SiO_2$ | Quartz | 4 | 30kb/600°C | Coesite | 4 | 10 |
| $SiO_2$ | Quartz | 4 | 130kb/1200°C | Rutile | 6 | 38 |
| $CrVO_4$ | Own structure type | 4.6 | 60kb/750°C | Rutile | 6.6 | 13.7 |

In ionic compounds, too, the inter-ionic distance is dependent on the co-ordination, as was found experimentally by GOLDSCHMIDT[4] and later put on a theoretical basis by ZACHARIASEN.[5] The effect of a higher co-ordination is in turn opposed by the greater distance, but the overall effect is a clear decrease of the volume with increasing co-ordination. A schematic diagram of the space-filling percentage as a function of the cation-to-anion-radius ratio of some AX compounds, as given by NEUHAUS,[6] is shown in Figure 2. It is clear from this picture that, for a given co-ordination number, the percentage falls down if the value of $r_+/r_-$ increases.

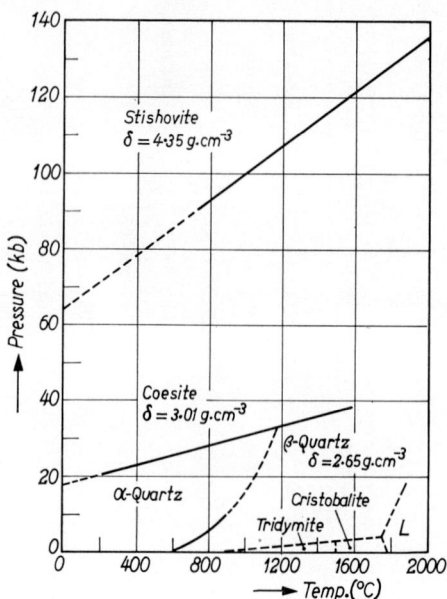

FIGURE 1.—Pressure/temperature phase diagram of $SiO_2$. (After data reported by several authors).

An example of a pressure-induced phase transition, in which the arrangement of the atoms has not been modified, but a change occurs in their radius, is given by the element cerium. This element, which normally crystallizes in the f.c.c. arrangement, undergoes a phase transition at room temperature and at about 7 kb to another f.c.c. arrangement, the two structures differing in their lattice constant: $a = 5 \cdot 16$ Å and $a = 4 \cdot 85$ Å respectively.[7] This implies a volume decrease of 17%, including the normal compressibility of the material

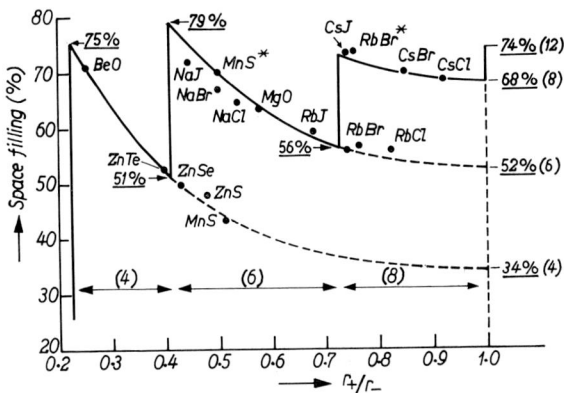

FIGURE 2.—Dependence of the space-filling percentage on the cation-to-anion radius ratio of some AX compounds (After reference 6).
* High-pressure phases.

up to the transition point. Measurements of the magnetic susceptibilities of both phases have proved that this change is caused by the shift of an electron from the 4f shell to 5d or 6s. The co-ordination number has been retained in this case. Similar phenomena have been observed in other elements, including cesium and barium.

Quite recently an analogous effect was found in the rare-earth monotellurides SmTe and EuTe.[8] These compounds which, like the other rare-earth monochalcogenides, crystallize in the rock-salt type of structure, show a phase transformation at pressures

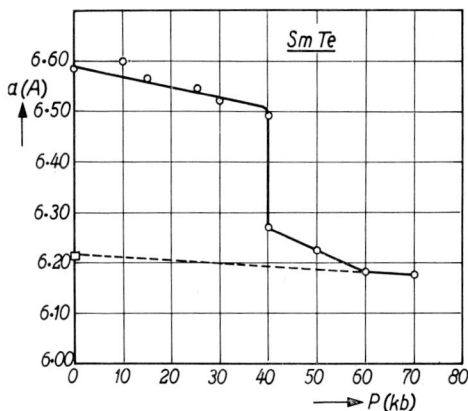

FIGURE 3.—Change in the lattice constant of SmTe as a function of pressure (After reference 2).

10

of about 30 kb to a high-pressure phase, again with the rock-salt type of structure. The variation of the lattice constant of SmTe with pressure is given in Figure 3. The transition can be described by the reaction:

$$Sm^{2+}.Te^{2-} \rightarrow Sm^{3+}.Te^{2-} + 1e$$

This reaction looks very uncommon at first sight. Some more details on the properties of the rare-earth chalcogenides will clear up the picture.

It was earlier observed by IANDELLI[9] that two groups of compounds can be distinguished in the monochalcogenides of the rare-earth elements, one consisting of the compounds with Eu, Sm and Yb as the cation, the other comprising the other rare-earth compounds. The differences between the two groups are summarized in Table 3.

**Table 3**

| Compound | Lattice constant | Magnetic properties | Electrical properties |
|---|---|---|---|
| SmX, EuX,* YbX $(R^{2+}X^{2-})$ | In accordance with radius divalent cation + radius divalent anion | Indicating divalent cation | Insulating or semi-conducting |
| Other RX compounds $(R^{3+}X^{2-} + 1e)$ | In accordance with radius trivalent cation and radius divalent anion | Indicating trivalent cation | Metallic |

*X = O, S, Se, or Te.

A graphical representation of the lattice constants as a function of the atomic number, as given in Figure 4, clearly shows this difference.

Without going into much detail, it can be said that the difference is caused by the strong tendency of the rare-earth elements to form trivalent cations. Only in those cases where the divalent stage leads to a favourable electronic configuration, e.g. a completely filled 4f shell as in $Yb^{2+}$, a half-filled 4f shell as in $Eu^{2+}$, or a nearly half-filled 4f shell as in $Sm^{2+}$, are the "normal" compounds encountered. Pressure causes here a shift of one electron from the 4f shell to the conduction band. The volume decrease of this isostructural transition is about 16%.

As already mentioned, a combination of both a different arrangement of the ions and a change in the charge of the constituent ions is also possible. The best example is given by the compound InTe.

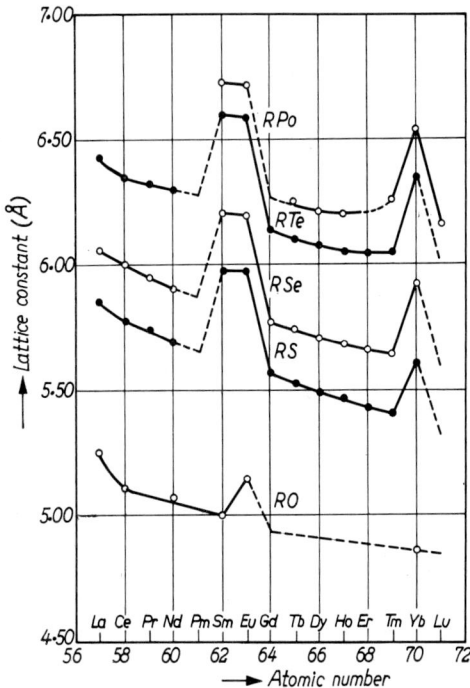

FIGURE 4.—Lattice constants of the rare-earth monochalcogenides as a function of the rare-earth atomic number (after reference 2).

Under normal conditions, InTe crystallizes in the tetragonal TlSe structure, in which half of the indium ions are four-co-ordinated by tellurium, the other half being in eight-co-ordination. In fact two valencies of the indium ions are present, and a better formula would therefore be $In^I.In^{III}.Te_2^{II}$. In accordance with this formula the low-pressure modification is diamagnetic. At pressures of about 30kb a transition takes place to a rock-salt type of structure.[19] The originally semiconducting compound has now become metallic. The charge distributions have changed from $In^I + In^{III}$ to $2In^{II}$.

## 3. THE INFLUENCE OF VERY HIGH PRESSURES ON CRYSTAL DEFECTS

Every material contains defects of thermodynamic and kinetic origin. These defects are, for instance, vacancies, interstitials, dislocations, clusters of a second phase, grain boundaries, and so on. Their concentration is not as a rule equal to the equilibrium value, but is determined by the state of the starting materials and from the heat or mechanical treatment to which the material has been subjected. It is clear, however, that the concentration of such defects will depend on both temperature and pressure. The creation for example, of interstitials or vacancies will be associated with a difference in volume. If $V$ is the change in partial molar volume involved in the creation of such defects, $C_0$ the concentration at atmospheric pressure, and $T$ the absolute temperature, then the equilibrium concentration at a pressure $p$ will be given by

$$C_p = C_0 \exp\left(-pV/RT\right).$$

Thus, for example, for a value of $V = 5$ cm$^3$ per mole and $T = 500°$K, $C_p/C_0$ will be 0·87 at 100 bars, but a value of $4·10^{-6}$ will be the equilibrium ratio at $T = 500°$ K and 100 kilobars (for $V$ independent of pressure). Cooling of the sample from 500° K to room temperature, or lower, in high-pressure conditions, and releasing the pressure afterwards, leads to non-equilibrium situations and makes it possible to investigate compounds with extremely low vacancy-concentrations. In general it can be said that those defects that involve an expansion of the volume will tend to decrease in number upon the application of pressure. On the other hand, defects which are connected with a negative change in volume are favoured under high-pressure conditions. An example of the last kind is possibly given by the high-pressure experiments on zinc ferrite by GOTO.[11]

The fact that it is very difficult to maintain reasonable hydro-static conditions at very high pressures causes sound experiments in this area to be rather scarce. It is obvious that the presence of uniaxial stresses introduces new defects. This can lead to high-pressure phases with abnormally high concentrations of defects, which can partially disappear by annealing under high-pressure conditions. ROY and co-workers[12] showed that the resistance of CdS pellets with rock-salt structure, measured under pressure, could be enhanced four to five orders of magnitude by such an annealing procedure. In a recent paper GALE and KULP[13] make mention of the preparation of a single crystal of CdS in the rock-salt modification. It was possible for these authors to recover this sample at ambient conditions without any structural change by freezing it at liquid-nitrogen temperature before release of the pressure. The optical characteristics of this crystal were markedly different from earlier measurements performed under non-hydrostatic conditions. Recent developments[14] in the techniques of maintaining quasi-hydrostatic pressures up to 60 kb make it probable that valuable studies on the effect of high pressures on defects can be expected. It would, for instance, be quite interesting to grow single crystals of a material under both normal and very-high-pressure conditions and to compare their physical characteristics afterwards.

## 4. THE INFLUENCE OF VERY HIGH PRESSURE ON REACTIVITY

The application of very high pressures during the reaction of ceramic materials may lead to a considerable enhancement of the reaction rate. Striking examples of such a behaviour, evidently caused by closer contact between the grains of the reacting materials, have been found by POCH,[15] among others. The formation of spinels from the constituent oxides under pressure may take place at considerably lower temperature than normally necessary. As this effect is discussed more extensively in the paper by Poch, it is mentioned here only in passing. Some other effects must also be taken into consideration: the influence of very high pressure on idiffusion, on the reaction mechanism, and on the sign of the difference n molar volume between the initial and final state of the compounds involved.

If the rate of the reaction is determined by the diffusion of one compound into the other, the activation volume for diffusion, which has a positive value, will cause a decreasing reaction rate. This effect is therefore opposite in sign to the one of closer contact mentioned first. One must be very careful, however, in interpreting the activation volumes for diffusion, and even more the activation volumes for transformation, the latter being the difference in molar volume between the starting material and the activated complex (the intermediate state or transition state) in the case of a polymorphic transformation. This applies especially in heterogeneous systems, in which the course of the reaction is not accurately known. Phase transitions at very high pressures proceed in many cases at extremely high rates, indicating that the normal concepts of reaction and transformation rate theory are no longer applicable. The graphite–diamond conversion or the wurtzite–rock-salt transition of CdS, both of which can be realized by shock waves, and therefore within $10^{-6}$ sec., make this clear. The resulting increase in temperature gives no sufficient explanation.

The actual reaction mechanism is also of considerable importance. REIJNEN[16] gives a detailed account of the interplay between starting materials and the surrounding gas atmosphere during the formation of some ferrite materials. In many cases the reaction path is insufficiently known, which makes interpreting of the results of high-pressure experiments on such reactions a difficult affair.

It is also necessary to compare the volumes of the reacting compounds. Let us for instance consider the reaction:

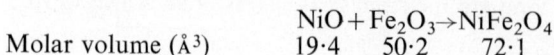

$$NiO + Fe_2O_3 \rightarrow NiFe_2O_4$$

Molar volume ($Å^3$)       19·4       50·2       72·1

Comparison of the molar volume change reveals that the compound to be formed has a lower density than the starting products. Very high pressures will therefore lead to a decomposition of the ferrite compound instead of to its formation. This effect has been observed by MAREZIO, REMEIKA and JAYARAMAN[17] on lithium ferrite. Here a pressure of about 35 kb at temperatures near 1000°C gives the reaction:

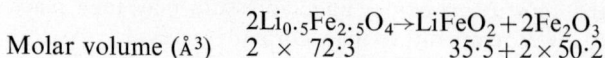

$$2Li_{0.5}Fe_{2.5}O_4 \rightarrow LiFeO_2 + 2Fe_2O_3$$

Molar volume ($Å^3$)       2 × 72·3       35·5 + 2 × 50·2

This reaction leads to a decrease in volume of 6%. Whereas in the spinel phase one third of the cations have a tetrahedral surrounding, the decomposition products $LiFeO_2$ and $Fe_2O_3$ have the cations in six co-ordination. The increase in volume going from NiO and $Fe_2O_3$ to $NiFe_2O_4$ can also be easily explained by the change in cation co-ordination—here from only six co-ordination in the starting materials to mixed four, six, in spinel.

In the reaction:

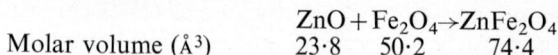

$$ZnO + Fe_2O_4 \rightarrow ZnFe_2O_4$$

Molar volume ($Å^3$)       23·8       50·2       74·4

it is seen that the change in volume is only very slight, the cation co-ordination remaining equal in both situations.

The conclusion must therefore be that the effect of very high pressures on reaction rates in ceramic systems cannot be simply reduced to the same denominator. Both the reaction mechanisms, and the structure and volume of the phases involved, may play a decisive role.

## 5. APPLICATIONS OF VERY-HIGH-PRESSURE TECHNIQUES TO CERAMIC SYSTEMS

Turning, finally, to the existing or prospective applications of very high pressures, a few topics merit some attention apart from the moulding techniques of hydrostatic extrusion and isostatic pressing.

As far as new materials are concerned, the synthesis of large single crystals of quartz by hydrothermal means at 1500 atm. and 400°C is certainly worth mentioning.[18] Crystals have been grown as large as a few hundred cubic centimetres. Diamond crystals up to about a millimetre in size, grown at pressures near 60 kb and temperatures above 1500°C,[19] are being produced now in increasing quantities up to several millions of carats a year to satisfy the demands of industrial drilling and polishing. Recently new superconducting materials, $Nb_3In$, $Nb_3Bi$ and $Mo_3Sn$, all with the $\beta$-W structure, have been synthesized.[20] Borazon, too, the boron nitride compound with diamond structure and corresponding properties, deserves to be mentioned.[21] Apart from these new modifications of sometimes already known compounds, completely new compounds have been synthesized, like the boron oxide, $B_2O$, called the "unsymmetrical" analogue of graphite, with hexagonal crystal structure.[22] Some new dichalcogenides of bismuth and lead have been prepared at high pressures and temperatures by SILVERMAN.[23]

The greater insight into the stabilities of crystal structures gained from the study of pressure- and temperature-induced polymorphism, is of considerable importance to our understanding of chemical reactions in the solid state. In Table 4 some thermodynamic data on inorganic materials obtained in this way have been compiled.

It may be remarked at this point that the systems given are treated as quasi-unary, allowing application of the Clausius–Clapeyron rule. Moreover, the $\Delta V$ values do not take into account either the thermal expansion or the compressibility of the two phases, except in the case of the $\alpha$- to $\beta$-quartz transition. In this approximation the corrections for these effects do not seem necessary. The difference in lattice energy between different modifications of one substance is a few kcal per mole, which is only a few percent of the lattice energy itself. The existence of different polymorphs of one and the same compound makes it possible to study the dependence of physical properties, such as electrical and thermal conduction, optical and magnetic behaviour, on structural characteristics such as co-ordination and inter-atomic distance. Examples are the spinel and the $Cr_3S_4$ polymorph of $FeCr_2S_4$ and related compounds,[24,25] or the phenacite and the spinel polymorph of $Li_2MoO_4$.[26] The effects of temperature on physical properties can now be investigated at constant volume.

High-pressure research has also made an important contribution to the understanding of geochemical and geophysical phenomena. The same is true of our knowledge about the origin and history of the meteorites. BERNAL's[27] hypothesis that the abnormally steep density-gradient in the earth's mantle between about 200 and 900 km depth could be explained by an olivine–spinel transition in the system $Mg_{2-x}Fe_xSiO_4$ has been strongly supported by high-pressure laboratory experiments, first on $Fe_2SiO_4$,[28] and quite

Table 4

| Compound | N.T.P. structure | Density | High-pressure structure | Density | $\Delta V$ $(cm^3 mol^{-1})$ from X-ray data | $dP/dT$ $(atm.\ deg.^{-1})$ | $\Delta S$ $(cal.\ mol.^{-1}$ $deg.^{-1})$ calculated | $P\Delta V$ | $T\Delta S$ | $\Delta U$ |
|---|---|---|---|---|---|---|---|---|---|---|
| | | | | | | | | (kcal.mol⁻¹ at 300°K) | | |
| $SiO_2$ | Quartz, $\alpha$ | 2·65 | Quartz, $\beta$ | 2·52* | 0·11* | 34·5 | 0·1 | 0·0 | 0·1* | 0·1* |
| | Quartz, $\alpha$ | 2·65 | Coesite | 2·92 | −2·0 | 10·7 | −0·5 | −0·9 | −0·2 | 0·7 |
| | Coesite | 2·92 | Stishovite | 4·28 | −6·6 | 22 | −3·4 | −14·7 | −1·0 | 13·7 |
| $ZnO$ | Wurtzite | 5·68 | Rock salt | 6·91 | −2·55 | 42·5 | −2·5 | −5·4 | −0·8 | 4·6 |
| $CdTiO_3$ | Ilmenite | 5·80 | Perovskite | 6·32 | −2·9 | −48·8 | 3·4 | −2·8 | 1·0 | 3·8 |
| $Mg_2GeO_4$ | Olivine | 4·04 | Spinel** | 4·38 | −3·5 | 40 | −3·3 | — | −3·6 | −3·6 |
| $CdS$ | Wurtzite | 4·83 | Rock salt | 6·37 | −7·2 | 10 | −1·7 | −3·0 | −0·5 | 2·5 |
| $FeCr_2S_4$ | Spinel | 3·83 | $Cr_3S_4$ type | 4·19 | −6·5 | −36·0 | 5·5 | −5·6 | 1·7 | 7·3 |

*At the transition temperature of 573°C.
**Spinel is the stable low-temperature modification; $T_{tr} \approx 810°C$ at one atmosphere; values applicable to this temperature.

recently on the whole system, $Mg_2SiO_4$ being found to crystallize in a spinel structure at pressures above 155 kb and temperatures of about 800°C.[29] At still higher pressures and temperatures, transitions to MgO and $SiO_2$, the latter one in the rutile modification, can be predicted.

Shock-wave experiments on metals and on silicates at pressures up to those to be expected in the earth's core have made it highly improbable that silicates are the main constituents. The experiments are in excellent agreement with a model in which the core consists mainly of iron, alloyed possibly with some 10% of silicon. Some of these results are shown in Figure 5. Extrapolation of the dependence

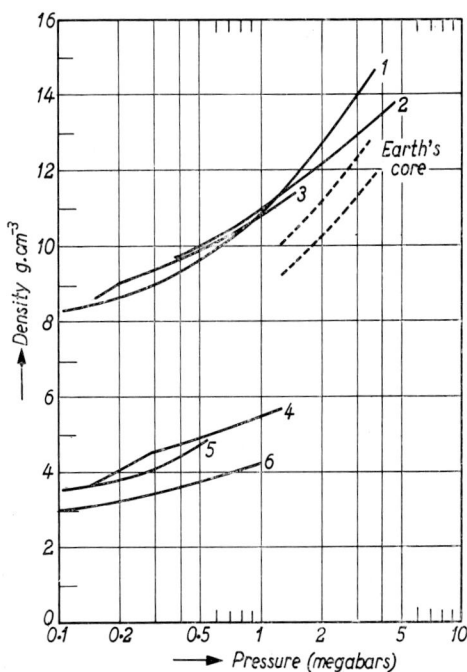

FIGURE 5.—Density *versus* pressure (logarithmic scale):
(1) Murnaghan equation for isothermal compression of iron.
(2) and (3) Shock compression of iron.
(4) Density of the earth's mantle.
(5) Shock compression of olivine, $Mg_2SiO_4$.
(6) Shock compression of aluminium.
The broken curves are the limiting densities for the earth's core. After F. Birch in "Solids under Pressure" (Ed. W. Paul and D. M. Warschauer) (McGraw Hill Book Company Inc.: 1963).

148          ROOYMANS:

of the melting point of iron and iron alloys on pressure gives rather
rough estimates of the temperature present on the mantle–core
boundary. Some of these results are shown in Figure 6.

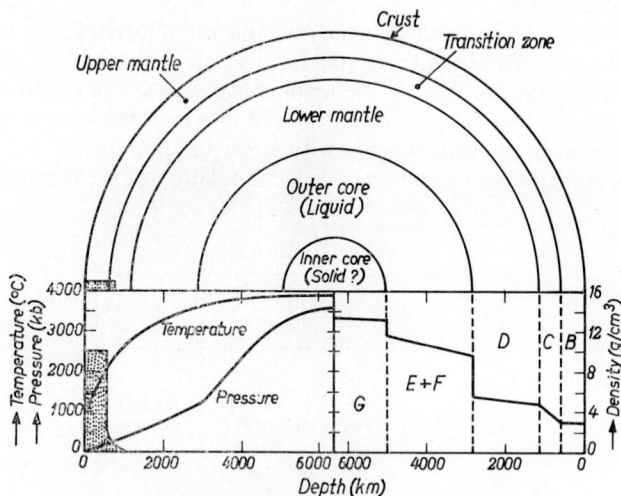

FIGURE 6.—Characteristics of the earth's interior. Stippled areas indicate
pressure/temperature capabilities of present-day apparatus. After R. C.
Newton in "Advances in High Pressure Research" (Ed. R. S. Bradley) Vol. I
(Academic Press: London and New York 1966).

## 6. CONCLUSION

In conclusion it can be said that the study of the effects of very
high pressures offers new prospects of deepening our understanding
of ceramic materials, both as regards synthesis, physical properties,
and their occurrence in nature.

**REFERENCES**

1. DRICKAMER, H. G., LYNCH, R. W., CLENDENEN, R. L., and PEREZ-ALBUERNE.
   E. A., "Solid State Physics" Vol. 19, 135, 1966 (Academic Press;
   New York and London).
2. ROOYMANS, C. J. M. Thesis, Amsterdam 1967. *Philips Res. Repts Supplements.*
   1968, No. 5.
3. RUDMAN, P. S., *Trans. metall. soc. A.I.M.E.*, **233**, 864, 1965.
4. GOLDSCHMIDT, V. M., Skrifter utgitt avdet Norske Vidensk. Ak. Oslo I.M.N.-
   Kl. 1926, p. 57.
5. See, for instance; KETELAAR, J. A. A., "The Chemical Constitution" (Elsevier
   Publish. Co.; Amsterdam, 1958).
6. NEUHAUS, A., BALLHAUSEN, C., MEYER, H. J., and STEFFEN, R., "Jahrbuch
   1965 des Landes Nordrhein-Westfalen/Landesamt für Forschung"
   (Westdeutscher Verlag, Köln, 1966).
7. JAYARAMAN, A., *Phys. Rev.*, **137**, A179, 1965

8. ROOYMANS, C. J. M., *Ber. Bunsenges. phys. Ch.*, **70**, 1036, 1966.
9. IANDELLI, A., "Rare Earth Research" (Ed. E. V. Kleber) (MacMillan Co.: New York, 1961), p. 135.
10. GELLER, S., JAYARAMAN, A., and HULL, G. W., *J. Phys. Chem. Sol.*, **26**, 353, 1965.
11. GOTO, Y., *Japan J. Appl. Phys.*, **3**, 309, 1964.
12. MILLER, R. O., DACHILLE, F., and ROY, R., *J. Appl. Phys.*, **37**, 4913, 1966.
13. GALE, K. A., and KULP, B. A., *J. Phys. Chem. Sol.*, **28**, 1233, 1967.
14. BARNETT, J. D., and BOSCO, C. D., *Rev. Sci. Instr.*, **38**, 957, 1967.
15. POCH, W., "Science of Ceramics", Volume 4 (Stoke-on-Trent: British Ceramic Society, 1968).
16. REIJNEN, P. J. L., "Science of Ceramics" Volume 4 (Stoke-on-Trent: British Ceramic Society, 1968).
17. MAREZIO, M., REMEIKA, J. P., and JAYARAMAN, A., *J. Chem. Phys.*, **45**, 1821, 1966.
18. LAUDISE, R. A., *J. Amer. Chem. Soc.*, **81**, 562, 1959.
19. BUNDY, F. P., *J. Chem. Phys.*, **38**, 631, 1963.
20. BANUS, M. D., REED, T. B., GATOS, H. C., LAVINE, M. C., and KAFALAS, J. A., *J. Phys. Chem. Sol.*, **23**, 971, 1962; KILLPATRICK, D. H., *J. Phys. Chem. Sol.*, **25**, 1499, 1964.
21. BUNDY, F. P., and WENTORF, R. H., *J. Chem. Phys.*, **38**, 1144, 1963.
22. HALL, H. T., and COMPTON, L. A., *Inorg. Chem.*, **4**, 1213, 1965.
23. SILVERMAN, M. S., *Inorg. Chem.*, **3**, 1041, 1964; **5**, 2067, 1966.
24. ALBERS, W., and ROOYMANS, C. J. M., *Sol. St. Comm.*, **3**, 417, 1965.
25. BOUCHARD, R. J., *Mat. Res. Bull.*, **2**, 459, 1967.
26. LIEBERTZ, J., and ROOYMANS, C. J. M., *Sol. St. Comm.*, **5**, 405, 1967.
27. BERNAL, J. D., *Observatory*, **59**, 268, 1936.
28. RINGWOOD, A. E., *Geochim. cosmochim. Acta*, **15**, 18, 1958. *Amer. Miner.*, **44**, 659, 1959.
29. RINGWOOD, A. E., and MAJOR, A., *Earth Planet. Sci. Letters*, **1**, 241, 1966; AKIMOTO, S., and IDA, Y., *Earth Planet. Sci. Letters*, **1**, 358, 1966.

# 11.—The Sintering of Fine-particle Oxides under High Pressure

By W. POCH

*Max Planck Institut für Silikatforschung, Würzburg, Germany*

*ABSTRACT* C525

*Spherical particles of Ag, W, and $Al_2O_3$ and MgO crystals of irregular shape have been "hot"-pressed at 20 kbar between 20° and 1250°C. Depending on the material and the pressing conditions, the particles deform plastically or by fracture, but in each case a relative density exceeding 0·9 is attained in 2 min. MgO with a crystallite size of 140 Å was sintered between 700° and 1000°C and at 40 kbar. During the first, rapid, stage of compaction, a density 0·95 is reached; during the following 3 min., e.g. at 1000°C and 20 kbar, 0·995 is attained. Strongly discontinuous crystal growth of the particles occurs, resulting in a structure of 25-μm particles in a sub-microscopic matrix. With MgO of 1000 Å as starting material, rapid densification also occurs at relatively low temperatures, but without discontinuous grain growth. $BaTiO_3$ and $NiFe_2O_4$ of 0·99 density and crystal sizes of 500 Å and 800 Å respectively were obtained by high-pressure sintering for 5 min. at 800°C and 20 kbar.*

*Le frittage d'oxydes en particules fines sous des pressions élevées*

*Des particules sphériques de Ag, W, $Al_2O_3$ et des cristaux de MgO de forme irrégulière ont été pressés à "chaud" à 20 kbar entre 20 et 1250°C. Suivant le matériau et les conditions de pressage, les particules se déforment plastiquement ou par fracture, mais dans chaque cas une densité relative dépassant 0,9 est atteinte en 2 minutes. MgO, sous forme de cristallites de 140 Å a été fritté entre 700 et 1000°C et à 40 kbar. Au cours du premier stade, rapide, de la compaction, une densité de 0,95 est atteinte; au cours des trois minutes suivantes à 1000°C et 20 kbar, la densité atteinte est de 0,995. Il se produit une croissance cristalline fortement discontinue des particules qui a pour effet de donner un produit constitué de particules de 25 microns dans une matrice submicroscopique. En opérant avec des cristallites de MgO de 1000 Å, la densification est également rapide à des températures relativement basses, mais se fait sans croissance discontinue des grains. $BaTiO_3$ et $NiFe_2O_4$ à densité 0,99 ont été obtenus par* *frittage à haute pression pendant 5 minutes sous 20 kbar avec des dimensions de cristaux respectives de 500 Å et 800 Å.*

### Hochdruck-Sintern feinteiliger Oxyde

*Ag-, W- und $Al_2O_3$-Kugeln sowie unregelmäßig geformte MgO-Kristalle wurden bei 20 kbar zwischen 20° und 1250°C "heiß"-gepreßt. Je nach dem Material und den Preßbedingungen werden die Teilchen entweder plastisch verformt oder durch Bruch zerkleinert. In jedem Fall wird nach 2 min. eine relative Dichte größer 0,9 erreicht. MgO mit einer Kristallitgröße von 140 Å wurde zwischen 700° und 1000°C bei 20–40 kbar gesintert. Im ersten, schnell ablaufenden Verdichtungsschritt wird eine Dichte von 0,95 erreicht, die während der folgenden 3 min z.B. bei 1000°C und 20 kbar auf 0,995 ansteigt. Es findet dabei ein stark diskontinuierliches Kristallwachstum statt, das zu einem Gefüge aus 25 μm großen Teilchen in einer submikroskopischen Matrix führt. Mit MgO von 1000 Å als Ausgangsmaterial tritt ebenfalls bei relativ niedrigen Temperaturen eine schnelle Verdichtung ein, jedoch ohne diskontinuierliches Kristallwachstum. $BaTiO_3$ und $NiFe_2O_4$ mit Dichten von 0,99 rel. und Kristallitgrößen von 500 Å bzw. 800 Å wurden innerhalb von 5 min. durch Hochdruck-Sintern bei 800°C und 20 kbar erhalten.*

## 1. INTRODUCTION

The advantages of hot pressing over normal sintering are that almost completely dense samples with a fine microstructure can be obtained at relatively low temperatures in a short time and that "difficult compounds" can be dealt with. The trend goes to pressures in the range of $10^3$ atmospheres, under isostatic conditions if possible.[1]

By observing the changes in structure during preliminary experiments with coarse particles we were able to deduce the mechanisms that are involved in high-pressure sintering. We then investigated in detail the sintering behaviour of fine crystalline materials and, as a result, were able to prepare dense $BaTiO_3$ and $NiFe_2O_4$ of submicroscopic structure.

Pressures of 20 and 40 kbar were applied in a Belt apparatus of the usual type at temperatures between 700° and 1000°C without the samples being evacuated.

## 2. DENSIFICATION OF COARSE PARTICLES

MURRAY et al.[2] have put forward a mathematical description of hot pressing which is based on the Mackenzie–Shuttleworth flow model. Although all the relevant parameters are taken into account, this model gives a good description only for the density range from 0·8 to 0·95 because the theory, which is derived from hydrostatic pressure distribution, does not take into account that the yield strength is pressure-dependent. Even more important, materials which are brittle under normal conditions become very ductile under very high isostatic pressure (e.g. $Al_2O_3$ at 20°C and 20 kbar can be elongated 30% before fracture). On the other hand, it is clear that, in pressure sintering, isostatic conditions only occur in the final stage. In the early stage, peak pressures occur at the areas of contact between particles, which leads to grain-boundary movement or fragmentation as the first step in densification.

For the temperature range in which we are interested, no data are available that enable calculated sintering curves (e.g. according to the Murray model) to be compared with experimental values. Furthermore, the curves obtained experimentally show an unfavourable density/time relationship. In addition, SCHOLZ and LERSMACHER [3] warn against describing sintering by means of a single model. It is therefore advisable to seek information on densification direct from high-pressure sintering.

The substances chosen for these experiments—Ag, W, MgO and $Al_2O_3$, have very different flow properties. Except for MgO, in all cases the starting material consisted of spherical polycrystalline particles of 30–90 $\mu$m diameter.

A pressure of 20 kbar (at 20°C) applied to silver spheres results instantaneously in a body of 0·995 relative density, probably compacted by plastic flow. This concept is supported by Figure 1A, which shows that the pressure applied is considerably higher than the yield strength.

Tungsten spheres were densified to 0·98 relative density in 1 min. at 400°C and 20 kbar, this pressure being approximately in the range of the critical strain for plastic flow. The resultant structure (Figure 1B) is in accordance with plastic deformation of the particles.

When heated to 800°C before the pressure is applied, coarse crystalline MgO (powdered single crystal, grain size 0·5–0·9 mm) can be densified to 0·97. The mechanism is a flow or slip of the original particles (Figure 1C) which can be concluded from the densities obtained, grain arrangement and the unchanged grain size of samples compacted at pressures below 20 kbar. If, however, the pressure is applied first, there is a reduction in particle size, and a density of 0·9 results; on heating at 800°C for 2 min., this value increases to 0·98.

Fragmentation is clearly observed when $Al_2O_3$ spheres are cold-pressed at 20 kbar (Figure 1D). Here (in contrast to MgO) the size

A

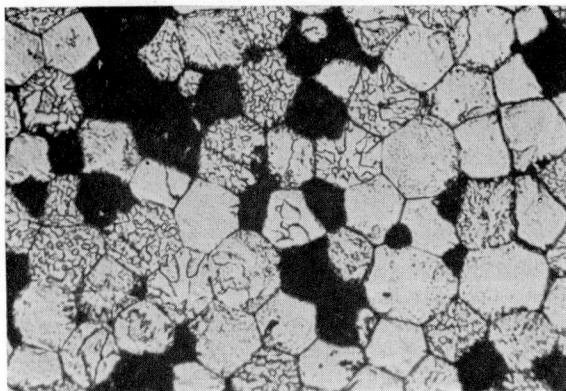

B

FIGURE 1
A. Ag spheres pressed at 20 kbar and 20°C.
B. W spheres pressed at 20 kbar and 400°C.

C

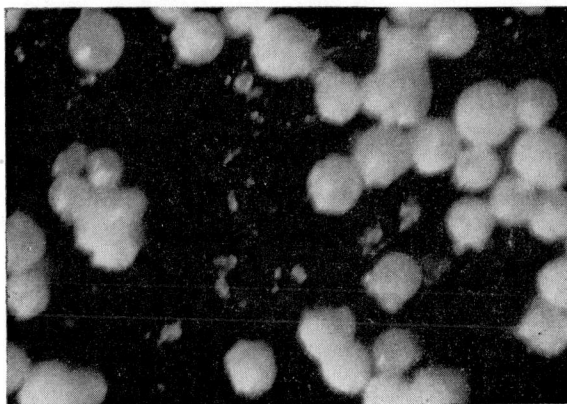

D

FIGURE 1—*continued*

C. MgO crystallites, 0·5–0·4 mm, pressed at 20 kbar and 800°C.
D. $Al_2O_3$ spheres pressed at 20 kbar and 20°C.

11

of the particles also decreases if the sample is heated to 1000°C before pressure is applied. In this case the pressure applied, in relation to the actual pressure at areas of contact, is higher than the rupture strength, which is lower than the critical yield strength. However, the main densification also takes place within 2 min., at the end of which the relative density is approximately 0·9. With $Al_2O_3$ spheres, plastic deformation under 20 kbar is observed only when the temperature is raised to 1250°C before the pressure has been applied. The relative density after 2 min. is 0·97. The structure is analogous to that of pressed spheres of Ag or W.

These experiments have shown that, by hot pressing under very high pressures, densification to a relative density $> 0·9$ occurs in a few minutes. Depending on the material, the temperature and the pressing conditions, densification is achieved by plastic deformation and/or by rupture of the spheres; probably grain-boundary movement is also involved. When plastic deformation is involved, the first stage of densification apparently leads to a higher density than does densification by fracture. In the slow, second, stage of high-pressure sintering, densification may be governed by diffusion.

In the following experiments with MgO, it should be remembered that heating at 800°C followed by the application of pressure causes plastic deformation of the particles which, together with grain-boundary movement, leads rapidly to a relative density of nearly 1.

## 3. SINTERING OF FINE POWDERS AT 20 AND AT 40 kbar

### 3.1 Magnesium Oxide

MgO powder A, with an average particle size diameter of 140 Å,* and powder B, of 1000 Å, were prepared by calcining MgCO₃ (p.a. Merck) at 600° and 1250°C respectively for 3 h.[4] BUDNIKOV[5] in a recent paper has described the sintering behaviour of MgO under normal hot-pressing conditions.

### 3.11 Densification

The sample, at an intermediate pressure, is heated to the appropriate temperature; the pressure is then increased by 5–10 kbar/min. up to either 20 or 40 kbar, as appropriate, and after a certain time the treatment is reversed. Figure 2 gives the relative density d** of powder A for several pressure + temperature treatments.

In considering these curves it should be noted that, taking the time at which sintering starts ($\tau = 0$) as the time at which the pressure

*Determined from broadening of the X-ray diffraction lines. We are indebted to Dr A. Krauth for these grain-size determinations.

**MgO single-crystal of 3·567 density is unity.

first reaches the maximum value, introduces a certain discrepancy for short sintering-periods. The real $d_0$ value should exceed 0·9, since a rapid increase of pressure up to 20 or 40 kbar at room temperature followed by immediate release causes a density of $\sim 0·9$.

It is clear from Figure 2 that, by applying a high enough pressure at relatively low temperatures (0·35 of the melting temperature of MgO), densities approaching 1 can be obtained almost instantaneously. This extremely rapid densification is even more striking if compared with sintering behaviour under normal hot-pressing conditions (at 1·4 kbar;[6] Figure 3).

The curves in Figure 2 indicate principally the final stage of high-pressure densification, which we attribute to a diffusion process. The increasing density both for a set of isobars with increasing temperature and for isotherms with decreasing pressure indicate a diffusion mechanism.

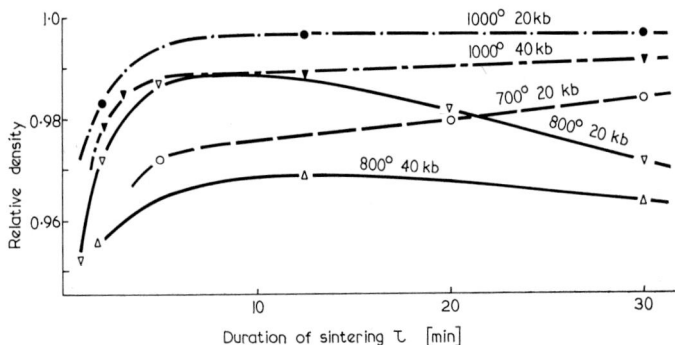

FIGURE 2.—Density/time curves of MgO A sintered at high pressures between 700° and 1000°C.

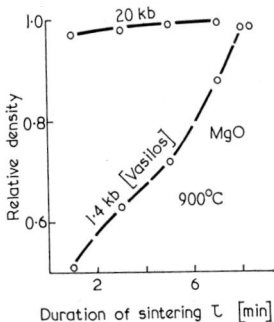

FIGURE 3.—Density of MgO during hot pressing and sintering under high pressure.

The decrease in density with time after $\sim 10$ min. may be due to interaction between the lattice and the gas within the sample, which is under high pressure. As the total surface of the particles decreases during crystal growth, the amount of entrapped gas will increase with the amount of gas absorbed, and at a certain level may cause the structure to explode if there are no large channels along which the gas can escape. The observation that the density decreases when the starting material has larger crystallites (MgO B, Figure 4) accords with this hypothesis.

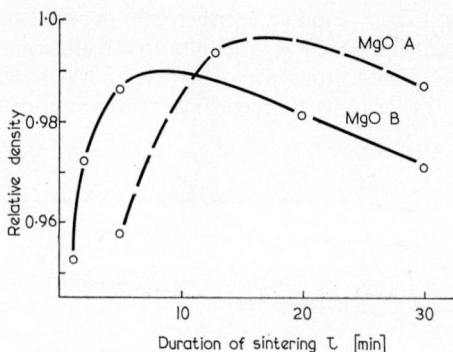

FIGURE 4.—Densification of MgO A and MgO B at 800° and 20 kb.

### 3.12 Microstructure, Strength and Crystal Growth

The photomicrographs of polished* samples of the 800°C, 20 kbar series with MgO A afford a good idea of the unusual structure obtained for this material (Figure 5,A–C). These spherical, nearly pore-free crystals appear in the amorphous matrix during the first 10 min. of sintering, the rate of growth being about 2 $\mu$m/min. With prolonged sintering the remaining space is mainly filled by the growth of newly nucleated crystals; a minor part is filled by growth of the crystals already present. Figure 5C shows various stages of this process. Two particles on the right-hand side seem to be an ideal example of the model of the sintering of two spheres, but in this case there is also mass transport from the neighbourhood to the spheres.

In order to characterize the "amorphous" phase and the "coarse" phase, we measured the microhardness, $H_V$. In one series of experiments the hardness of the embedded phase was found to be nearly independent of sintering time: the measured value is 450–500 kp/mm$^2$ (the hardness of a single crystal of MgO is 680 kp/mm$^2$), that of

*Pieces of material contained in plastic have been polished with diamond (3–0·25$\mu$m) etched in HNO$_3$ 1:1 during 15–30 sec. and lightly polished.

the fine-grained matrix lies between 270 and 400 kp/mm$^2$, the value depending slightly on the density. The measured "hardness" of the matrix seems to represent the strength of the structure rather than the real hardness of its constituent units.

If the temperature is increased to 1000°C (at 20 kbar), after a sintering period of 2 min. the two phases are no longer easy to distinguish (Figure 6A). In such cases, areas with similar hardness may have different porosities. After 30 min. sintering, a non-porous uniform structure is obtained (Figure 6B) similar to that formed by hot pressing a material of large crystallites. The cohesion of the particles, however, is not very good.

At 700°C and 20 kbar, almost no structure can be seen, the difference between the "crystalline" and the "amorphous" areas being even less well defined than in the 800°C experiment. The samples obtained at 700°C are susceptible to fracture.

Increasing the pressure to 40 kbar favours the formation of the large crystals and, in contrast to the experiments at 700°C and 20 kbar, the two phases are well separated, the coarse crystals are well formed and the growth surface is clearly distinguishable (compare Figures 7 and 5C). Figures 8 and 6B show that the well defined end-stages are retarded when the pressure is raised to 40 kbar. An increase in pressure increases the rate of nucleation. After 2 min. at 800°C and 40 kbar, about 5 times more macrocrystals have grown in the matrix than after 20 min. at 800°C and 20 kbar (Figures 9 and 5B).

As the crystals increase in size, the rate of crystallite growth in the initial stage slows down, possibly due to a retardation effect of the separated gas at the growth front. This decrease in growth rate, in conjunction with continuous nucleation of new crystals, leads to the fairly uniform structure of the final stage.

Discontinuous grain growth, which has not been observed to the same extent during normal sintering, only occurs when the grain size of fine raw materials is very small. It can be suppressed using MgO powder B ($D = 1000$ Å) which, when sintered at 800°C and 20 kbar for 30 min., produces a dense material without observable structure (compare Figures 10 and 5C).

The tensile strength of the samples was determined on tiny rods by the 3-point method, but as only one or two values for each sample were determined, accurate values cannot be given. Those of dense samples of which the fraction of the large crystallites does not exceed 50% have the highest strength. The samples with the highest strength were obtained from powder B at 800°C and 20 kbar for 12 min. Somewhat lower values have been measured on samples made

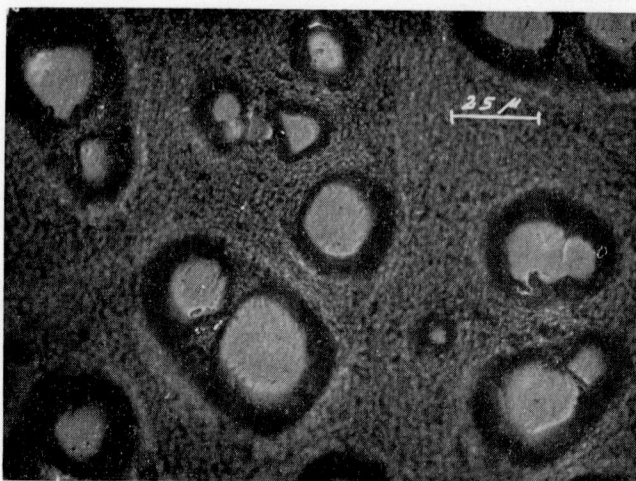

A

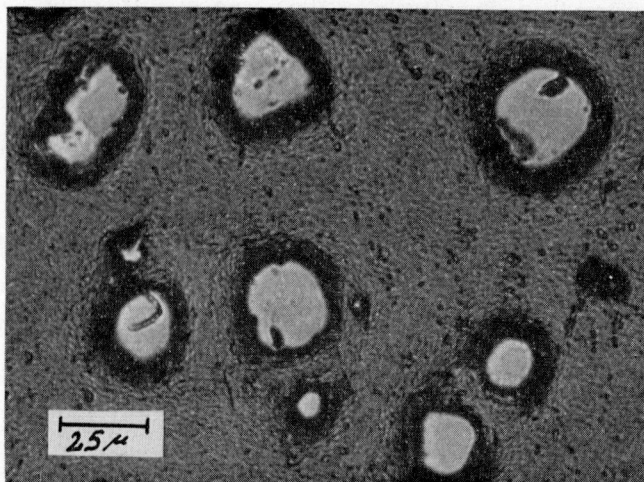

B

FIGURE 5.—Microstructure MgO A after sintering at 20 kbar, 800°C:
A. 12·5 min.    B. 20 min.

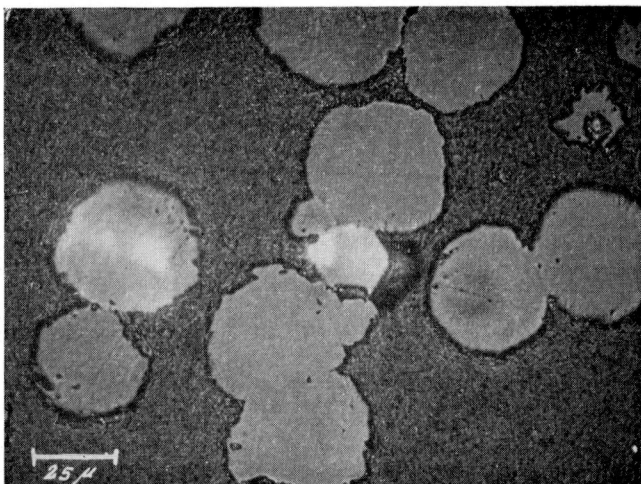

c
FIGURE 5—*continued*
c. 30 min.

of MgO A at: 800°C, 20 kbar, 12·5 min.; 1000°C, 20 kbar, 2 min.; 700°C, 40 kbar, 60 min.

Apart from densification, the problem of grain growth under pressure is of importance. The size of crystallites smaller than 1000 Å has been determined from broadening of the X-ray diffraction lines. The size distribution for particles larger than 1 $\mu$m was determined by means of a particle size analyser according to ENDTER. Conversion to real diameters of spheres seems neither necessary nor allowable.[7] The average particle diameter was estimated by taking 0·7 times the averaged counted diameter.[4]

Under high pressure there are opposite effects influencing crystal growth. The diffusion coefficient is lower, sintering time is short and temperature is low, and these conditions hamper crystal growth. On the other hand, the particles are in very close contact, which favours the transport of matter from one particle to the other. The latter effect apparently is the more important. In several cases grain growth is not proportional to the square root of time, and it seems impossible to describe the grain growth with a simple mathematical formula.

To sum up the investigations on MgO, a starting material with a grain size of 500–1000 Å and sintering at 800°C and 20 kbar for 10 min. are the optimal conditions for obtaining a dense material with uniform microstructure of small crystallites.

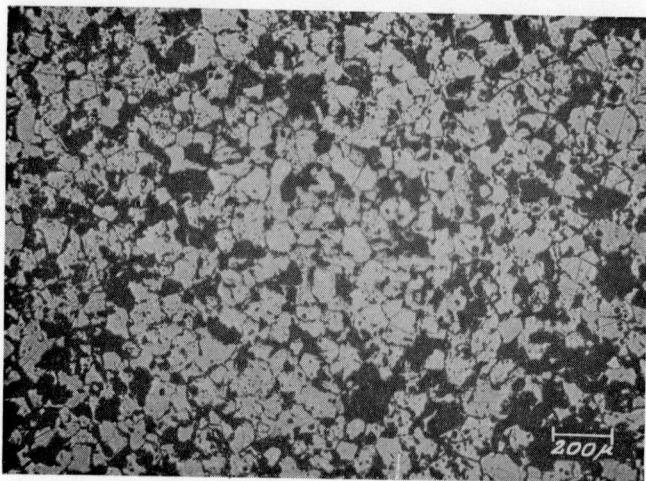

A

B

FIGURE 6.—Microstructure of MgO A after sintering at 20 kbar, 1000°C.
A. 2 min.    B. 30 min.

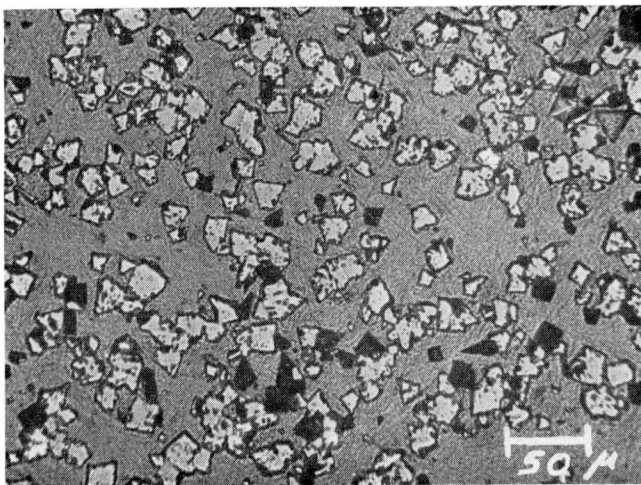

FIGURE 7.—MgO A. Sintering conditions: 30 min. at 700°C, 40 kbar.

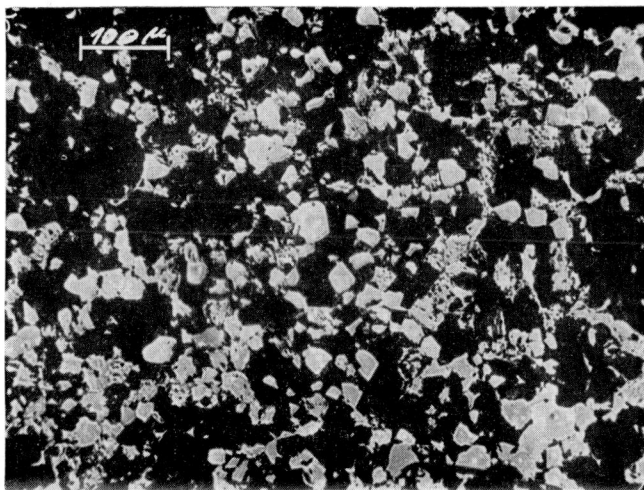

FIGURE 8.—MgO A. Sintering conditions: 30 min. at 1000°C, 40 kbar.

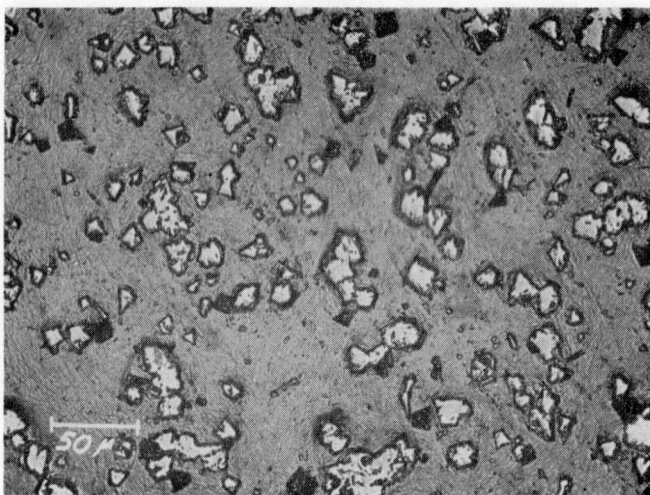

FIGURE 9.—MgO A. Sintering conditions: 2 min. at 800°C, 40 kbar.

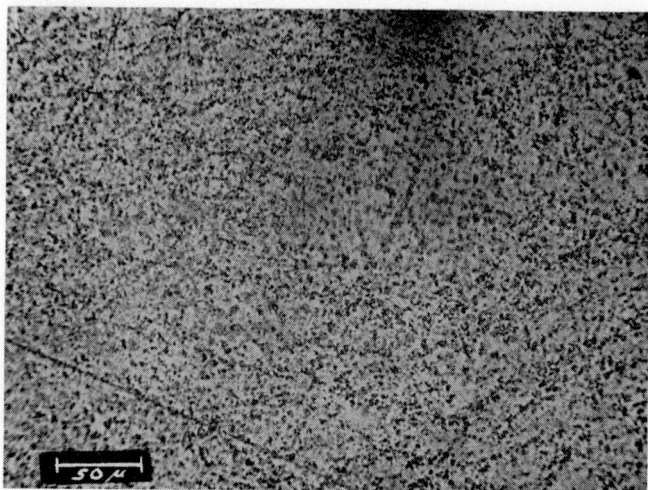

FIGURE 10.—MgO B. Sintering conditions: 30 min. at 800°C, 20 kbar.

**Table 1**
**Particle Size of Samples Sintered under Pressure**

| | Sintering conditions | | | Large crystallites | | Fine particles |
|---|---|---|---|---|---|---|
| Duration (min.) | | Temperature (°C) | Pressures (kbar) | $D(\mu m)$ | Vol. (%) | $D$ (Å) |
| MgO A | 30 | 700 | 20 | — | — | 500 |
| | 5 | 800 | 20 | 12 | 5 | 400 |
| | 12·5 | 800 | 20 | 18 | 10 | 500 |
| | 20 | 800 | 20 | 20 | 20 | 600 |
| | 30 | 800 | 20 | 25 | 30 | 800 |
| | 2 | 1000 | 20 | 21 | 40 | > 1000 |
| | 12·5 | 1000 | 20 | 17(?) | 60 | > 1000 |
| | 30 | 1000 | 20 | 22 | 100 | > 1000 |
| | 30 | 700 | 40 | 5·4 | 50 | 900 |
| | 60 | 700 | 40 | 4·5 | — | > 1000 |
| | 2 | 800 | 40 | 6 | 22 | 300 |
| | 12·5 | 800 | 40 | 9 | 40 | 800 |
| | 30 | 800 | 40 | 8 | 80(?) | > 1000 |
| | 2 | 1000 | 40 | 7 | (?) | 800 |
| | 12·5 | 1000 | 40 | 6 | 80 | > 1000 |
| | 30 | 1000 | 40 | 9 | 100 | > 1000 |
| MgO B | 5 | 800 | 20 | | Between 1000 Å and 1 $\mu m$ | |
| | 12·5 | 800 | 20 | | | |
| | 30 | 800 | 20 | | | |

## 4. PREPARATION OF DENSE, FINE-PARTICLE BaTiO₃ AND NiFe₂O₄

The dielectric and ferromagnetic properties of these materials are very dependent on the size of the crystallites. For instance, the initial permeability increases rapidly with decreasing grain size.[9,10] From the results for MgO one might expect that $BaTiO_3$ and $NiFe_2O_4$ ceramics with a relative density of nearly 1, a crystal size of about 0·1 $\mu m$, but free of inhibitors, could well be prepared by sintering under very high pressure.

The starting materials were prepared by thermal decomposition of Ba titanyl oxalate or Fe oxalate + $NiCO_3$. The crystal size of the $BaTiO_3$ powder was determined as about 500 Å, the size of the $NiFe_2O_4$ particles being 800 Å. Relative densities of 99·7% and 99·2% respectively were obtained by sintering these materials at 800°C and 20 kbar for 5 min. Practically no growth of the crystallites occurred. A polished and subsequently etched surface does not show any structure. Discontinuous grain growth, which is observed with fine particles of MgO, could not be detected.

The dielectric and the ferromagnetic properties of these bodies have not yet been determined.

## REFERENCES

1. BUGL, J., *Ber. deutsch. keram. Ges.*, **43**, 577, 1966.
2. MURRAY, P., LIVEY, T. D., and WILLIAMS, J.,"Ceramic Fabrication Processes" (London: Chapman and Hall Ltd. 1958), p. 147.
3. SCHOLZ, S., and LERSMACHER, B., *Ber. deutsch. keram. Ges.*, **41**, 98, 1964.
4. KRAUTH, A., and OEL, H. J., *Ber. deutsch. keram. Ges.*, **43**, 264, 1966.
5. BUDNIKOV, P. P., CHARITONOV, F. J., KERBE, F., and SACHIN, V. S., *Silikattechn.*, **17**, 375, 1966.
6. SPRIGGS, R. M., BRISETTE, L. A., and VASILOS, T., *J. Amer. Ceram. Soc.*, **46**, 508, 1963.
7. OEL, H. J., *Ber. deutsch. keram. Ges.*, **43**, 624, 1966.
8. BURKE, J. E., "Ceramic Fabrication Processes" (London: Chapman and Hall Ltd., 1958), p. 120.
9. KNIEKAMP, H., and HEYWANG, W., *Naturwiss.*, **41**, 61, 1954.
10. GMELINS "Handbuch der anorg. Chemie," "Magnetische Werkstoffe" (Weinheim: Verlag Chemie, Syst. Nr. 59, 8 Auf. 1., Part D, 1959).

*Part Three*

# REACTIONS DURING SINTERING

# 12.—Sintering Behaviour and Microstructures of Aluminates and Ferrites with Spinel Structure with Regard to Deviation from Stoicheiometry

By P. J. L. REIJNEN

*Philips Research Laboratories,*
*N.V. Philips Gloeilampenfabrieken*
*Eindhoven, Netherlands*

***ABSTRACT***                                  C525/E21

*The microstructures and densities of Ni aluminates and Ni ferrites have been studied in relation to their deviation from stoicheiometry. The highest densities are obtained when the sample is anion-deficient whereas cation-deficient samples are sintered to low densities. In anion-deficient aluminates pore growth is retarded, which provokes discontinuous grain growth. The sintering behaviour and microstructures of the Ni ferrites are strongly influenced by the ferric/ferrous–oxygen equilibrium, causing exaggerated pore growth in the cation-deficient region when the samples are sintered in an oxidizing atmosphere. Moreover the final density in this region depends on the oxygen partial pressure during firing, which is in agreement with theory.*

### *Comportement au frittage et microstructure d'aluminates et ferrites à structure de spinelle*

*La microstructure d'aluminates et de ferrites du type spinelle a été étudiée dans les domaines de composition $(NiO)_{1-x}(Al_2O_3)_x$ et $(NiO)_{1-x}(Fe_2O_3)_x$ pour 0, $4 \leqslant x \leqslant 0$, 6. Les ferrites ont été frittés à 1350°C dans diverses atmosphères: azote, azote+1% d'oxygène, oxygène et vide. Les aluminates ont été frittés à 1600°C dans une atmosphère d'oxygène. Lorsque le nombre de vacances cationiques est important, il en découle des densités faibles pendant le frittage par suite d'une forte dissolution de $Al_2O_3$ ou de $Fe_2O_3$ dans la phase spinelle. Ceci est expliqué par la diminution correspondante du nombre de vacances d'oxygène qui sont essentielles pour la diffusion d'ions oxygène en grandes quantités. L'azote emprisonné dans les pores inhibe la croissance des grains exactement comme une seconde phase dispersée. Des ferrites avec du $Fe_2O_3$ en excès et frittés dans une atmosphère d'oxygène, présentent des microstructures caractéristiques, avec un petit nombre de grands pores exclusivement aux joints des grains. Le gros du cristal est absolument exempt de pores. Une* *explication est donnée pour la concentration des grands pores aux joints des grains.*

## Sinterverhalten und Mikrostruktur von Aluminaten und Ferriten mit Spinellstruktur

*Die Mikrostrukturen von Aluminaten und Ferriten vom Spinelltyp wurden im Bereich der Zusammensetzung $(NiO)_{1-x}(Al_2O_3)_x$ und $(NiO)_{1-x}(Fe_2O_3)_x$ für $0,4 \leqslant x \leqslant 0,6$ untersucht. Die Ferrite wurden bei 1350°C in verschiedenen Atmosphären, wie Stickstoff, Stickstoff plus 1% Sauerstoff, Sauerstoff und Vakuum gesintert. Die Aluminate wurden bei 1600°C in Sauerstoff gesintert. Wenn der Kationen-Unterschuß groß ist, infolge der Auflösung von $Al_2O_3$ oder $Fe_2O_3$ in der Spinellphase, so ergeben sich beim Sintern niederige Dichten. Dies wird durch die entsprechende Abnahme der Zahl der Sauerstoff-leerstellen, die wesentlich für die Volumendiffusion der Sauerstoffionen sind, erklärt. In Poren eingeschlossener Stickstoff hemmt das Kornwachstum ebenso wie eine dispergierte zweite Phase. Ferritzus-ammensetzungen mit einem Überschuß an $Fe_2O_3$ und in Sauerstoff gesintert, zeigen charakteristische Mikrostrukturen, in welchen eine kleine Zahl großer Poren ausschließlich an den Korngrenzen gefunden wird. Das Innere des Kristalls ist absolut frei von Poren. Es wird eine Erklärung für die Konzentration großer Poren an den Korngren-zen gegeben.*

## 1. INTRODUCTION

The present investigation is concerned with the marked differences in microstructure between spinel-type ferrites having different deviations from stoicheiometry. Materials containing excess mono- or di-valent oxide have low porosities and many small pores within the grains (Figure 1) whereas the materials with excess iron oxide are less dense and have a small number of large pores preferentially at the intersections of three or four grain boundaries, the grains being absolutely pore-free (Figure 2). To see whether this has something to do with the ferric/ferrous–oxygen equilibrium in ferrites, the corresponding aluminates, with which no comparable effect occurs, were also investigated.

## 2. EXPERIMENTAL

A series of Ni ferrites and Ni aluminates were prepared from pure raw materials by normal ceramic techniques. The compositions of the samples are expressed by a composition parameter $x$: $(1-x)NiO + xFe_2O_3$ and $(1-x)NiO + x\ Al_2O_3$. The values of $x$ were chosen as follows: 0·60, 0·57, 0·55, 0·54, 0·53, 0·52, 0·51, 0·50, 0·49, 0·48, 0·47, 0·46, 0·45, 0·43, 0·40.

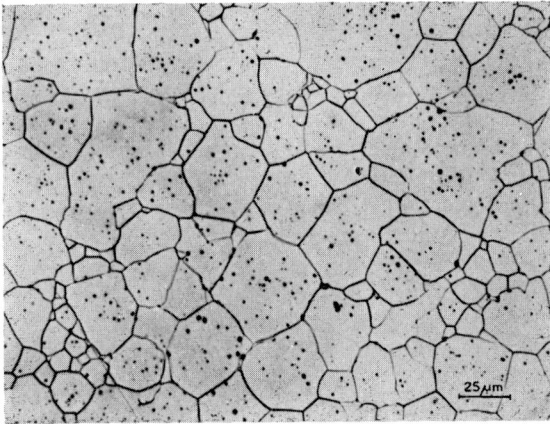

FIGURE 1.—Microstructure of spinel-type ferrite with excess mono- or di-valent oxide.

FIGURE 2.—Microstructure of spinel type ferrite with excess iron oxide sintered in an oxidizing atmosphere.

The oxide mixtures were ball-milled for 6 h; the Ni ferrite series was pre-fired at 950°C for 1 h and the Ni aluminate series at 1200°C for 1 h. Thereafter the reaction products were again ball-milled for 6 h, dried and isostatically pressed into discs at 1000 kg/cm². The weight of one of the discs was about 6 g. The aluminates were sintered for 24 h at 1600°C in oxygen and the ferrites for 24 h at 1350°C, one series in purified nitrogen, one in nitrogen $+1\%$ oxygen, and one in pure oxygen.

12

Densities of the samples were determined gravimetrically by suspension in water, with an accuracy of about $0.05\%$ for dense samples. The volumes of samples with open porosity were determined by immersing them in mercury and determining the volume of the mercury displaced. An apparatus for doing this quickly and accurately has been specially developed and constructed. The accuracy is about $0.5\%$.

The composition $NiFe_2O_4+0.5\ Fe_2O_3$ $(x=0.6)$ was studied in more detail. Isostatically pressed discs were sintered at 1350°C in various atmospheres:

12h in oxygen and then 12h in nitrogen.
12h in vacuum and then 12h in nitrogen.
24h in nitrogen.
12h in nitrogen and then 12h in oxygen.

These experiments were carried out in the same horizontal tube furnace so as to make a comparison between the various products possible.

A number of sintered specimens were polished and the microstructures were studied. Part of the polished surfaces were etched to make the grain boundaries visible: the aluminates by thermal etching at 1400°C, and the ferrites by chemical etching in boiling diluted $H_2SO_4$.

## 3. RESULTS

Determination of the porosity of the samples from experimental densities is possible when the corresponding theoretical densities are known. Theoretical densities were calculated on the basis of the following assumptions:

(1)   The solubility of NiO in $NiAl_2O_4$ and in $NiFe_2O_4$ is negligibly small.

(2)   The spinel compositions of the aluminates and ferrites are respectively:

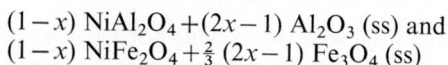

$$(1-x)\ NiAl_2O_4+(2x-1)\ Al_2O_3\ (ss)\ and$$
$$(1-x)\ NiFe_2O_4+\tfrac{2}{3}\ (2x-1)\ Fe_3O_4\ (ss)$$

where ss stands for solid solution.

On cooling, the excess $Al_2O_3$ remains dissolved in the spinel phase. In the sample with $x=0.6$, no $\alpha$-$Al_2O_3$ could be detected by X-ray analysis even after reheating for 6h at 1000°C. On cooling in oxygen, part of the dissolved $Fe_3O_4$ is oxidized to $Fe_2O_3$, mainly at the outer surface of the discs but, for the sake of convenience, theoretical densities were calculated on the assumption that all excess $Fe_2O_3$ is present as $Fe_3O_4$ (ss).

(3) The solid solutions have ideal behaviour. This means that there is no change in volume and enthalpy when the solid solutions are formed. The values of the cell parameters of $NiAl_2O_4$, $Al_{8/3}V_{1/3}O_4$, $NiFe_2O_4$ and $Fe_3O_4$ are taken from literature. The cell dimensions of spinels have been reviewed by G. BLASSE.[1]

In Table 1 the calculated X-ray densities, the experimental densities and the porosities are given. In Figure 3A and B the porosities are plotted *versus* the composition parameter $x$.

The microstructures of three aluminate samples ($x = 0.49$, $0.50$, $0.51$) are given in Figure 4. The first two samples, A and B, are dense and have a large number of small pores within the grains. The third sample is very porous and has such a small grain size that the individual grains cannot be detected microscopically. When the samples with excess $Al_2O_3$ are sintered at higher temperatures ($24$ h, $1800°C$), discontinuous grain growth takes place (Figure 5).

A decrease in density when $x > 0.5$ is also observed in the Ni ferrite series, that is to say when the samples are sintered in oxygen, but the decrease is not sharp and less drastic than in the corresponding aluminate case, which can be seen from the plot of the porosities *versus* composition in Figure 3B.

Moreover when the samples are sintered in low oxygen partial pressures, no significant difference in density is observed (Figure 3B).

There is a drastic difference in microstructure in the Ni ferrite series sintered in oxygen: samples with $x < 0.50$ have a second phase [$NiO$–$FeO$(ss)–$Fe_2O_3$(ss)] and pores within the grains while samples with $x > 0.50$ have no second phase and a small number of large pores exclusively on the intersections of grain boundaries. This is illustrated in Figure 6, where the microstructures of samples with $x = 0.49$, $0.50$ and $0.51$ are given.

The microstructures of the same compositions sintered in nitrogen (Figure 7) do not differ from each other in any respect. Second phase [$NiO$–$FeO$(ss)–$Fe_2O_3$(ss)] is found in all three samples, the amount being considerably larger than in the same compositions sintered in oxygen. No exaggerated pore growth has taken place, but a large number of small pores is found within the grains.

The microstructures of samples which have the composition $NiFe_2O_4 + 0.5 Fe_2O_3$ ($x = 0.6$) and which are sintered in different atmospheres are shown in Figure 8.

The sample sintered in oxygen for $12$ h and then in nitrogen for $12$ h has low density and the microstructure indicates that exaggerated pore growth has taken place (A).

REIJNEN:

Table 1

| x | Ni ferrite | | | | | | | Ni aluminate | | |
|---|---|---|---|---|---|---|---|---|---|---|
| | 24h $O_2$ 1350°C | | | 24h $N_2$+1%$O_2$ 1350°C | | 24h $N_2$ 1350°C | | 24h $O_2$ 1600°C | | |
| | $d_{calc}$ | $d_{exp}$ | Porosity (%) | $d_{exp}$ | Porosity (%) | $d_{exp}$ | Porosity (%) | $d_{calc}$ | $d_{exp}$ | Porosity (%) |
| 0·40 | 5·56 | 5·40 | 2·8 | 5·32 | 4·3 | 5·47 | 1·6 | 4·79 | 4·75 | 0·9 |
| 0·43 | 5·51 | 5·37 | 2·5 | 5·29 | 3·9 | 5·42 | 1·6 | 4·70 | 4·65 | 1·0 |
| 0·45 | 5·47 | 5·35 | 2·2 | 5·28 | 3·5 | 5·39 | 1·6 | 4·64 | 4·59 | 1·0 |
| 0·46 | 5·46 | 5·35 | 2·0 | 5·27 | 3·4 | 5·38 | 1·5 | 4·61 | 4·56 | 1·1 |
| 0·47 | 5·44 | 5·33 | 2·1 | 5·27 | 3·2 | 5·36 | 1·5 | 4·59 | 4·59 | 1·0 |
| 0·48 | 5·43 | 5·30 | 2·4 | 5·27 | 2·9 | 5·35 | 1·3 | 4·56 | 4·51 | 1·0 |
| 0·49 | 5·41 | 5·28 | 2·4 | 5·26 | 2·8 | 5·34 | 1·3 | 4·53 | 4·49 | 1·0 |
| 0·50 | 5·40 | 5·25 | 2·7 | 5·26 | 2·6 | 5·34 | 1·0 | 4·50 | 4·43 | 1·7 |
| 0·51 | 5·39 | 5·26 | 2·4 | 5·26 | 2·5 | 5·33 | 1·2 | 4·48 | 3·12* | 29·4 |
| 0·52 | 5·39 | 5·21 | 3·2 | 5·27 | 2·1 | 5·32 | 1·3 | 4·46 | 3·06* | 31·2 |
| 0·53 | 5·38 | 5·16 | 4·2 | 5·25 | 2·4 | 5·29 | 1·7 | 4·44 | 3·00* | 32·4 |
| 0·54 | 5·38 | 5·12 | 4·7 | 5·28 | 1·8 | 5·30 | 1·5 | 4·41 | 2·99* | 32·3 |
| 0·55 | 5·37 | 5·12 | 4·8 | 5·26 | 2·1 | 5·29 | 1·4 | 4·40 | 3·08* | 30·0 |
| 0·57 | 5·36 | 5·04 | 5·9 | 5·24 | 2·4 | 5·27 | 1·8 | 4·36 | 3·05* | 30·0 |
| 0·60 | 5·43 | 4·86 | 9·0 | 5·16 | 2·4 | 5·24 | 2·1 | 4·29 | 2·94* | 31·3 |

*Densities measured by immersing pellets in mercury

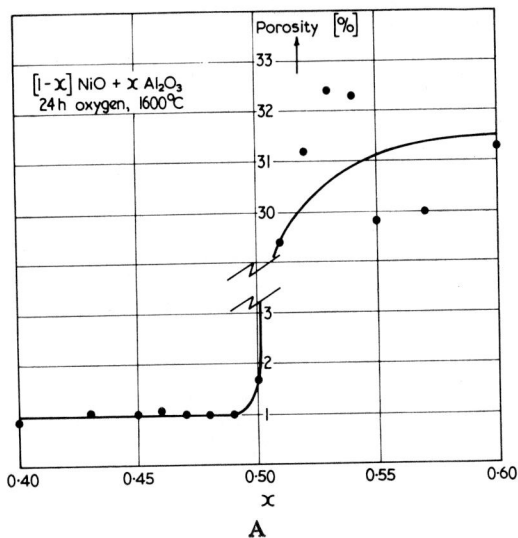

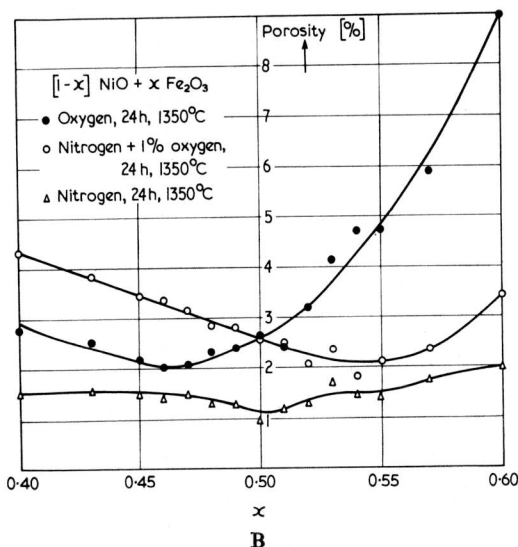

FIGURE 3
Densities and porosities of:

A. Ni aluminates sintered 24 h at 1600°C in oxygen as function of the composition parameter $x$, $[(1-x)NiO + xAl_2O_3]$.
B. Ni ferrites sintered 24 h at 1350°C in: oxygen; nitrogen + 1% oxygen; nitrogen; as function of the composition parameter $x$, $[(1-x)NiO + xPe_2O_3]$.

A

B

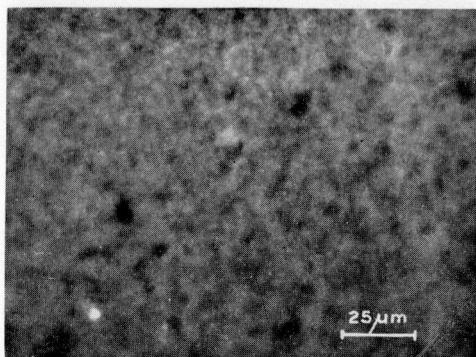

C

FIGURE 4

Microsructures of Ni aluminates with composition parameters:
A. $x=0.49$.  B. $x=0.50$.  C. $x=0.51$.

No individual grains can be detected in the sample with excess $Al_2O_3$.

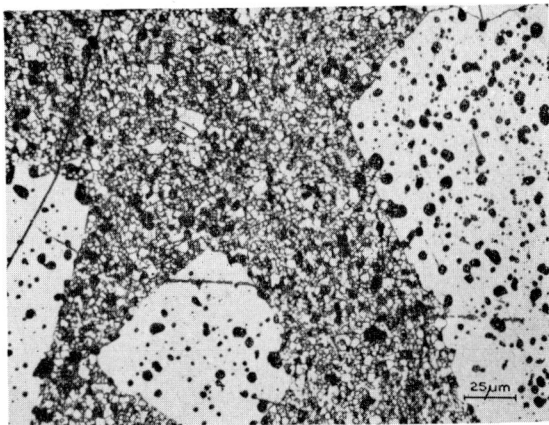

FIGURE 5.—Discontinuous grain growth in Ni aluminate with excess $Al_2O_3$ sintered for 24 h at 1800°C.

The sample sintered in nitrogen for 24 h (B) has high density and many pores within the grains.

The sample sintered for 12 h in vacuum and then for 12 h in nitrogen has high density and pores within the grains but the number of pores is smaller and the grains are larger than in the sample sintered in nitrogen only (C).

The sample sintered in nitrogen for 12 h and then for 12 h in oxygen has a typical microstructure in which exaggerated pore growth has taken place but there are still some pores left in the grains (D).

## 4. DISCUSSION

### 4.1 Density

The difference in density in the Ni aluminate series between samples with $x \leqslant 0.50$ and $x > 0.50$ is thought to be due to a difference in defect structure.

The exact stoicheiometric composition has a cation/anion ratio of 3/4 ($NiAl_2O_4$). By dissolving a slight amount of NiO in the stoicheiometric composition, extra oxygen vacancies and interstitial cations are introduced, and by dissolving $Al_2O_3$ in the stoicheiometric composition, extra cation vacancies are formed. As spinels consist of a close-packed oxygen lattice, volume diffusion of oxygen vacancies is thought to be the rate-determining step in sintering. It can be shown theoretically that the highest densification rates are obtained when the material is slightly anion-deficient, that is to say, when sintering takes place by volume diffusion of vacancies. This explains the

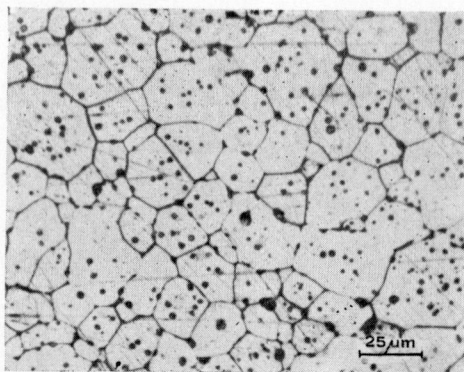

A

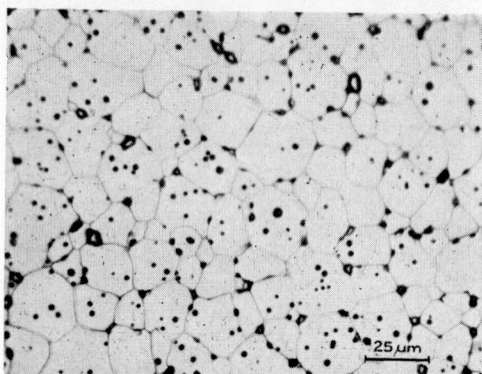

B

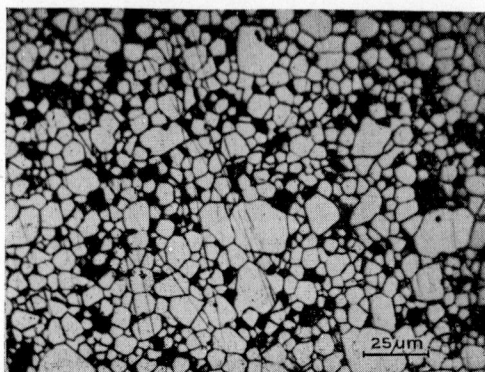

C

FIGURE 6.
Microstructures of Ni ferrites sintered in oxygen with composition parameters:
A. $x=0.49$.    B. $x=0.50$.    C. $x=0.51$.
In sample C exaggerated pore growth has taken place.

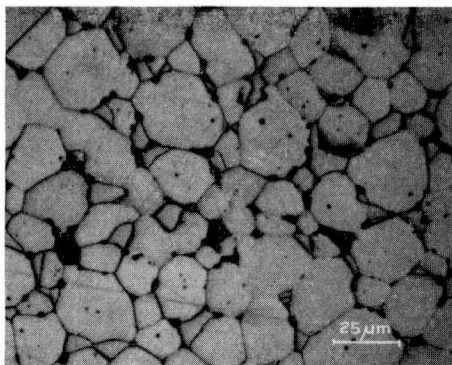

A

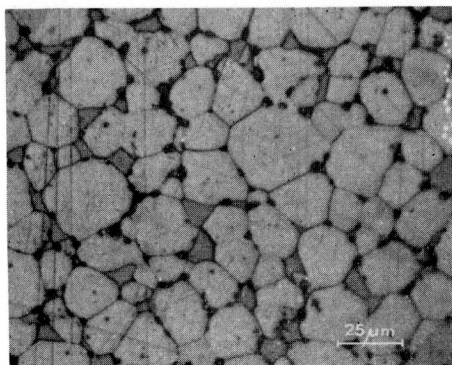

B

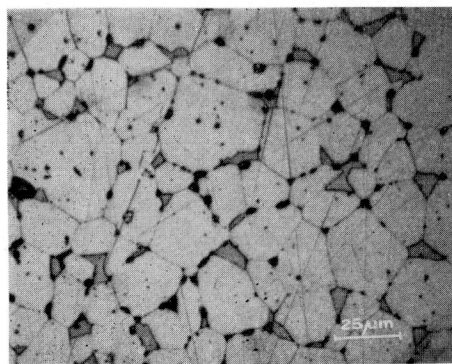

C

FIGURE 7

Microstructures of Ni ferrites sintered in nitrogen with composition parameters:
A. $x=0.49$.　B. $x=0.50$.　C. $x=0.51$.
There is no essential difference in the three microstructures.

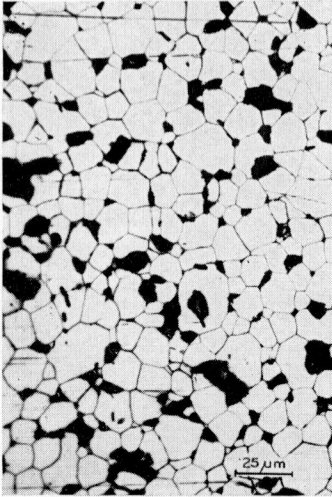

A

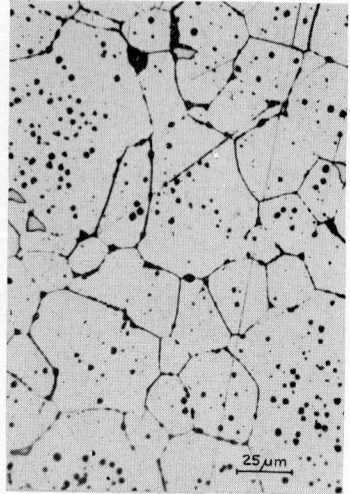

B

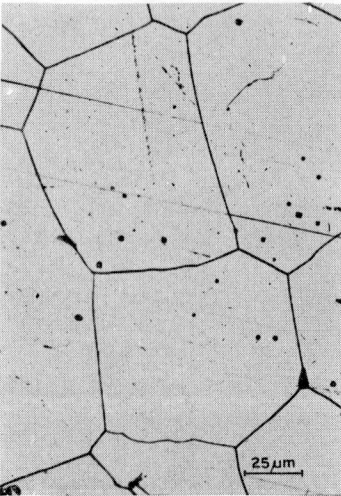

C

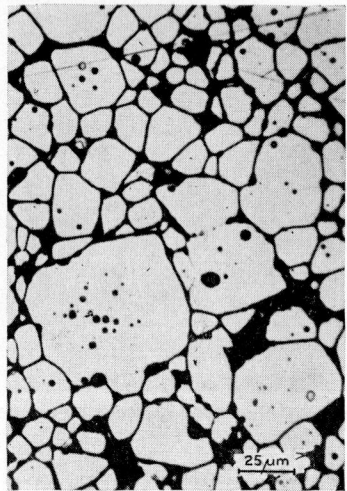

D

FIGURE 8

Samples of composition $NiFe_2O_4 + 0.5Fe_2O_3$ $(x=0.6)$ sintered at $1350°C$:

A. 12h in oxygen + 12h in nitrogen.
B. 24h in nitrogen.
C. 12h in vacuum + 12h in nitrogen.
D. 12h in nitrogen + 12h in oxygen.

high densities of the samples with $x < 0\cdot50$, which at the sintering temperature consist of a spinel phase saturated with NiO and a second phase, NiO. The sample for which $x = 0\cdot50$ has a somewhat lower density, presumably because the NiO content of the solid solution is not optimal. In cation-deficient samples which are solid solutions of $Al_2O_3$ in $NiAl_2O_4$, the diffusion of oxygen is much lower, resulting in very low densification rates. The full theoretical background is somewhat outside the scope of this paper.

The lower densities in the Ni ferrite series when $x > 0\cdot50$ is explained in the same way. The decrease in density is, however, less drastic than in the corresponding aluminates.

The reason is that excess $Fe_2O_3$ is mainly dissolved as $Fe_3O_4$ and only partly as $Fe_2O_3$. Sintering at higher temperatures or at lower oxygen partial pressures causes a further decrease of the cation vacancy concentration:[2]

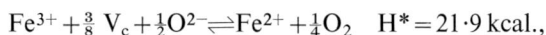

$$Fe^{3+} + \tfrac{3}{8} V_c + \tfrac{1}{2}O^{2-} \rightleftharpoons Fe^{2+} + \tfrac{1}{4}O_2 \quad H^* = 21\cdot9 \text{ kcal.,}$$

and consequently higher densities are found when the atmosphere during sintering contains less oxygen. When the samples are fired in low oxygen partial pressures, the evolution of oxygen during sintering when closed porosity is developed and the formation of large amounts of second phase $[NiO + FeO(ss) + Fe_2O_3(ss)]$ have an unpredictable influence on the final density. The density *versus* composition curve is therefore less unequivocal than with $NiAl_2O_4$.

## 4.2 Microstructure and Pore Growth

The difference in microstructures of anion-deficient and cation-deficient samples in the $NiAl_2O_4$ series is also seen as a consequence of the decreased volume diffusion of oxygen ions in the cation-deficient region, because pore growth and grain growth also take place by the volume diffusion of ions. Along with low densities, the material will have many small pores and small grains. The effect of small grain size is, however, magnified owing to the presence of small pores, which retard grain growth. This can even lead to discontinuous grain growth (Figure 5), which might sound somewhat contradictory. The reason is that the driving force for grain growth is proportional to the reciprocal of the radius of the curved grain boundary, which determines the vacancy gradient over the grain boundary. For the boundary between small grains and a large grain this radius is equal to the radius of the small grains, whereas the grain boundary between grains of almost the same size is practically straight. If nucleation of a somewhat larger grain in a matrix of very small grains has taken place, the driving force for grain growth becomes suddenly very large, causing the phenomenon of discontinuous grain growth.

The microstructures of anion-deficient and cation-deficient ferrites when sintered in an oxidizing atmosphere also show a marked difference.

As mentioned before, pore growth is *retarded* in cation-deficient aluminates, but in the corresponding ferrites *exaggerated* pore growth has taken place. In any material, pore growth takes place by volume-diffusion and grain-boundary diffusion of vacancies, because a large pore acts as a vacancy sink for micropores within a radius of about twice its own radius. Other mechanisms for pore growth require the interaction of grain boundaries. When a moving grain-boundary meets a pore it is hampered in its progress and an energy barrier must be surmounted before the pore can be left behind in the grain. The grain boundary exerts a pull on the pore, which then acquires two radii of curvature in order that the surface tensions remain in equilibrium (Figures 9 and 10). The pore can

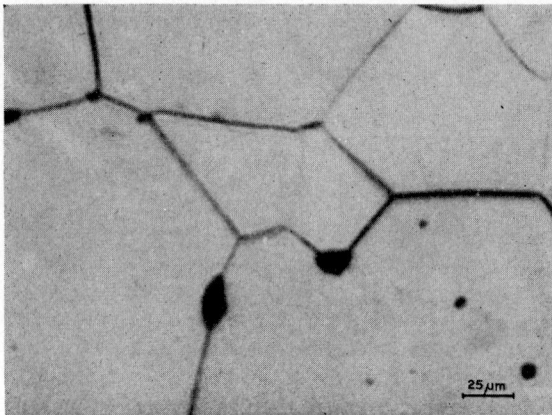

FIGURE 9.—A pore on a boundary which tends to move obtains two different radii of curvature

move along with the grain boundary if material can be transported in a sufficiently short time from one side of the pore to the other. Owing to different curvatures of the surface of the pore, vacancy gradients are set up which are the driving forces for material transport by volume or by surface diffusion. For very small pores these gradients can be sufficiently large for the pores to be swept

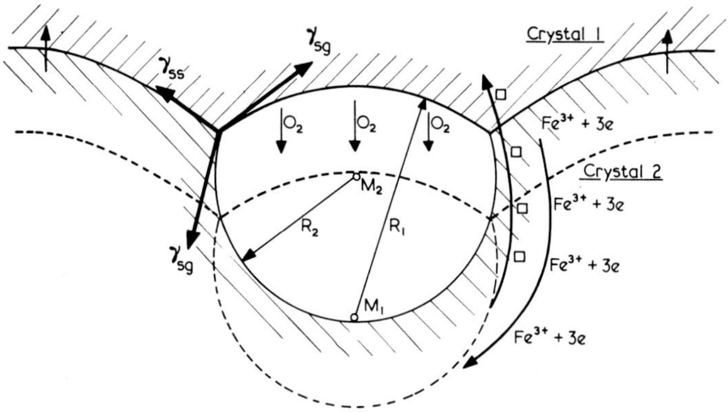

FIGURE 10.—Mechanism of pore migration in ferrites with excess $Fe_2O_3$ sintered in an oxidizing atmosphere. Cations are transported by volume- or by surface-diffusion. No space charge is built up, owing to an equivalent transport of electrons. Oxygen is transported via the gas phase. Formation of oxygen from oxygen ions, and the reverse reaction, form a continuous source of and sink for electrons.

by the grain boundary. For a detailed analysis of this phenomenon see SPEIGHT and GREENWOOD.[3]

The movement of pores along with grain boundaries, whatever the mechanism is, leads to exaggerated pore growth, as has been pointed out by KINGERY and FRANCOIS.[4] The microstructure will be characterized by pore-free crystals and a small number of large pores will be found at the intersections of three or four grain-boundaries.

Transport of material from one side of the pore to the other is possible via the gas atmosphere, depending on the type of material and the atmosphere. In the present materials and under the given experimental conditions, transport of both metal and oxygen from one pore side to the other via the gas phase is not likely.

The phenomenon of exaggerated pore growth as observed in ferrites containingexcess iron oxide and which are sintered in oxygen is therefore thought to be due to transport of oxygen via the gas phase and cations by volume or surface diffusion.

Diffusion of, say, cations must be accompanied by an equivalent flux of anions or electrons, and in materials like alumina and aluminates, which are good insulators, therefore it is impossible for the cations to be transported by volume or surface diffusion, and oxygen via the gas phase.

In materials which contain cations of the same element in different valency states, a flux of cations can be maintained without a space

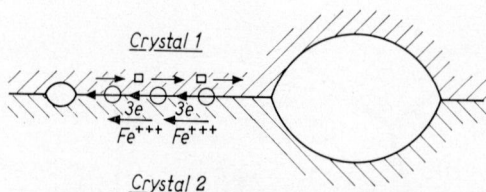

FIGURE 11.—Pore-growth mechanism in ferrites with excess $Fe_2O_3$ sintered in an oxidizing atmosphere. Oxygen atoms diffuse along the grain boundary, owing to different oxygen partial pressures in the two pores. Cations are transported by volume diffusion. Electrons have the same role as in the mechanism of Figure 10.

charge being built up, owing to an equivalent flux of electrons. This situation arises in cation-deficient ferrites sintered in an oxidizing atmosphere.

The following equilibrium exists between a ferrite and the gas atmosphere:

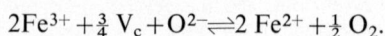

$$2Fe^{3+} + \tfrac{3}{4} V_c + O^{2-} \rightleftharpoons 2 Fe^{2+} + \tfrac{1}{2} O_2.$$

The value of the equilibrium constant $K$ has slightly different values at the pore surfaces, and concentration gradients are thus set up:

$$\frac{\Delta K}{K} = 2\frac{\Delta[Fe^{2+}]}{[Fe^{2+}]} + \tfrac{1}{2}\frac{\Delta p_{O_2}}{p_{O_2}} - 2\frac{\Delta[Fe^{3+}]}{[Fe^{3+}]} - \tfrac{3}{4}\frac{\Delta[V_c]}{[V_c]}$$

The gradient of ferrous ions is opposite to the gradient of ferric ions, as a result of which a volume flux of cations only can be maintained without transport of charge.

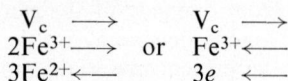

$$\begin{array}{ccc} V_c \longrightarrow & & V_c \longrightarrow \\ 2Fe^{3+} \longrightarrow & or & Fe^{3+} \longleftarrow \\ 3Fe^{2+} \longleftarrow & & 3e \longleftarrow \end{array}$$

The total process is indicated schematically in Figure 10. The pressure gradient that is required to maintain an oxygen flux which is equivalent to the bulk diffusion of cations and cation vacancies is negligibly small.

As $p_{O_2}$ is large, the term $\Delta p_{O_2}/p_{O_2}$ can be neglected in the expression above.

Morever as $\Delta[Fe^{2+}]$, $\Delta[Fe^{3+}]$ and $\Delta[V_c]$ are of comparable magnitude and $[Fe^{2+}]$ and $[Fe^{3+}]$ are much larger than $[V_c]$, it follows that

$$\frac{\Delta K}{K} \approx -\frac{\Delta[V_c]}{[V_c]} \approx \frac{2\gamma V_m}{RT}\left(\frac{1}{\rho_1} - \frac{1}{\rho_2}\right) \approx \frac{10^{-7}}{\rho_1}$$

$\gamma$ = Free energy of the pore surface per
cm$^2 \approx 500$ ergs/cm$^2$
$V_m$ = Volume of $\frac{1}{4}$ mole ferrite $\approx 10$ cm$^3$
$RT$ = Usual meaning $\approx 10^{11}$ ergs

Suppose that the grain boundary has a velocity of $10^{-7}$cm/sec ($10\mu$m in 3 h), then the vacancy flux per cm$^2$ must be about $10^{-8}$mole/ cm$^2$sec. With a volume diffusion coefficient of $10^{-6}$cm$^2$/sec, the magnitude of which is indicated further on, it follows from the first law of Fick

$$\frac{\Delta[V_c]}{\rho} \cdot 10^{-6} = 10^{-8} \text{ thus } \Delta[V_c] = 10^{-2} \rho$$

The material is cation-deficient—let us say that 1 cation site in $10^3$ sites is empty—then 10 cm$^3$ material contains $10^{-3}$ mole $V_c$, which makes $[V_c] = 10^{-4}$.

It now follows that $\Delta[V_c] = 10^{-2}\rho = 10^{-11}/\rho$ and thus $\rho \approx 0.3\mu$m Pores with a diameter of about $0.5 \ \mu$m can thus be swept by a moving grain boundary with a velocity of $10\mu$m in 3 h. The outcome of this calculation can easily be in error by a factor 2 or 3, but it shows in any case that quite large pores can move along with the grain boundary according to the proposed mechanism.

The diffusion coefficient $D_c$ is correlated to the self-diffusion coefficient $D_c^*$ by $D_c^* = D_c f_c$, in which $f_c$ is the fraction of empty vacancy sites.

The self-diffusion coefficient of iron ions has been determined by SCHMALZRIED[5] as being $\sim 10^{-9}$cm$^2$/sec for a composition Mg$_{0.71}$ Fe$_{2.29}$O$_4$ in air at 1365°C. According to our work[6] on the MgFe$_2$O$_4$ system, the fraction of empty cation sites is $4 \times 10^{-3}$ under these circumstances, and thus $D_c \approx 10^{-6}$ cm$^2$/sec. ($D_c$ is the diffusion coefficient of cation vacancies).

Another mechanism for pore growth operates under the same circumstances as the preceding one. Consider two pores, a large one of radius $\rho_1$, and a small one of radius $\rho_2$. The large pore will grow at the expense of the small one if the latter is close enough to the larger —i.e., within twice the radius of the large pore. When the material is cation deficient, diffusion of vacancies is strongly decreased, slowing down both densification and pore growth.

The situation is again different when the material contains cations of the same element in different valency states.

The oxygen partial pressure in equilibrium with the solid phase is somewhat higher in the large pore than in the small pore, which makes the transport of oxygen *atoms* along a grain boundary possible in principle. The equivalent transport of cations can take place by volume diffusion, as the material contains free electrons. (Figure 11). The formation of oxygen in the large pore is a continuous source of electrons and the formation of oxygen ions in the small pore a continuous sink for electrons. The grain-boundary diffusion of oxygen atoms is probably the rate-determining step of the process. Let us therefore try to calculate the pressure gradient between the two pores.

The same correlation between the gradients exists as has derived before:

$$\frac{\Delta K}{K}=2\frac{\Delta[Fe^{2+}]}{[Fe^{2+}]}+\frac{1}{2}\frac{\Delta p_{O_2}}{p_{O_2}}-2\frac{\Delta[Fe^{3+}]}{[Fe^{3+}]}-\frac{3}{4}\frac{\Delta[V_c]}{[V_c]}$$

It is assumed that the gradients which are required to maintain the equivalent flux of cations through the bulk are negligible small, thus

$$\frac{\Delta p_{O_2}}{p_{O_2}}\approx\frac{10^{-7}}{\rho_2}$$

If $\rho_2=10^{-5}$ cm and $p_{O_2}=1$ atm., $\Delta p_{O_2}\approx10^{-2}$ atm.

The length of the grain boundary is say $10^{-3}$ cm and the pressure gradient between the pores then becomes 10 atm/cm.

Whether this value is high enough to allow appreciable diffusion of oxygen atoms along grain boundaries is difficult to say. The amount of material which must be transported is, however, very small.

Anyhow, the grain boundaries in such materials are very clean (Figure 2), which is experimental evidence for this mechanism of pore growth.

### 4.3 Miscellaneous

From the foregoing considerations it is clear that ferrites with excess $Fe_2O_3$ sintered in nitrogen will not exhibit exaggerated pore growth (Figures 7 and 8B). Owing to the low oxygen partial pressure, the cation vacancy concentration is low and therefore both $\Delta p_{O_2}$ and $\Delta[V_c]$ have low values, which means that transport of material from one side of the pore to the other is not rapid enough for the pore to move along with the moving grain-boundary. The trapped nitrogen gas in the pores somewhat retards the disappearance of micropores which, in its turn, has a slight effect on grain growth, as can be seen by comparison with the sample that has been sintered in vacuum (Figure 8C). In the sample which had been sintered first in nitrogen and then in oxygen (Figure 8B), exaggerated pore growth took place only in the latter stage of sintering, by which time the grain size is already rather large and the grains are therefore not rendered absolutely pore-free by grain-boundary movement. Re-oxidation of the material in the second part of the firing cycle takes place by grain-boundary diffusion of oxygen and formation of new oxygen layers on grain boundaries. The bulk of a crystallite obtains equilibrium composition by cation diffusion.

The re-oxidation therefore leads to enormous tensions in the material, causing cracks (Figure 12) and poor magnetic performance.

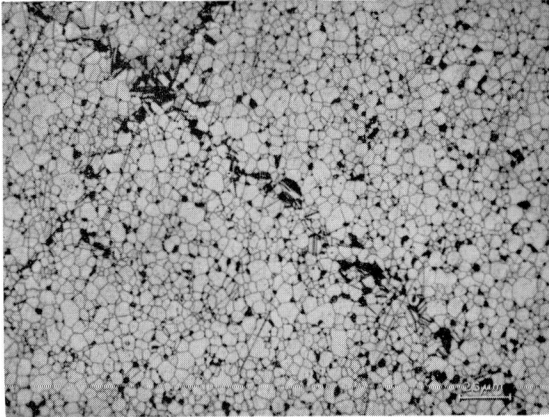

FIGURE 12.—A cation-deficient ferrite sintered first in nitrogen and then in oxygen is reoxidized by grain-boundary diffusion of oxygen. New oxygen layers are formed on grain boundaries, causing tensions in the material. $\alpha$-Fe$_2$O$_3$ is precipitated along the resulting cracks.

## 5. CONCLUSIONS

(1) In materials which contain free electrons, separate volume diffusion of cations can take place because the charge transport can be compensated by a flux of electrons in the same direction. The equivalent amount of oxygen is transported via the gas phase or along grain boundaries. The formation of oxygen from oxygen ions is a continuous source of electrons and the reverse reaction is a continuous sink. Such types of transport are shown to exist in cation-deficient ferrites sintered in an oxidizing atmosphere, causing the phenomenon of exaggerated grain growth.

(2) As in a lattice consisting of close-packed anions the diffusion constant of cation vacancies is much larger than that of anion vacancies such materials are sintered to highest densities when they are slightly anion-deficient, and sintered to low densities when they

13

188 REIJNEN:

are cation-deficient. The concentration of cation vacancies in cation-deficient ferrites depends strongly on the oxygen partial pressure. Such materials are therefore sintered to high densities in low oxygen partial pressures, in full agreement with theory.

## ACKNOWLEDGMENT

The author is grateful to G. Aarts for his assistance in collecting the experimental data.

## REFERENCES

1. BLASSE, G. *Philips Res. Repts. Suppl.* **3**, 20, 1964.
2. REIJNEN, P. "Reactivity of Solids" (Ed. G. M. Schwab) (1965) p. 562.
3. SPEIGHT, M. V., and GREENWOOD, G. W., *Phil. Mag.*, **6.**, 683, 1964.
4. KINGERY, W. D. and FRANÇOIS, B., *J. Amer. Ceram. Soc.*, **48**, 546, 1965.
5. SCHMALZRIED, H. Private communication.
6. REIJNEN, P. To be published in *Philips Res. Repts*, **23**, 151, 1968.

# 13.—The Influence of Atmosphere on the Sintering of High-surface-area Uranium Dioxide

By D. KOLAR, B. S. BRČIĆ, B. ZATLER and V. MARINKOVIĆ

*Nuclear Institute J. Stefan, Ljubljana, Yugoslavia*

*ABSTRACT* C525/E21

*Experiments have shown that the influence of reducing and of oxidizing atmospheres on the sintering of the uranium dioxide powders with different specific surface areas depends on the activity of the powder. The combination of excess oxygen in non-stoicheiometric $UO_2$ and high surface area of the initial powder may limit the sintered density and lead to the formation of intergranular porosity. In such cases, higher densities may be achieved by sintering at lower temperatures or by sintering in hydrogen.*

*Influence de l'atmosphère sur le frittage de bioxyde d'uranium à surface spécifique élevée*

*Des expériences ont montré que l'influence d'atmosphères réductrices et oxydantes sur le frittage de poudres de $UO_2$ de surfaces spécifiques différentes dépend de l'activité de la poudre. La combinaison de l'oxygène en excès, dans $UO_2$ non stoechiométrique, et de la surface spécifique élevée de la poudre initiale peut limiter la densité du produit fritté et conduire à la formation de porosité intergranulaire. Dans ce cas, on peut obtenir des densités plus élevées par frittage à de plus basses températures ou par frittage dans l'hydrogène.*

*Einfluss der Atmosphäre auf das Sintern von $UO_2$ mit grosser Oberfläche*

*Untersuchungen haben gezeigt, daß der Einfluß reduzierender und oxidierender Atmosphäre auf das Sintern von $UO_2$ mit verschiedenen spezifischen Oberflächen von der Aktivität der Pulver abhängt. Die Kombination von überschüssigem Sauerstoff in nicht stöchiometrischem $UO_2$ und großer Oberfläche der Ausganspulver kann die zu erreichende Dichte begrenzen, da sie zur Bildung von intergranularer Porosität führt. In diesen Fällen können höhere Dichten durch Sintern bei niederen Temperaturen oder durch Sintern in Wasserstoff erreicht werden.*

# 1. INTRODUCTION

Factors affecting the sinterability of uranium dioxide powders have been investigated by many authors. Since $UO_2$ oxidizes to higher oxides, the sintering must be carried out in reducing, neutral or slightly oxidizing atmospheres. The favourable influence of excess oxygen in non-stoicheiometric $UO_2$ on sintering was recognized early and it was reported[1,2] that, to achieve comparable densities, sintering non-stoicheiometric powders in hydrogen required temperatures approximately 350°C higher than those in argon. However, all the published data are not in accord, for it was also reported that, at the same temperature, lower densities were achieved in an argon atmosphere than in a dry hydrogen atmosphere.[3] BEL et al.[4] obtained the same densities when sintering in hydrogen and in argon, and water vapour has been found to have a beneficial effect when used alone[5] or in hydrogen gas (moist hydrogen)[3] but again, this influence was not confirmed by all investigators.[6]

As differences in the results might be at least partially ascribed to differences in the starting materials, we tried to resolve these apparent disagreements by investigating the influence of reducing or slightly oxidizing atmosphere on the sintering of uranium dioxide powders with different specific surface areas.

# 2. EXPERIMENTAL

The $UO_2$ powders used were prepared by decomposing and reducing active ammonium diuranate (surface area, 19 m²/g) at 400°–1000°C to give $UO_2$ powders having surface areas varying from 19 to 6 m²/g. After reduction, the powders reoxidized when exposed or on standing at room temperature in air, and became non-stoicheiometric. Characteristics of the powders are given in Tables 1–3. Surface areas were measured by the B.E.T. method.

**Table 1**
**Impurities in $UO_2$ (p.p.m./U)**

| Al | B | Ca | Cu | Fe | Mg | Si | C |
|----|----|----|----|----|----|----|----|
| 7 | 0·3 | 12 | 29 | 63 | 30 | 26 | 50 |

The powders were die-compacted at ~ 25 t.s.i., usually without any additions, into cylinders 6 mm diam. by 10 mm high. The die wall was lubricated with a light coating of stearic acid in acetone, and to some samples 0·2% PVA in water was added as binder. The resulting green densities were between 4·4 and 5·7 g/cm³ (40–52% of theoretical density.)

Specimens were sintered in a Pt–Rh resistance alumina-tube laboratory furnace. Temperature was controlled by Pt–Rh thermocouple and held constant within $\pm 20°C$. The sintering atmosphere was dry hydrogen or dry nitrogen (carefully purified with active Cu and dried over magnesium perchlorate and in traps cooled with liquid nitrogen), wet hydrogen or nitrogen (the gases were bubbled through room-temperature water, moisture content $10 \, g/m^3$). To achieve stoicheiometry, some samples were kept in hydrogen at $600°C$ for 3 h prior to further sintering. The rate of rise of temperature during sintering was $300° \, C/h$, and the specimens were held at temperature for 3 h, except for a few experiments, when they were held for 24 h. All samples were cooled in hydrogen.

Density was measured by mercury displacement under vacuum. The reported densities are mean values taken on three compacts sintered in separate runs. Reproducibility was generally within $\pm 0.1 \, g/cm^3$ ($\pm 1\%$ theoretical density).

For electron microscope examinations, carbon replicas of fractured and/or polished and etched samples were prepared. Etching was with $H_2O_2 + H_2SO_4$.

## 3. RESULTS

Powders from the first series (Table 2) were sintered for 3 h at $1400°C$ in wet nitrogen or hydrogen. Values of density measured are plotted *versus* surface area in Figure 1. Green pellets from this series contained a small amount of binder ($0.2\%$ PVA in water).

### Table 2
#### Characteristics of the $UO_2$ Powders used in Series 1

| Specific surface area $(m^2/g) \pm 10\%$ | $O/U$ ratio $\pm 0.01$ | Green density $(0.2\% \, PVA \, added)$ $(\% \, theor.) \pm 2\%$ |
|:---:|:---:|:---:|
| 19·0 | 2·15 | 44 |
| 18·1 | 2·04 | 44 |
| 15·4 | 2·10 | 45 |
| 14·3 | 2·01 | 45 |
| 13·1 | 2·07 | 44 |
| 10·5 | 2·03 | 47 |
| 5·9 | 2·01 | 52 |

Figure 1 shows that the "beneficial" effect of a wet hydrogen or wet nitrogen atmosphere depends on the specific surface area of the particular $UO_2$ powder. High-surface-area powder, sintered

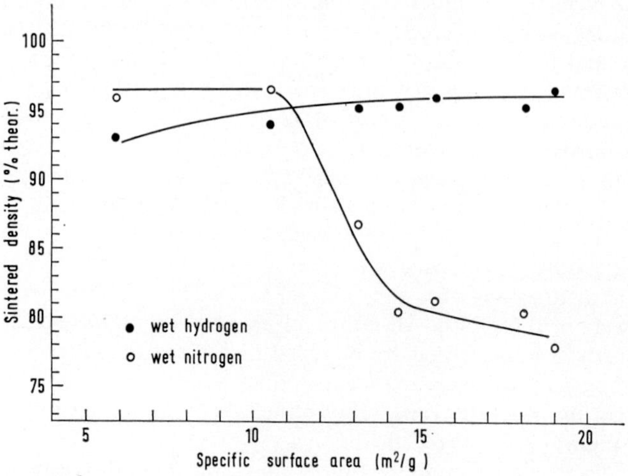

FIGURE 1.—Effect of furnace atmosphere on sintering (3 h at 1400°C) $UO_2$ samples (with 0·2% binder) with different specific surface areas.

for the same time and at the same temperature, attained a much higher density in hydrogen than in nitrogen (96·5% TD compared with 78% TD). On the other hand, a wet nitrogen atmosphere was better for sintering low-surface-area samples (96% TD compared with 93% TD). Different results reported in the literature regarding the influence of various atmospheres on the sintering of uranium dioxide powders might be at least partly due to differences in the $UO_2$ powders used.

To verify the behaviour described above, a new series of powders was prepared (Table 3). Pellets were pressed without any binder and were sintered for 3 h at 1400°C in dry hydrogen, in wet hydrogen and in nitrogen. To check the influence of initial non-stoicheiometry,

Table 3

Characteristics of the $UO_2$ Powders used in Series 2

| Specific surface area $(m^2/g) \pm 10\%$ | $O/U$ ratio $\pm 0·01$ | Green density (no binder added) (%theor.) $\pm 2\%$ |
|---|---|---|
| 17·2 | 2·11 | 40 |
| 14·0 | 2·08 | 42 |
| 10·7 | 2·07 | 45 |
| 7·2 | 2·06 | 51 |

in some experiments the samples were reduced at 600°C in hydrogen for 3 h and subsequently sintered in wet nitrogen. The results are shown in Figure 2.

The general effect of atmosphere on sintered density is not influenced by the absence of binder. It is again shown that the influence of the furnace atmosphere on sintered density depends on the activity of the powder (surface area).

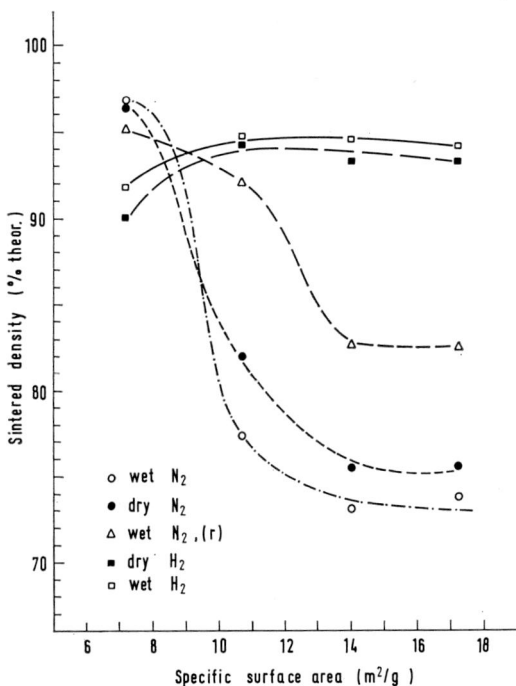

FIGURE 2.—Effect of wet and dry furnace atmospheres on sintering (3 h at 1400°C). $UO_2$ samples (without binder) with different specific surface areas. (r) Sample reduced at 600°C prior to sintering.

## 4. DISCUSSION

COBLE [2] described two possible important atmospheric effects in the sintering of ceramics: changes in atmosphere may affect the densification rate or the limiting density, and for $UO_2$ it appears that both influences are valid.

Table 4 lists the sintered densities for two limiting powders, i.e. with surface area of 17·2 and 7·2 m²/g, and shows that the furnace atmosphere which is beneficial for the sintering of low-surface-area powder, adversely affects the sintering of high-surface-area powder. A $UO_2$ powder with a specific surface area of 7·2 m²/g meets the specifications for the "ceramic grade" $UO_2$ used by most investigators reporting on the beneficial influence of non-stoicheiometric oxygen in $UO_2$ during sintering. Our experiments confirm this influence, since the best results were achieved during the sintering of non-stoicheiometric $UO_2$ in oxidizing (due to decomposition of water vapour at high temperatures) atmospheres. When the atmosphere became more reducing, the resulting densities were lower, the lowest being observed with dry hydrogen.

Low densities with high-surface-area samples sintered in an oxidizing atmosphere might be explained as follows. An oxidizing atmosphere preserves or causes the non-stoicheiometry of $UO_2$. Excess oxygen assists the sintering of $UO_2$. Since the predominant mechanism in the first stage appears to be movement of surface ions,[8] the presence of oxygen or hydroxyl ions which create a defect lattice accelerates this process. On the other hand, increasing the surface area of the powder increases the rate of densification at intermediate temperatures. Therefore both effects lead to more rapid sintering and the formation of closed pores, with the entrapment of furnace gases and other gases, which may be evolved from $UO_2$ at high temperatures—for example CO from carbon contamination.[6] It was actually observed[8] that very large pores might be closed at as low as 75% of theoretical density and that in such cases high final sintered densities are difficult to achieve, since densification depends on solubility and diffusion of entrapped gases.

Microscope observations of our pellets support the above explanation. It has been reported[4,12-14] that, when $UO_2$ is sintered in hydrogen, two different types of porosity may develop. The first type consists of small round pores, distributed at random and mainly inside the grains. This intragranular porosity is usual for sintered $UO_2$ and normally may be reduced with increasing time and temperature of sintering. Under special conditions, a second type of porosity may develop, consisting of large pores located at grain boundaries and irregular in shape. Such pores remain even after prolonged heating at very high temperatures and are typical for $UO_2$ pellets with pores closed during the early stage of sintering.[6]

Micrographs of pellets sintered in hydrogen and in wet nitrogen are shown in Figures 3–8. Hydrogen-sintered pellets (Figures 3, 5 and 7) exhibit intragranular porosity, with pores much smaller than grains (approximate ratio of grain size to pore size, 6 : 1, and even 10:1 in pellets with the lowest density). Pellets made from high-surface-area $UO_2$ develop some intergranular porosity (Figure 3), but there are still many small pores within the grains, indicating that the density achieved at 1400°C (94% TD) is not the ultimate.

Micrographs of pellets sintered in wet nitrogen (Figures 4, 6 and 7) show intergranular porosity predominant, especially pronounced in pellets made from high-surface-area powders (Figure 4) which sintered to low density. The ratio of grain size to pore size is approximately 3:1.

The closure of porosity at 26% porosity in sintered bodies is generally unusual. However, it has already been reported in $UO_2$[8] and it seems to be connected with the special type of intergranular porosity as shown in Figure 4. The porosity of some low-density pellets was measured by mercury pressure porosimeter, and the total amount of porosity measured by this technique was 5–8%, which was not comparable with the calculated porosity from density measurements (20–25%). However, when the pellets were crushed into pieces of approximately 1 mm, the porosity as measured by porosimeter increased to some 15%, which may be regarded as additional evidence for closed porosity.

The difference in pore type between nitrogen-sintered and hydrogen-sintered $UO_2$ is well seen in Figures 5 and 6, which are micrographs of pellets made from the same powder and sintered approximately to the same density. With decreasing specific surface area and increasing density of pellets sintered in wet nitrogen, an increasing amount of small pores is found within the grains (Figure 6), showing approach to "normal" $UO_2$ structure which enables high sintered densities to be obtained.

Such behaviour of $UO_2$ is an exception in general sintering practice. It is generally accepted that the essence of ideal sintering is to keep the pores at the grain boundaries until they are eliminated, since the elimination of pores entrapped within the grains is too slow. In $UO_2$ almost the opposite behaviour is noted, for the pores, at least under special conditions of sintering, prefer to segregate at grain boundaries. BELLE[15] has pointed out that pores in $UO_2$ may act as anchoring points for grain boundaries, since the latter cannot move without first changing the pore shape.

FRANCOIS and KINGERY[6] have also reported that the residual porosity located at junctions between $UO_2$ grains practically stops the densification. This type of microstructure and the density remain stable over a wide range of time and temperature.

The influence of water vapour (Table 4) does not seem to be significant unless to introduce the excess oxygen in $UO_2$ at higher temperatures or to help in preserving initial non-stoicheiometry during the sintering of $UO_2$ in wet nitrogen. Since lower sintering temperature decreases the sintering rate, better results may be

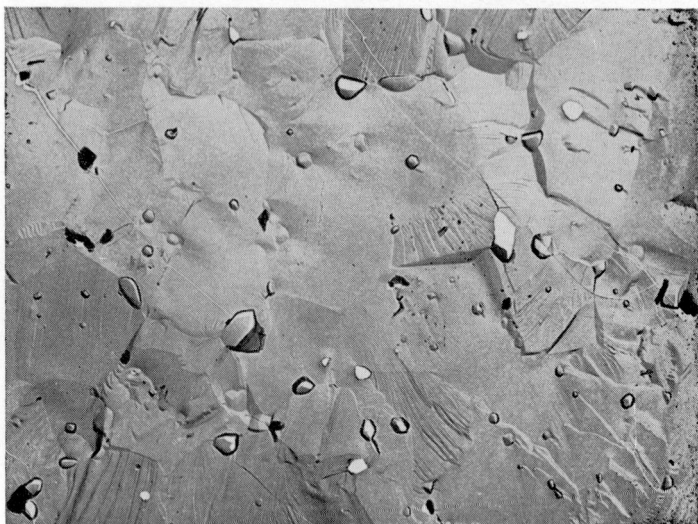

FIGURE 3.—Fractured surface of hydrogen-sintered pellet (powder surface area, 17·2 m²/g). Density, 94% of theoretical (× 7,000).

FIGURE 4.—Fractured surface of nitrogen-sintered pellet (powder surface area, 17·2 m²/g). Density, 74% of theoretical (× 7,000).

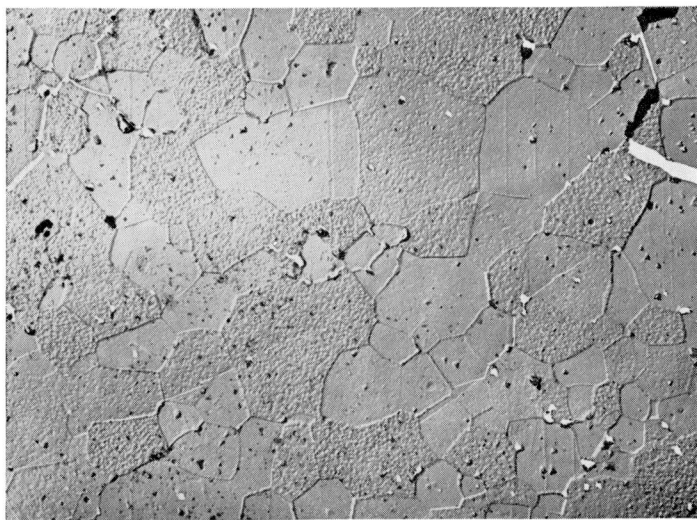

FIGURE 5.—Polished and etched section of hydrogen-sintered pellet (powder surface area, 10·7 m²/g). Density, 93% of theoretical. (× 7,000).

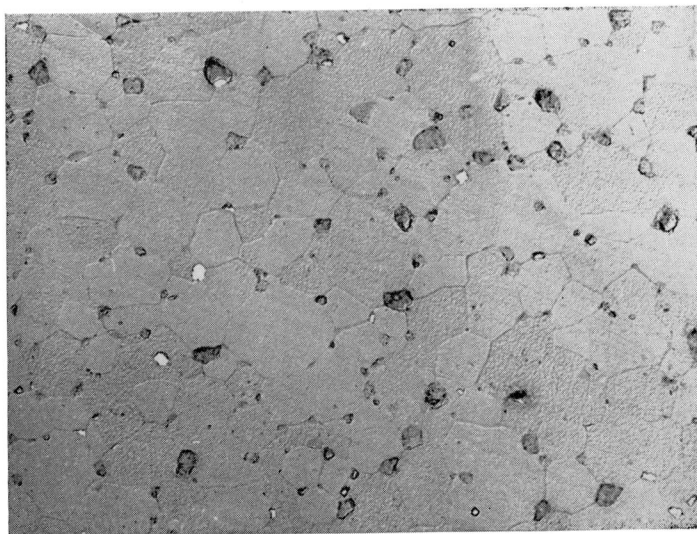

FIGURE 6.—Polished and etched section of nitrogen-sintered pellet (powder surface area, 10·7 m²/g). Density, 92·5% of theoretical (× 7,000).

FIGURE 7.—Fractured surface of hydrogen-sintered pellet (powder surface area, 7·2 m²/g). Density, 92% of theoretical. (×7,000).

Table 4

Influence of Furnace Atmosphere on the Sintered Densities of Powders with Different Specific Surface Areas

| Furnace atmosphere | Starting pellets | Sintered density (% theor.) | |
|---|---|---|---|
| | | UO₂ 17·2 m²/g | UO₂ 7·2 m²/g |
| Wet N₂ | Non-stoicheiometric | 73·7 | 96·7 |
| Dry N₂ | Non-stoicheiometric | 75·6 | 96·4 |
| Wet N₂ | Reduced | 82·5 | 95·2 |
| Wet H₂ | Reduced | 94·1 | 91·8 |
| Dry H₂ | Reduced | 93·2 | 90·0 |

expected with high-surface-area non-stoicheiometric $UO_2$ sintered at lower temperatures.

To check this deduction, $UO_2$ powder was used from series I, with specific surface area of 15·4 m²/g. The results are shown in Figure 9. Starting with non-stoicheiometric $UO_2$, highest densities were achieved at 1000°C. Removal of non-stoicheiometric oxygen prior to sintering (3 h in hydrogen at 600°C) increases the optimum temperature, since the oxidizing influence of water vapour becomes significant at higher temperatures. Similar optimum temperature ranges for $UO_2$ have been observed by other authors, for example on sintering in hydrogen,[9] in nitrogen[10] or in $CO_2$.[11]

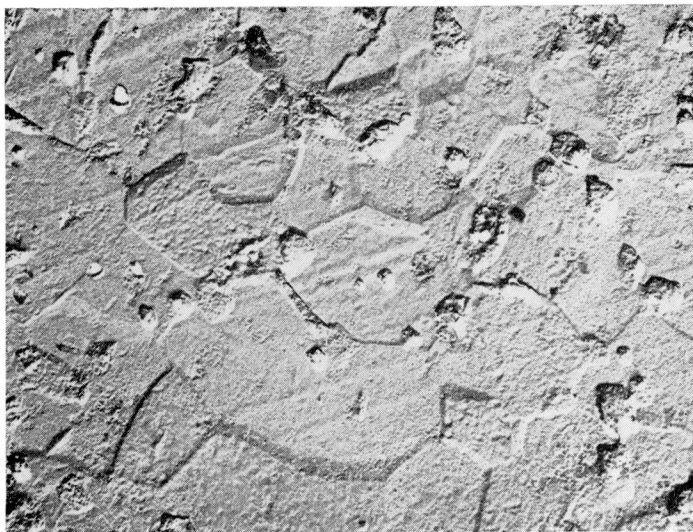

FIGURE 8.—Polished and etched section of nitrogen-sintered pellet (powder surface area, 7·2 m²/g). Density, 96% of theoretical (× 17,000).

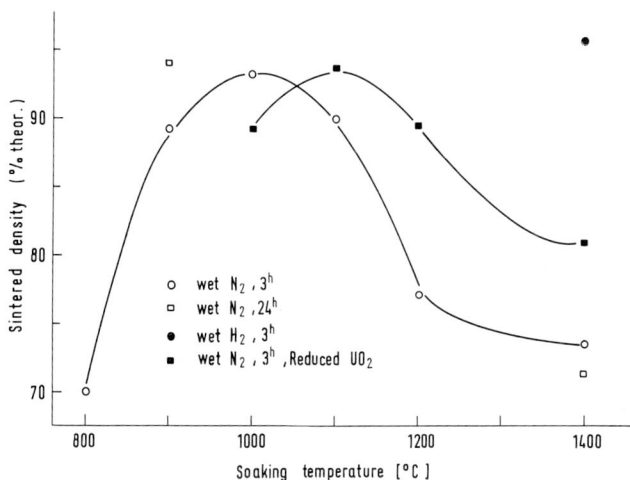

FIGURE 9.—Sintered density of high-surface-area UO₂ as a function of soaking-temperature.

Whereas prolonged soaking (24 h) at 900°C resulted in increased density, as would be normally expected, prolonged soaking at 1400°C even showed a reduction in final density. The phenomenon has been explained as a "bloating" effect[7] and in certain cases was ascribed to CO generation in closed pores.[6,4] Again, sintering in hydrogen at 1400°C gave much higher final densities, comparable with those obtained in nitrogen at 1000°C.

## 5. CONCLUSION

The beneficial influence of a reducing or slightly oxidizing atmosphere depends on the reactivity (specific surface area) of the $UO_2$ powder used. The combined effects of excess oxygen in non-stoicheiometric $UO_2$ and high surface area of the initial powder may limit sintered density and lead to the formation of large intergranular pores. This may explain some conflicting results in the literature regarding the influence of atmosphere on the sintering of $UO_2$.

### ACKNOWLEDGMENT

This work was performed under the auspices of the Yugoslav Federal Nuclear Energy Commission.

### REFERENCES

1. MURRAY, P., PUGH, S. F., and WILLIAMS, J., Fuel Elements Conference, Paris 1957, US AEC Report TID–7546, Vol. 2, p. 433.
2. WILLIAMS, J., BARNES, E., SCOTT, R., and HALL, A., *J. Nucl. Mat.*, **1**, 28, 1959.
3. RUNFORS, U., SCHÖNBERG, N., and KIESSLING, R., Proceedings of the 2nd United Nations International Conference on the Peaceful Uses of Atomic Energy, Geneva, 1958, Vol. 6 (UN: Geneva, 1958) pp. 612–619.
4. BEL, A., DALMAS, R., and FRANCOIS, B., *J. Nucl. Mat.*, **1**, 259, 1959.
5. WEBSTER, A. H., and BRIGHT, N. F. H., Research Report R–2, Department of Mines and Technical Surveys, Ottawa (1958).
6. FRANCOIS, B., and KINGERY, W. D., CEA Report DM-1469 (1965). (Paper submitted at the International Conference on sintering and related phenomena, Notre Dame 1965).
7. COBLE, R. L., *J. Amer. Ceram. Soc.*, **45**, 123, 1962.
8. JOHNSON, J. R., and SOWMAN, H. G., "Uranium Dioxide" (Ed. J. Belle) (US AEC: Washington, 1961) p. 339.
9. FRANCOIS, B., DELMAS, R., CAILLAT, R., and LACOMBE, P., *J. Nucl. Mat.*, **15**, 105, 1965.
10. FUHRMAN, N., HOWER, L. D., and HOLDEN, R. B., *J. Amer. Ceram. Soc.*, **46**, 114, 1963.
11. MÜLLER, N., *Ber. Dtsch. Keram. Ges.*, **40**, 140, 1963.
12. KIESSLING, R., and RUNFORS, U., US AEC Report TID 7546, p. 411.
13. FRANCOIS, B., DELMAS, R., CAILLAT, R., and LACOMBE, P., *C. R. Acad. Sci.*, **256**, 925, 1963.
14. FRANCOIS, B., KURKE, G., and DELMAS, R., 2nd International Powder Metallurgy Conference, Prague 1966, Vol. 1, (Academia: Prague 1966) p. 177.
15. BELLE, J., Fuel Elements Conference, Paris, 1967. US AEC Report TID 7546/2, p. 477.

# 14.—The Kinetics of Recrystallization during Solid State Sintering

By P. P. BUDNIKOV and F. JA CHARITONOV

*Mendeleev Institute for Chemical Technology, Moscow*

*ABSTRACT*                                                    C525/A43

*Factors such as growth of the area of contact between particles, the coalescence of pores, particle size distribution and temperature during densification have been investigated for $Al_2O_3$, $MgO$ and $ZrO_2$. Equilibrium conditions at particle surfaces in relation to surface free energy, and the coalescence of pores during sintering, are examined. Under certain firing conditions small pores inhibit the development of larger pores and a fine-grained material develops (barrier effect). The extent of recrystallization, given by the ratio of crystal size at the end to that at the beginning of sintering, is inversely proportional to the square of the grain size before sintering. Density is proportional to the extent of recrystallization, as there is a linear relationship between the size of recrystallized grains and density. It is assumed that particles above an optimal size do not take part in recrystallization during sintering. A dense product with a fine crystalline structure will be obtained as the surface free energy and the molar volume of the starting material are lowered by processing and sintering. Investigations on pure oxides showed that recrystallization of the grains is not caused by sintering and intergrowth of the grains but by displacement of the barriers between adjacent crystals and between crystals and pores. Recrystallization rates determined for $Al_2O_3$ particles between 10 and 40 μm showed a maximum for particles between 20 and 25 μm. Activation energies connected with the sintering and recrystallization of $Al_2O_3$ and $MgO$ were calculated.*

## La cinétique de la recristallisation au cours du frittage en phase solide

*Les facteurs suivants: croissance de la surface de contact entre les particules, coalescence des pores, répartition granulométrique et température sont étudiés, au cours de la densification, pour $Al_2O_3$, $MgO$, et $ZrO_2$. Les conditions d'équilibre aux surfaces des particules sont examinées en relation avec l'énergie libre de surface ainsi que la coalescence des pores pendant le frittage. Dans certaines conditions de cuisson, les pores de petites dimensions empêchent le développement* *de pores plus grands et il se forme un matériau à grains fins (effet de barrière). L'importance de la recristallisation, donnée par le rapport de la dimension du cristal à la fin du frittage à celle du début, est inversement proportionnelle au carré de la dimension du grain avant le frittage. La densité est proportionnelle à l'importance de la recristallisation puisque la relation entre la dimension des grains recristallisés et la densité est linéaire. On admet que les particules situées au-dessus d'une dimension optimale n'interviennent pas dans la recristallisation au cours du frittage. Un produit dense à fine structure cristalline est obtenu, quand l'énergie libre de surface et le volume molaire du matériau initial ont été abaissés par un traitement préalable et un frittage. Des recherches sur des oxydes purs montrent que la recristallisation des grains n'est pas due au frittage et à l'intercroissance des grains mais au déplacement des barrières entre les cristaux adjacents et entre les cristaux et les pores. Les vitesses de recristallisation déterminées pour des particules de $Al_2O_3$ situées entre 10 et 40 μm présentent un maximum pour les particules situées entre 20 et 25 μm. Les énergies d'activation associées au frittage et à la recristallisation de $Al_2O_3$ et MgO sont calculées.*

### Die Kinetik der Rekristallisation während des Sinterns im festen Zustand

*Faktoren wie das Wachsen der Kontaktflächen zwischen Teilchen, das Zusammengehen von Poren, Teilchengrößenverteilung und Temperatur während der Verdichtung wurden an $Al_2O_3$, MgO und $ZrO_2$ untersucht. Es werden die Gleichgewichtsbedingungen zwischen den Teilchenoberflächen und der Porenabnahme in Beziehung zur feien Oberflächenenergie während des Sinterns erforscht. Unter bestimmten Brennbedingungen verhindern kleine Poren die Entwicklung größerer Poren, und ein feinkörniges Material bildet sich aus (barrier effect). Das Ausmaß der Rekristallisation als Ergebnis aus dem Verhältnis der Kristallitgröße am Ende zur Kristallitgröße zu Beginn des Sinterns ist umgekehrt proportional zum Quadrat der Korngröße vor dem Sintern. Die Dichte ist dem Ausmaß der Rekristallisation proportional, weil eine lineare Beziehung zwischen der Größe der rekristallisierten Körner und der Dichte besteht. Es wird vermutet, daß die Teilchen oberhalb einer optimalen Größe nicht an der Rekristallisation während des Sinterns teilnehmen. Ein dichtes Produkt mit feinkristalliner Struktur wird erhalten, weil die freie Oberflächenenergie und das molare Volumen des Ausgangsmaterials durch die Herstellung und das Sintern erniedrigt werden. Untersuchungen an reinen Oxiden zeigten, daß Rekristallisation nicht durch Sintern und Einwachsen der Körner sondern durch Versetzung der Hindernisse* *zwischen angrenzenden Kristalliten und zwischen Kristalliten und Poren verursacht wird. Rekristallisationsverhältnisse an Al$_2$O$_3$-Teilchen zwischen 10 und 40 μm zeigten ein Maximum für Teilchen zwischen 20 und 25 μm. Bs wurden Aktivierungsenergien von Sintern und Rekristallisation an Al$_2$O$_3$ und MgO berechnet.*

## 1. INTRODUCTION

Recent advances in the theory of sintering of ceramics, especially of pure oxide materials, have solved a series of important problems. Polycrystalline materials consisting practically of a single phase have been prepared that have properties closely resembling those of single crystals. The kinetics of sintering and recrystallization, however, has not been sufficiently investigated, and it therefore seemed desirable to study the sintering and recrystallization of some refractory oxides such as Al$_2$O$_3$ and MgO, investigating the importance of factors such as particle size distribution, purity of the original components, growth of the area of contact between particles and coalescence of pores during these processes.

## 2. EXPERIMENTAL

### 2.1 Raw Materials
In this investigation the following raw materials were used.

### 2.11 Technical Alumina (99·1% Al$_2$O$_3$)
The alumina was fired at 1450°C after the addition of 1% of boric oxide. Batches of the compositions given in Table 1 were prepared by wet grinding. Mixtures K2 and MK were prepared under such conditions that, after sintering, finely crystalline corundum was obtained.[1]

### 2.12 Spectroscopically Pure MgO with Additions of Hafnium
Spectroscopically pure MgO with additions of hafnium was prepared from the hydroxide by ammonia precipitation from a solution containing spectrally pure magnesium chloride and hafnium sulphate p.a. The precipitates were filtered off, dried and fired for 1 h at 600°C. The particle size distribution of the primary material was measured by sedimentation analysis and the grain size of the sintered products was measured by microscope.

### 2.2 Experimental Methods
An important property for characterizing the kinetics of sintering and recrystallization is the linear rate of displacement of phase

14

### Table 1
### Preparation of Technical Alumina

| Code No. | Wet grinding | | Remarks |
|---|---|---|---|
| | Type of mill | Duration of grinding (h) | |
| K1 | Flint mill | 10–100 | Hydrochloric acid treatment to remove iron impurities |
| K2 | Corundum mill | 50 | Addition of 0·1 % MgO |
| MK | Corundum mill | 50 | Addition of MgO as $MgCl_2.6H_2O$ |

boundaries in the sintering particles. This value can be measured approximately by several methods with different degrees of accuracy.

Solid-state sintering of oxides is characterized by the change in dimensions of particles of different sizes at different rates not only in magnitude but in sign, i.e. there are different rates of particle growth and resorption.

Bearing this in mind we measured the rates of recrystallization of grains within a given size range. In order to determine the rate at which the diameter changes we also needed to know how particle size distribution changed with time. When a grain size distribution is estimated, a fraction showing a certain spread is always observed in diameter. The diameters of two neighbouring fractions differ in size by an amount $\Delta x$; as sintering is measured at finite intervals of time, $\Delta t$, the recrystallization, can be calculated as follows:[2]

$$v(xt) = \frac{\delta \Delta N_x}{2\bar{n}\Delta t} \quad . \quad . \quad . \quad . \quad (1)$$

$\delta$ = Fraction width ( = distribution interval)
$\Delta N_x$ = Change in total number of grains in a fraction (changes caused by growth and resorption)
$\bar{n}$ = Mean number of grains of a fraction during the time considered ($\Delta t$).

For measurements of the distribution functions and to obtain information about the role of the grain-to-grain contacts during sintering, pieces of the sintered materials were ground and polished, etched with concentrated NaOH and observed under the microscope. Pores which sometimes occupied up to 10% of the areas inspected were ignored.

## 3. RECRYSTALLIZATION OF CORUNDUM-TYPE CERAMICS

Figure 1 shows distribution curves for material K1 after being heated at 1750°C for 3 h. Based on this grain size distribution, the rates of recrystallization have been calculated for a series of adjacent fractions between 10 and 40 $\mu$m using Equation (1). The results show certain limits of grain size at which the rate of recrystallization is maximum and for corundum this maximum coincides with diameters of 20–25 $\mu$m. Larger and smaller particles recrystallize slower, which leads us to assume the existence of a diameter limit above which grains do not take part in recrystallization.

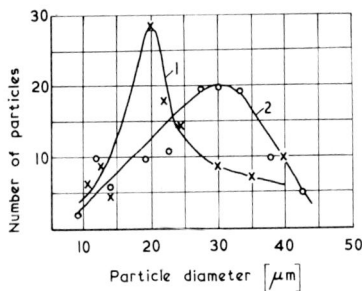

FIGURE 1.—Grain size distribution for corundum ceramic K1 sintered at 1750°C: (1) 1 h. (2) 3 h.

Table 2 shows (for alumina of various particle sizes) the density and grain size of corundum ceramic K1 for various firing temperatures, from which it can be concluded that a finer starting material would increase the apparent density and the crystal size of the product.

### Table 2
### Bulk Density and Particle Dimensions of Corundum Ceramic K1

| Firing temperature (°C) | Mean particle diameter of alumina ($\mu$m) | Bulk density of body (g/cm³) | Mean particle dimensions of corundum ($\mu$m) |
|---|---|---|---|
| 1450 | 0·85 | 3·32 | 2×7 to 5×7 |
| 1450 | 0·73 | 3·75 | 5×7 to 10×40 |
| 1450 | 0·44 | 3·86 | 7×14 to 14×300 |
| 1750 | 1·95 | 3·53 | 2×7 |
| 1750 | 1·65 | 3·78 | 10×6 |

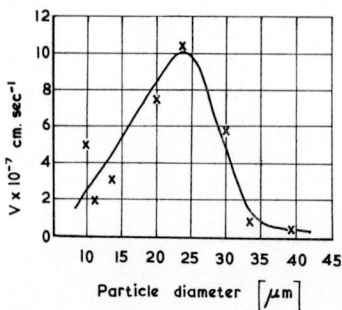

FIGURE 2.—Recrystallization rate of K1 as a function of grain diameter.

## 4. MICROSTRUCTURE

Figures 3–6 show that in most cases the crystals arrange themselves in such a way that at any point three grains meet. The average angle between any two grain-boundaries is 120°, which is in accord with the 6-fold symmetry of the crystals. This becomes more evident from the electron micrograph of spectroscopically pure MgO with 0·25 $^a/_o$ Hf (Figure 6). Details of the microstructure can be understood if attention is paid to the energy conditions at grain boundaries. Both sintering and recrystallization are exhibited by systems trying to reach a state of equilibrium, which is characterized by minimum free energy. For such equilibrium, minimum grain-boundary surface is necessary. In an attempt to minimize the total surface area, the grain boundaries move from one particle into another, thereby increasing the grain contacts; in other words they migrate towards their centre of curvature.

To obtain minimum boundary surface in this way, the crystal surfaces should be flat or have equal radii of curvature which are perpendicular to each other and of opposite sign.

The observations made on Figures 3–6 that in pure one-phase ceramics the angles between grain-boundaries are ±120° supports the supposition that the phase boundaries possess equal boundary energy. When a starting material of high purity is used, this equilibrium state is approached more closely, as is shown by the structures of polycrystalline and transparent corundum ceramics.[3–6] In K2 and MK (Figures 4 and 5) which were sintered after addition of MgO, grain growth has been hindered owing to the formation at the corundum boundaries of a layer of new reaction product, mainly $MgAl_2O_4$, causing a decrease in the free boundary energy of the system. In addition the high proportion of small particles in MK

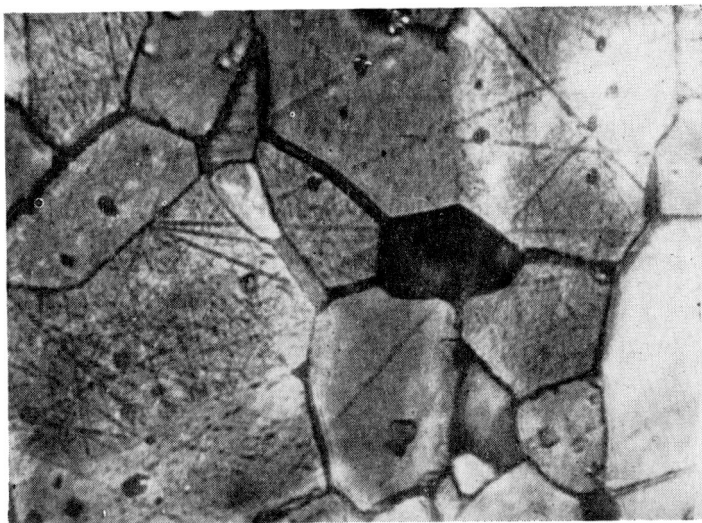

FIGURE 3.—Microstructure of K1 sintered at 1750°C for 1 h (etched with NaOH; ×900).

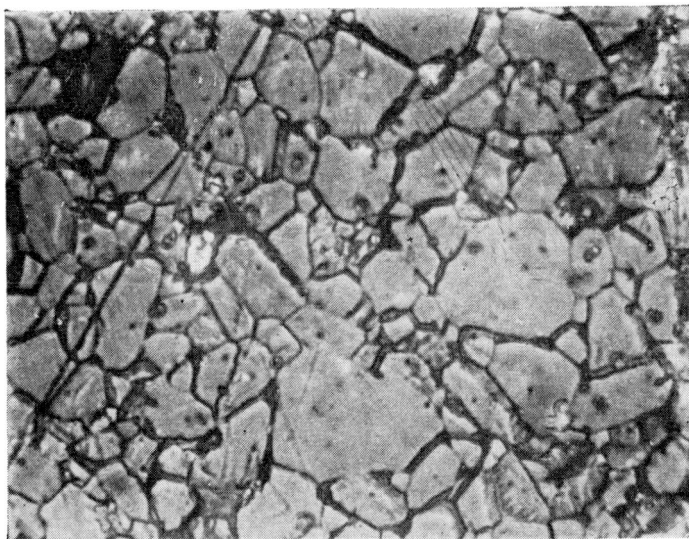

FIGURE 4.—Microstructure of K2 sintered at 1750°C, for 1 h (etched with NaOH; ×900).

FIGURE 5.—Microstructure of fine crystalline MK sintered for 1 h at 1750°C (etched with NaOH; ×900).

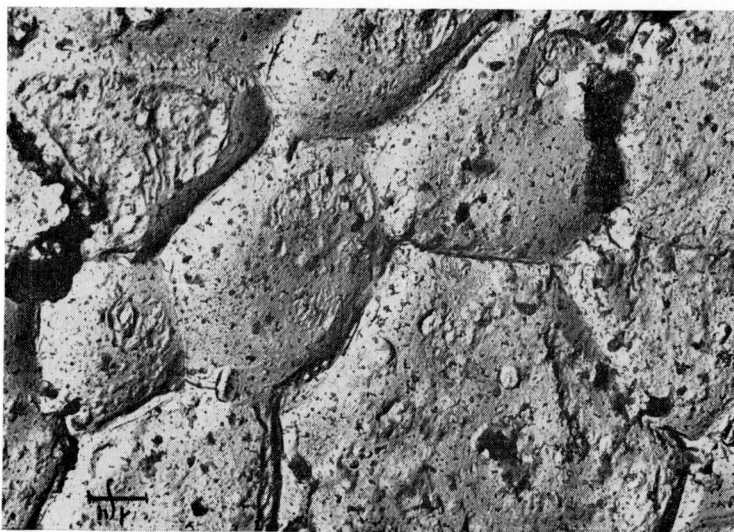

FIGURE 6.—Microstructure of spectrally pure MgO+0·25% Hf sintered at 1400°C for 1 h. (Electron micrograph; ×10,000).

retards the growth of larger grains by forming a homogeneous, fine-grain structure (barrier effect). The coarser crystalline structure of K2 may indicate that the proportion of MgO added (0·1%) was too small and that secondary recrystallization of the corundum has taken place.

## 5. RECRYSTALLIZATION OF SPECTRALLY PURE MgO WITH HAFNIUM ADDITIONS

The influence of small additions of cation on sintering and re-crystallization in very pure oxides is of great interest. When, for example, the purest magnesium oxide is used, it is possible to study the influence of an addition in "undiluted" form, namely under conditions in which the quantities of the additions are many times larger than the total of all the impurities present and at the same time small enough to find a site in the "host lattice".

Figure 7 refers to the sintering and recrystallization of spectrally pure magnesium oxide, to which 0·25ª/ₒ Hf has been added. The apparent density and grain size change in different ways. About 1100°C there is an important increase in relative density to 0·95, after which the density increases only slightly, but grain growth

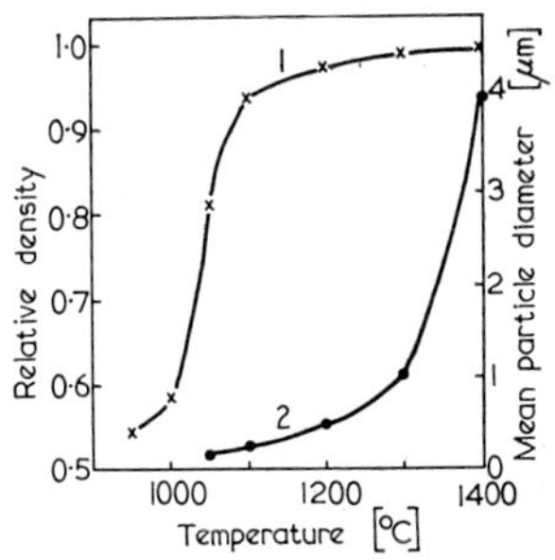

FIGURE 7.—Sintering and grain growth of spectrally pure MgO + 0·25ª/ₒ Hf as a function of temperature:
(1) Relative density; (2) average grain diameter.

becomes fairly rapid so that at 1400°C and at 1500°C the grains are 4 μm and 20 μm respectively. It can be concluded that densification and grain size growth start at temperatures 0·4 to 0·5 of the melting temperature of MgO. Under isothermal conditions the rate of grain growth of polycrystalline ceramics is inversely proportional to the grain diameter:

$$\frac{\mathrm{d}D}{\mathrm{d}t} = \frac{b}{D} \qquad . \qquad . \qquad . \qquad . \qquad . \qquad . \qquad (2)$$

When the diameter of the primary grain is small compared with that of the recrystallizing grain, its value can be neglected and Equation (2) can be written:

$$D = Bt^{\frac{1}{2}} \qquad \qquad (3)$$

$D$ being the grain diameter at time $t$.

It follows from Equation (3) that the average grain diameter is proportional to the square root of the sintering time, which has been confirmed by isothermal experiments on spectrally pure MgO to which $0.25^{a}/_{o}$ Hf had been added (Figure 8).

For considering the influence of temperature on the kinetics of recrystallization, Equation (3) can be written:

$$D = a.e^{-Q/2RT}t^{\frac{1}{2}} \qquad \qquad (4)$$

or

$$\log D = \log a - \frac{0.434Q}{2RT} + \tfrac{1}{2} \log t \qquad . \qquad . \qquad (5)$$

Earlier published results,[7] on the recrystallization at 1100°C and at 1400°C of pure MgO with the same proportion of Hf, have shown that the activation energy calculated with Equation (5) is 89 kcal/mol; this value is in good agreement with values obtained earlier[8,9] on spectrally pure MgO in experiments concerned with sintering and electrical conduction.

Experimental results on the sintering kinetics of MgO have shown the applicability of the following relation:

$$\log \gamma = \log A - \frac{0.434Q}{2RT} + \tfrac{1}{2} \log t \qquad . \qquad . \qquad (6)$$

where $\gamma$ is the relative density at time $t$.

The slope of the straight line $\gamma = f (1/T)$ corresponds to $0.217$ $Q/R$ from which the activation energy at 1500° to 1600°C is calculated as 86 kcal/mol.[10]

Substituting Equations (5) and (6); $\log D = \log \gamma - \log A + \log a$ or $D/\gamma = a/A = K$.

This relation indicates that the diameter of the recrystallizing grains is proportional to the relative density of the sintered product, and earlier experiments on spectrally pure MgO[7] confirm this conclusion.

Figure 9 shows the influence of the addition of Hf on the sintering and recrystallization of spectrally pure MgO at 1400°C. Very small additions (0.05–0.1 %) produce a more rapid crystal growth and higher density; with a higher proportion of Hf, a limit is reached.

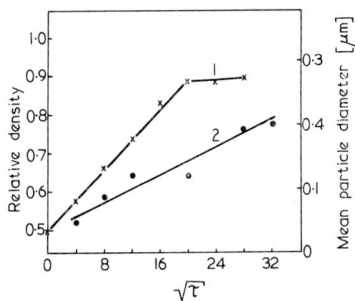

FIGURE 8.—Sintering and recrystallization at 1100°C of spectrally pure MgO+
0·25$^a$/$_o$ Hf as a function of the sintering time:
(1) Relative density; (2) average grain diameter.

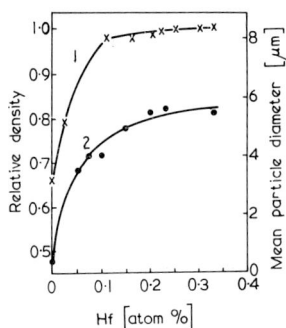

FIGURE 9.—Sintering and recrystallization of spectrally pure MgO at 1400°C
as a function of the proportion of Hf added:
(1) Relative density; (2) average grain diameter.

## 6. PORE COALESCENCE

Densification is accompanied by pore coalescence, smaller pores
being absorbed by larger ones. Atoms at the surface of a small
pore contribute less to the total free energy than those at the surface
of a larger pore. Under conditions of increased mobility of the lattice
components, solid-state diffusion is initiated, resulting in an increase
in the volume of the larger pores at the expense of the smaller
ones, which may result ultimately in complete resorption of the
latter. With interconnected pores, coalescence can be obtained by
surface migration of solid particles; with closed pores, migration by
a process of volume diffusion using lattice defects is possible.

In the latter case the presence of phase boundaries in the sintering material is of great importance. For diffusion along a phase boundary, the diffusion coefficient is many times larger than for diffusion through the crystal, and it is to be expected that the rate of pore coalescence is higher when the phase boundary network is denser. The structure of fine crystalline corundum is practically pore-free; whereas in coarse-grained corundum inter-crystalline pores preponderate (Figures 4 and 5), in sintered MgO and $ZrO_2$, however, most of the pores are at the grain boundaries (Figures 6 and 10). To eliminate this residual porosity, sintering in high vacuum, as is usual in the production of pure transparent corundum and magnesia is recommended.

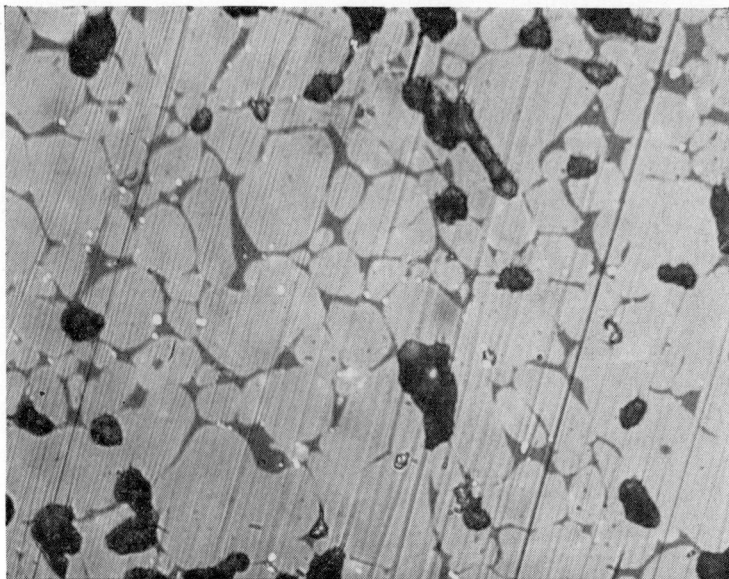

FIGURE 10.—Microstructure of $ZrO_2$ stabilized with $7^w/_o$ CaO sintered $\frac{1}{2}$ h at 1750°C (etched with NaOH; ×550).

## 7. SUMMARY

With corundum and spectrally pure magnesium oxide to which Hf has been added, the rate of recrystallization of corundum grains measured for a series of adjacent fractions in the grain diameter range 10–40 $\mu$m attains a maximum which for corundum is between 20 and 25 $\mu$m.

An important densification and the beginning of grain growth takes place in spectrally pure magnesium oxide with $0.25 \,^a/_o$ hafnium at temperatures $0.4$ to $0.5$ of the melting temperature of MgO. In isothermal sintering of pure MgO the average grain diameter is proportional to the square root of the sintering time. The activation energy of the grain growth of MgO at $1100°$ and at $1400°C$ is calculated to be 89 kcal/mol. The diameter of the recrystallizing grains is proportional to the apparent density. Very small additions ($0.5$–$1.0 \,^a/_o{}^2$) accelerate grain growth and promote the formation of a dense product.

## REFERENCES

1. PAWLUSKIN, N. M., "Speceni Korund", (Gosstroiisdat, Moskau, 1961).
2. CHARITONOV, F. J., *Steklo i Keramika*, **17**, (1), 18, 1960.
3. DEGTIAREWA, A. W., *Neorganitscheskije materialy*, **1**, (2), 281, 1965.
4. DEGTIAREWA, A. W., *Neorganitscheskije materialy*, **2**, (2), 2058, 1966.
5. PHILIPPE, D. S., "Process for producing transparent polycrystalline alumina", *U.S. Pat.*, 3,026,177, 20/3/1962.
6. COBLE, R. L., "Transparent alumina and method of preparation", *U.S. Pat.* 3,026, 210, 20/3/1962.
7. BUDNIKOV, P. P., MATWEJEW, M. A., JANOWSKIJ, W. K., and CHARITONOV, F. J., *Neorganitscheskije materialy*, **1**, (8), 1349, 1965.
8. BUDNIKOV, P. P., MATWEJEW, M. A., and JANOWSKIJ, W. K., *Ogneupori*, **30**, (4), 32, 1965.
9. BUDNIKOV, P. P., and JANOWSKIJ, W. K., *Shurnal prikladnoj chimij*, **37**, (6), 1247, 1964.
10. BUDNIKOV, P. P., MATWEJEW, M. A., and JANOWSKIJ, W. K., *Dokladi Akad. Nauk SSSR*, **159**, (4), 872, 1964.

# Part Four

# REACTIONS IN MULTIPHASE CERAMICS DURING FIRING

# 15.—Investigations of Structure Transformations in Refractories above 2000°C

By M. Foex

*Laboratoire des Ultra Réfractaires du Centre National de la Recherche Scientifique, Montlouis, 66, France*

***ABSTRACT***                                                    E21d/C526

*By combining the use of a solar or plasma centrifuge with thermal analysis, the high-temperature crystal forms of refractory oxides which normally do not persist after quenching can be investigated under the conditions of their formation. Arrests in the cooling curves of zirconia and the rare-earth oxides have been detected and the crystal conversions identified. The effect of additives on the cubic–tetragonal transformation of zirconia about 2300°C and the ranges of the hexagonal (A), monoclinic (B) and the cubic (C) modifications of the rare-earth sesquioxides have been determined. Two new modifications of the latter oxides which are stable above 2000°C have been identified: one, a hexagonal (H), which relates to all the sesquioxides except the last of the series; the other, a cubic form (X) which relates to the first few oxides only. The conversion from one form to the other has been studied and similarities noted between the A⇌H transformation in these oxides and the α⇌β transformation in quartz.*

### Recherches sur la transformation des oxydes réfractaires

*En combinant l'utilisation d'un four solaire ou d'un four à plasma centrifuges avec l'analyse thermique, les formes cristallines haute-température des oxydes réfractaires qui ne persistent pas normalement après le refroidissement brusque peuvent être étudiées dans les conditions de leur formation. Des irrégularités sont décelées dans les courbes de refroidissement de la zircone et des oxydes de terres rares et les transformations cristallines sont identifiées. L'effet d'ajouts sur la transformation cubique tétragonale de la zircone vers 2300°C est déterminé ainsi que le domaine des modifications hexagonale (A), monoclinique (B) et cubique (C) des sesquioxydes à l'exception du dernier de la série, l'autre, une forme cubique X, pour quelques-uns des premiers oxydes seulement. La transformation d'une forme en une*

217

*autre est étudiée et on observe des similitudes entre les transformations $\overline{A}\rightleftharpoons H$ de ces oxydes et la transformation $\alpha \rightleftharpoons \beta$ du quartz.*

### Untersuchung von Umwandlungen von feuerfesten Oxiden

*Unter kombinierter Benutzung eines Sonnen- oder Plasma-Zentrifugenofens und der thermischen Analyse können die Hochtemperatur-Kristallformen von feuerfesten Oxiden, die normalerweise nach dem Abschrecken nicht mehr existieren, unter ihren Entstehungsbedingungen untersucht werden. Unregelmäßigkeiten der Abkühlkurve von Zirkoniumoxid und der Seltenen Erd-Oxide wurden entdeckt und die Kristallumwandlungen identifiziert. Der Einfluß von Zusätzen auf die Umwandlung tetragonal in kubisch von Zirkoniumoxid bei ca. 2300°C und der Bereich der hexagonalen (A), der monoklinen (B) und der kubischen (C) Modifikationen der Sesquioxide der Seltenen Erden wurde ermittelt. Zwei neue Modifikationen der letzteren Oxide, die oberhalb 2000°C stabil sind, wurden identifiziert: eine hexagonale Form H, die bei allen Sesquioxiden mit Ausnahme letzten Oxide der Serie auftritt, und eine kubische Form X, die nur bei einigen der ersten Oxide der Reihe auftritt. Die Umwandlung von der einen Form in die andere wurde untersucht, wobei Ähnlichkeiten zwischen der A–H-Umwandlung dieser Oxide und der α–β-Umwandlung des Quarzes bemerkt wurden.*

## 1. INTRODUCTION

The rate at which structural conversions take place depends mainly on temperature, and so it is interesting to consider the difference between the equilibrium temperature and the temperature of the melting point of a substance. If these temperatures are widely separated, which normally means that the conversion is at a relatively low temperature, it is difficult to recover the original phase on cooling.

For a long time several of these transformations had been considered irreversible, but it is now possible to effect them by prolonged treatment by, for instance, grinding and pressure.

However, if the conversion temperature is close to that of the melting point, a change in crystal structure takes place rapidly as soon as conditions deviate slightly from equilibrium. In these cases, it is often not possible by rapid quenching to preserve the high-temperature modification at room temperature, and it is necessary to use methods which enable the high-temperature modifications to be studied in the conditions in which they are formed. Thermal analysis is the most convenient method, and for these measurements a recording optical pyrometer is essential.

## 2. THERMAL ANALYSIS USING A SOLAR FURNACE

This apparatus, which has been described in detail elsewhere,[1,2] incorporates a 2-kW solar furnace. The product is heated in air. A parabolic mirror of 2 m diam. and with a focal length of 0·85 m is set up in the laboratory (Figure 1), and this stationary mirror receives the solar radiation via a movable mirror outside the laboratory.

The samples are placed in a rotating cylindrical metal container having a volume of 10 to 25 cm$^3$ and an opening of 11 mm diam.; it is water-cooled. The material that is nearest the centre of the open end is in focus and will melt, and a cavity is formed by centrifugal force. After the radiation has been screened off, the cooling curve of the bottom of the cavity is measured with a pyrometer equipped with cesium cell and a rapid-response recorder. With this technique, changes in crystal structure at very high temperatures can be detected, but when the cooling rate is very rapid, there is a lag, especially when the conversion temperature is well below the melting point of the oxide concerned. However, much slower rates of cooling have been achieved by means of a pyrometer that enables measurements to be made in the solar beam, the beam being increasingly obscured by a screen.

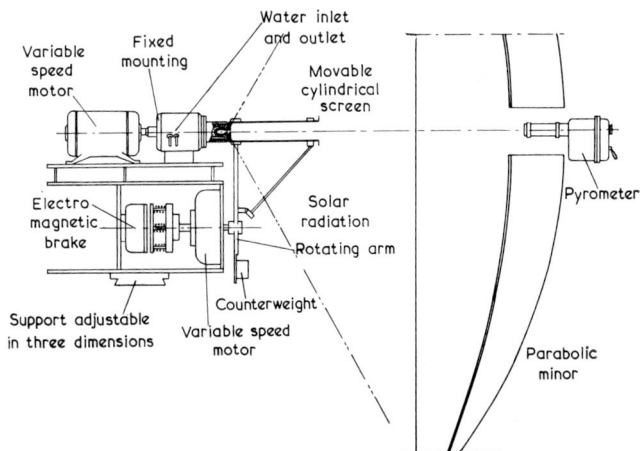

FIGURE 1.—Schematic representation of the solar furnace used for thermal analysis at high temperature.

## 3. THERMAL ANALYSIS USING A PLASMA FURNACE

The sample is centrifuged in a rotating metal cylinder which has two axial openings and the product is preheated with a conventional

15

plasma torch; in this way a cylindrical cavity lined with the hot material is formed into which is started an intense long arc, the electrodes being the preceding plasma and a second plasma on the outside of the furnace (Figure 2). This furnace operates with high efficiency with several gases; contamination by material from the walls of the cylinder and from the electrodes is small, and there may be none at all.

Figure 2 shows the "centrifuge furnace" (5) fed by the two plasma torches (1,3) which receive their current through (2) and (4) respectively. A third supply (8) feeds the arc discharge heater. The powdered sample (10) melts in the centre (9). A cylinder (6) enclosing the furnace has a gas-tight connection (7) to one of the surfaces; which enables the desired atmosphere to be introduced after the current to the torch (1) has been cut off. An automatic pyrometer (11) working in the red or infra-red region records after the temperature the plasma torches and the arc discharge have been extinguished.

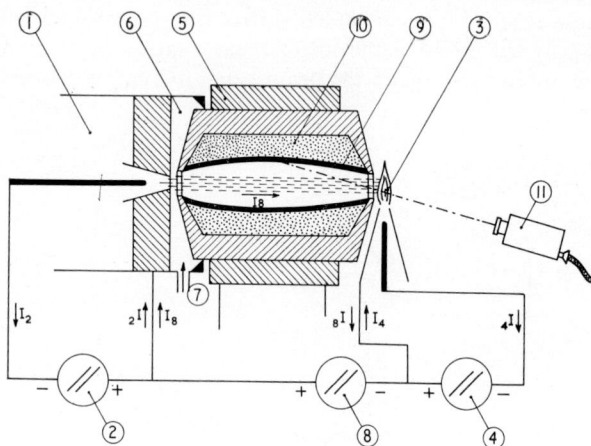

FIGURE 2.—Apparatus for thermal analysis at high temperature using a plasma furnace.

The first plasma tube is oriented along the axis of the apparatus. The second plasma nozzle is placed in front of the opening in such a way that the plasma does not penetrate the furnace. The arc in the centre of the furnace may easily use 50 kW in a length of 25 cm, and all refractory oxides can be easily melted in this furnace. This apparatus enables the thermal analysis of approximately 1000 g of oxide.

With the two preceding apparatus we found several anomalies in the temperature/time curves, and so in order to determine the nature of the transformations and the crystal structures involved we submitted the materials concerned to X-ray diffraction under the conditions described below.

## 3.1 High-temperature X-ray Camera

The high-temperature X-ray camera was a Philips-Norelco, with a goniometer. The sample holder consists of a metal strip furnace. The camera has beryllium or mylar windows. The electrical leads and the camera itself are water-cooled. The height of the sample can be monitored from the outside of the camera. The heating-strip is either of tungsten or of rhenium, and the atmosphere is helium, or hydrogen, or a helium-rich mixture of the two gases. Some experiments have been carried out in a slightly oxidizing atmosphere, and for these an iridium strip was used.

In order to ensure that the powder sample is not affected by reaction with the supporting strip or with the atmosphere, all the experiments were carried out with two different types of heating-element and in two different atmospheres. All the samples were re-examined after the heating.

## 4. TRANSITION OF CUBIC TO TETRAGONAL ZIRCONIA AT HIGH TEMPERATURES

Some time ago we determined a well pronounced arrest in the cooling curve of zirconia somewhat above 2300°C (Figure 3). The part at 2710°C corresponds to the solidification of the oxide. The cooling rate does not seem to affect the temperature of these anomalies (using various methods we obtained cooling-rates 10 times slower and 4 times faster). The arrest in the cooling curve indicates a change in crystal structure at that temperature, but even with rapid quenching we were unable to obtain at room temperature another crystal structure, so examination with a high-temperature X-ray camera was essential.

Measurements by SMITH and CLINE,[7] WOLTEN[8] and BOGANOV et al.[9] confirm this transition at 2300°C. Above this temperature, the tetragonal structure of zirconia changes to cubic. WEBER[10] has expressed doubt about this phase transition, and he discusses the possibility of contamination from the support and deviation from the stoicheiometric composition. The experiments were carried out in vacuum, mostly on a tungsten support. According to the three authors mentioned above, the transition is reversible, the hysteresis, if present at all, being less than 30°C. In some experiments WOLTEN[8] observed the two phases to co-exist in a narrow temperature region. BOGANOV et al.[9] have shown that there is an abrupt change from tetragonal to cubic structure; no progressive change has been observed by these authors in $c$ and $a$ parameters of the tetragonal structure near the transition temperature.

The conversion described has nothing to do with the well known monoclinic⇌tetragonal transition of zirconia, which takes place at 1100°C on heating and at 900°C on cooling.

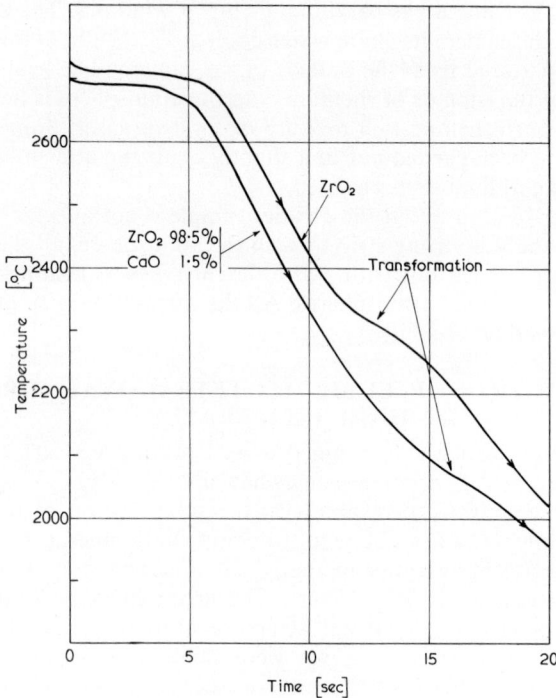

FIGURE 3.—Cooling curve in air of pure zirconia and of $ZrO_2$ containing 1·5% CaO.

In order to get a better insight into the processes by means of which the cubic⇌tetragonal transition of zirconia takes place, we carried out high-temperature X-ray measurements. Figure 4 consists of several diagrams of samples heated on a rhenium support in helium gas. The distance between the 002 and 200, 022 and 220, 113 and 311 lines decreases as the temperature approaches the conversion temperature.

Near the transition temperature there is a marked increase of $a$ and a decrease of $c$, and about 2310°C there occurs a drastic change

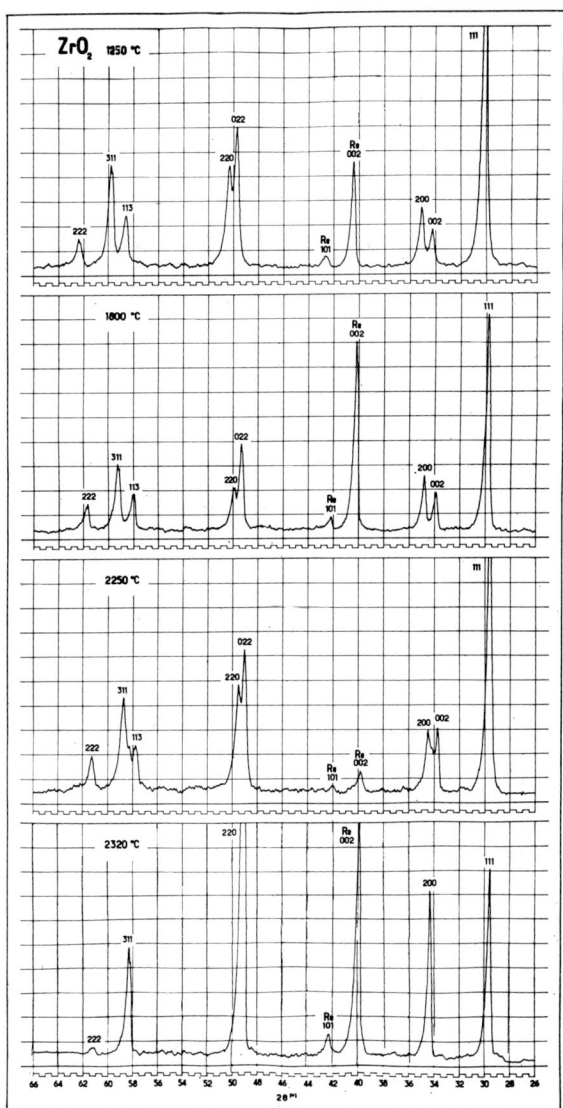

FIGURE 4.—X-ray powder pattern (CuKα) of $ZrO_2$ at different temperatures (rhenium support, He atm.):

Tetragonal form at 1250°C
Tetragonal form at 1800°C
Tetragonal form at 2250°C
Cubic form at 2320°C.

in structure which is preceded by a zone in which *c* and *a* change rapidly. It is difficult to detect the presence or absence of two phases in a narrow range of temperature. Our results show that the transformation of zirconia at high temperature is not caused by chemical attack or dissociation, and it is reversible; after the completion of every experiment the X-ray diagrams were identical with those of the starting material.

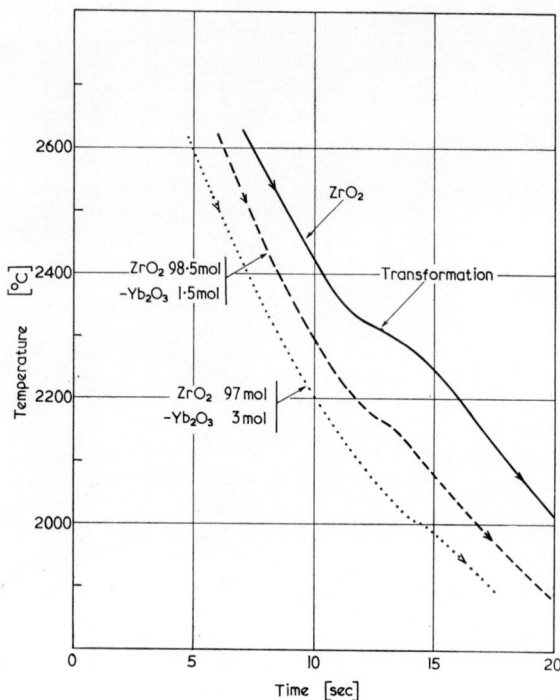

FIGURE 5.—Cooling curves of zirconia: the influence of $Yb_2O_3$ additions.

The temperature of conversion to the cubic phase decreases when other oxides, particularly calcia and rare-earth oxides, are added (Figure 5). When sufficient oxide is added, the cubic phase can be obtained at room temperature. However, this does not necessarily mean that this structure is stable at room temperature, because in several cases it disappeared after prolonged heating. When the transition temperature is below $\sim 1800°C$, the high-temperature modification can easily be obtained by quenching.

## 5. TRANSFORMATION OF THE LANTHANIDE SESQUI-OXIDES AT HIGH TEMPERATURES

In 1925 Goldschmidt, Ulrich and Barth reported the existence of three allotropic modifications of the sesquioxides of the lanthanide elements: A, hexagonal; B, monoclinic; C, cubic; and they determined the respective regions of stability. For the whole series of oxides, when the temperature is increased the conversions occur in the sequence C > B > A. At constant temperature the sequence is A > B > C, the changes occurring according to the increase in atomic number. WARSHAW and ROY[12] recently reviewed the situation. Forms A and B are more dense than C, the co-ordination number of the A and B structures being 7 whereas that of the C structure is 6.

Recent experiments based on thermal analysis and carried out in the Ultra Réfractaires laboratory indicate that other conversions occur at high temperature, and this has been verified by high temperatures X-ray investigations.[13,14] The results for several sesquioxides of lanthanide follow.

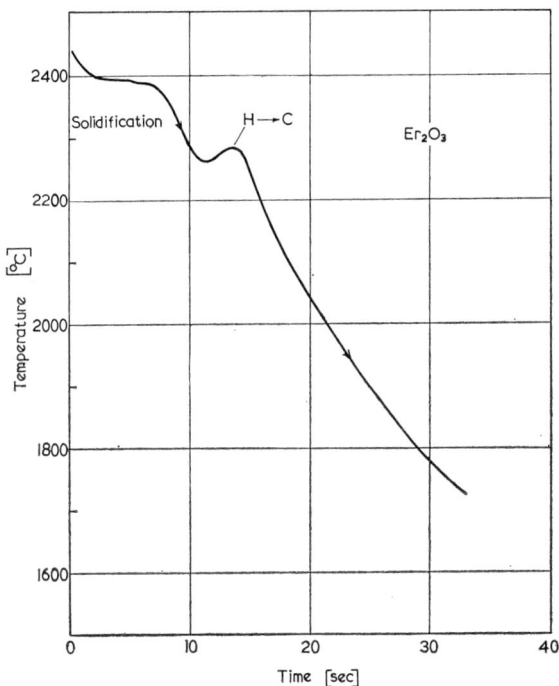

FIGURE 6.—Thermal analysis of $Er_2O_3$.

## 4.1 Erbium Oxide

The cooling curve of $Er_2O_3$ (Figure 6) shows, in addition to the arrest associated with the melting point, another pronounced anomaly near 2300°C, corresponding to a structural transformation, which has been confirmed by high-temperature X-ray diagrams. Figure 7 enables the conversion from one phase to the other to be followed. The hexagonal structure, referred to as H, is not the same as structure A, although very similar. The C⇌H conversion spreads over a wide range of temperature, with considerable thermal hysteresis.

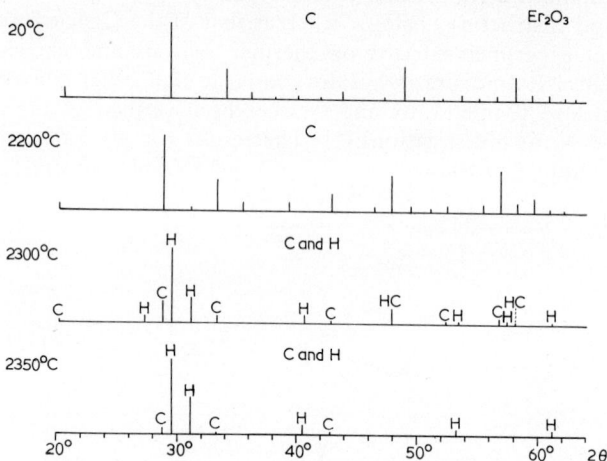

FIGURE 7.—Schematic representation of X-ray powder patterns (Cu Kα) of $Er_2O_3$ at various temperatures (rhenium support, He atm):

Form C room temperature
Form C at 2200°C
Form C and H at 2300°C
Form C and H at 2350°C.

## 4.2 Holmium Oxide

The thermal analysis of $Ho_2O_3$ indicates two changes in crystal structure, at 2240° and at 2180°C, which X-ray diagrams (Figure 8) confirm. Holmium oxide is cubic at room temperature and hexagonal at high temperature, like erbium oxide. An intermediate phase, with the monoclinic B structure, occurs at about 2250°–2300°C.

## 4.3 Neodymium Oxide

The behaviour of $Nd_2O_3$ is similar to that of $La_2O_3$ and $Pr_2O_3$. The thermal analysis shows two arrests, one at 2200°C and the other

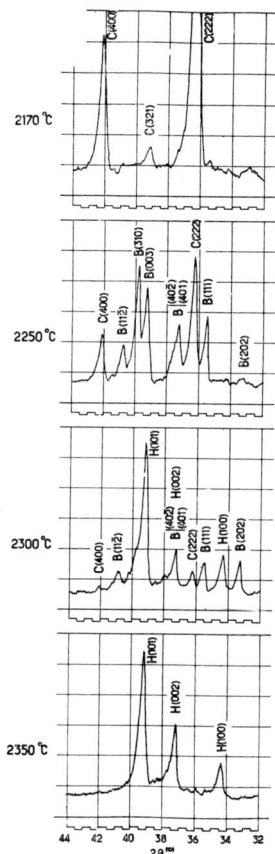

FIGURE 8.—X-ray diagram of $Ho_2O_3$ (Fe K$\alpha$) at different temperatures (rhenium support, He–$H_2$ atm.)

Form C at 2170°C
Forms C and B at 2250°C
Forms C, B and H at 2300°C
Form H at 2350°C.

at 2100°C (Figure 9). High-temperature X-ray investigation indicates the hexagonal room-temperature modification A persisting until about 2050°C. A considerble expansion of both $c$ and $a$ occurs (Figure 10) at 2100°C and a change to another hexagonal form (referred to as H) appears (Figure 11); $a$ and $c$ increase suddenly but $c/a$ decreases. At 2200°C another structure occurs; a strong line at $d = 3\cdot11$ appears near the 002 line of the H form, sometimes accompanied by three additional lines (indicated by dash at the bottom of

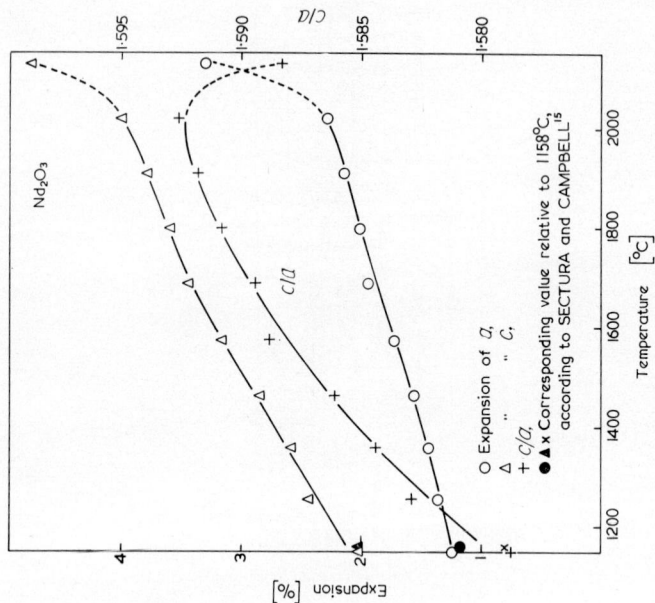

FIGURE 10.—Thermal expansion of $a$ and $c$ of hexagonal $Nd_2O_3$ between 1150°C and 2150°C on the basis of the values at 25°C.

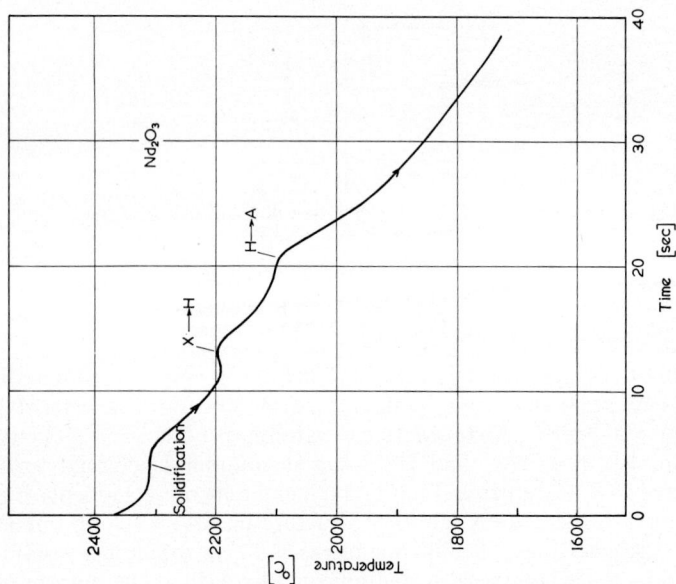

FIGURE 9.—Thermal analysis of $Nd_2O_3$.

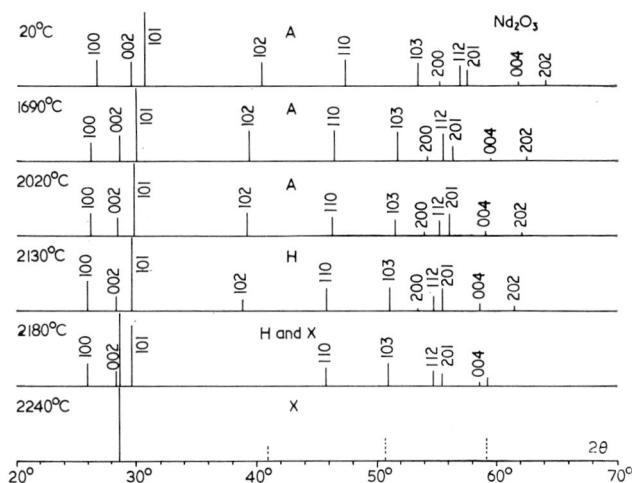

FIGURE 11.—X-ray diagrams of $Nd_2O_3$ (Cu K$\alpha$) at different temperatures (rhenium support, He + $H_2$ atm.):

Form A at 20, 1690 and 2020°C
Form H at 2130°C
Form H and X at 2180°C
Form X at 2240°C.

Figure 11); the $\sin^2\theta$ values are in the ratio of 1, to 2,3,4, possibly indicating that a body-centred cubic structure is formed.

## 5. REGIONS OF STABILITY OF THE STRUCTURES OF THE LANTHANIDE OXIDES

New results are presented in Figure 12. The curve at the top gives the melting points of the lanthanide sesquioxides.

The two curves at the bottom of the figure separate the regions of existance of A, B and C. Our results extend the curve determined by other authors[11,12,5] to higher temperatures. The curves in the middle indicate the regions of stability of the new structures H and X.

The conversion temperatures as determined by thermal analysis and from X-ray diagrams are in good agreement.

The A⇌H conversion involves a rearrangement similar to that for quartz at 573°C, except that with quartz the $c/a$ ratio increases instead of decreases. As there is no breaking in the cell, little activation energy is required. Conversion should occur at a well defined temperature, and no hysteresis phenomena are observed. The co-existence of the two phases was not detectable and the thermal effect which accompanies the conversion is small.

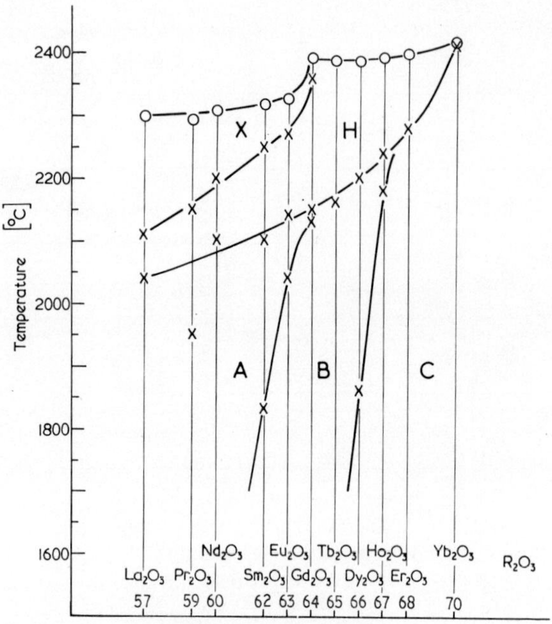

230                    FOEX:

FIGURE 12.—Regions of stability of the crystalline forms of the lanthanide sesquioxides at high temperature.

As with quartz, the high-temperature modification would be expected to have the highest degree of symmetry, which may cause the formation of twinned crystals after cooling to room temperature. The microtwinning found in lanthanum and neodymium oxide by MULLER-BUSCHBAUM and VON SCHNERING[16] seems to support this view.

With regard to form X, the results obtained for $Nd_2O_3$ and $Gd_2O_3$ seem to indicate that this X structure is cubic, although the diagrams, except the first reflection, are difficult to reproduce.

The influence of the difference between melting point and conversion temperature on the rate of conversion is nicely illustrated by the $B \rightleftharpoons C$ transformation of the lanthanide oxides. With $Sm_2O_3$, on heating, C converts to B at 1100°C,[15] which is 1200°C below the melting point, and it is nearly impossible to reobtain the form C by cooling. With $Eu_2O_3$ and $Gd_2O_3$ it is possible, though difficult, when they are heated for a long time at 950°C. The higher the atomic number of the lanthanide ion, the higher the conversion temperatures and the easier it is to obtain the low-temperature modification. The conversion temperatures of $Tb_2O_3$ and $Dy_2O_3$ are 1400°C and 1900°C respectively. The behaviour of the next oxide, $Ho_2O_3$ is quite different. The conversion temperature is about 2200°C which is some 200°C below the melting point. As soon as conditions depart from equilibrium, the structure converts so quickly that it is nearly impossible to retain the monoclinic form, despite rapid quenching.

## REFERENCES

1. FOEX, M., *Bull. Soc. Chim. Fr.*, p. 137, 1962.
2. FOEX, M., *Rev. Hautes Temp. Réfract.*, 3, 309, 1966.
3. FOEX, M., and DELMAS, R., "Four Centrifuge à Arc comportant des Chalmeaux à plasma comme Électrodes". Notice éditée par le Centre National de la Recherche Scientifique, Paris (1964).
4. INTRATER, J., and HURWITT, S., *Rev. Sci. Instrum.*, 32, 905, 1961.
5. SMITH, D. K., *Norelco Reporter*, 10, 19, 1963.
6. FOEX, M., and ROUANET, A. "Etude au four solaire des diagrammes de solidification de quelques systèmes formés par les oxydes réfractaires". Colloquium organized by the Solar Energy Society, Boston, March 1966.
7. SMITH, D. K., and CLINE, C. F., *J. Amer. Ceram. Soc.*, 45, 249, 1962.
8. WOLTEN, G. M., *J. Amer. Ceram. Soc.*, 46, 418, 1963.
9. BOGANOV, A. G., RUDENKO, V. S., and MAKAROV, L. P., *J. Doklad. Akad. URSS*, 160, 1065, 1965.
10. WEBER, B. C., *J. Amer. Ceram. Soc.*, 45, 614, 1962.
11. GOLDSCHMIDT, V. M., ULRICH, F., and BARTH, T., "Geochemische Verteilungsgesetze der Elemente", Skrifter Norske Videnskaps-Akademi, Oslo Mat. Maturv. kl. 5, 1924.
12. WARSHAW, I., and ROY, R., *J. Phys. Chem.*, 65, 2048, 1961.
13. FOEX, M., and TRAVERSE, J. P., *Bull. Soc. Franc. Min. Crist.*, 89, 184, 1966.
14. FOEX, M., and TRAVERSE, J. P., *Rev. Hautes Temp. Réfract.*, 3, 429, 1966.
15. STECURA, S., and CAMPBELL, W. J., "Thermal Expansion and Phase Inversion of Rare-earth Oxides" U.S. Bur. Mines. Rep. Invest. 5847, 1961.
16. MULLER-BUSCHBAUM, H. K., and VON SCHNERING, H. G., *Z. anorg. Chem.*, 340, 232, 1965.

# 16.—The Influence of Additions of Rare-earth Oxides on the Polymorphism of Zirconia and Hafnia

By E. K. Köhler and V. B. Glushkova

*Institute of Silicate Chemistry of the Academy of Sciences of the U.S.S.R. (Leningrad)*

**ABSTRACT**                                               E18-E21/A511

*Additions of oxides of the rare-earth elements to zirconia lead to the development of solid solutions of these elements in zirconia and considerably decrease the temperature of the monoclinic–tetragonal transformation. The method of preparation of the initial mixture influences the properties of the product obtained. In the co-precipitation method a metastable solid solution develops as an intermediate product. The mechanism of separation of the metastable solid solution into stable phases has been studied. Investigation of a number of $ZrO_2$–$M_2O_3$ systems (where M is a rare-earth element) has enabled the phase relationships in the regions rich in zirconia, and also the minimum additions of rare-earth elements necessary to stabilize zirconia, to be defined. The $HfO_2$-rich and $ZrO_2$-rich regions of the $ZrO_2$–$M_2O_3$ and $HfO_2$–$M_2O_3$ systems are compared.*

*L'influence d'additions d'oxydes de terres rares sur le polymorphisme de la zircone et de l'oxyde de hafnium*

*Des additions d'oxydes d'éléments de terres rares à la zircone y provoquent le développement de solutions solides de ces éléments et abaissent considérablement la température de la transformation monoclinique–tétragonale. La méthode de préparation du mélange initial influe sur les propriétés des produits obtenus. La méthode de la co-précipitation amène le développement d'une solution solide métastable comme produit intermédiaire. Le mécanisme du passage de la solution solide métastable en phases stables est étudié. L'étude d'un certain nombre de systèmes $ZrO_2$–$M_2O_3$ (où M est un élément de terres rares) a permis de définir les relations de phases dans les régions riches en zircone et également les additions minimales d'éléments de terres rares nécessaires à la stabilisation de la zircone. Les régions des systèmes $ZrO_2$–$M_2O_3$ et $HfO_2$–$M_2O_3$ riches en $HfO_2$ et ZrO sont comparées.*

*Der Einfluß von Zusätzen der Seltenen-Erd-Oxide auf die Poly-
morphie von Zirkonium- und Hafniumoxid*

*Zusätze von Oxiden der Seltenen Erden zu ZrO₂ und HfO₂ führen
zur Entwicklung von festen Lösungen dieser Elemente in ZrO₂ und
HfO₂ und senken die Umwandlungstemperatur monoklin–tetragonal
beträchtlich. Die Herstellungsweise der Ausgangsmischung beeinflußt
die Eigenschaften des Produkts. Bei der gemeinsamen Ausfällung ergab
sich eine metastabile feste Lösung als Zwischenprodukt. Der Mechanismus der Trennung der metastabilen festen Lösung in stabile Phasen
wurde untersucht. Durch Untersuchung einer Anzahl ZrO₂–M₂O₃
Systeme (wobei M das seltene Erdmetall ist) konnten die Phasen-
beziehungen im ZrO₂–reichen Gebiet und die notwendigen Minimal-
zusätze zur Stabilisierung des ZrO₂ definiert werden. Die ZrO₂-
reichen und die HfO₂-reichen Gebiete der ZrO₂–M₂O₃ und HfO₂–
M₂O₃ Systeme werden verglichen.*

## 1. ZIRCONIUM DIOXIDE AND HAFNIUM DIOXIDE

Zirconium dioxide and hafnium dioxide are highly refractory
oxides and so the investigation of their properties at high temperatures, and of the effect of minor additions of other oxides on these
properties, is of interest in various fields of engineering.

The polymorphism of zirconium dioxide and the influence on it
of additions of rare-earth oxides are being studied in considerable
detail,[1–6] and a number of $ZrO_2$–rare-earth-oxide phase diagrams
have also been investigated.[7] But so far, there are some problems
that have not been definitely solved. For example, there is no agreed
view concerning the existence of heavy rare-earth zirconates. Investigation into the interaction of hafnia and rare-earth oxides is
still in its early stages, and the investigation of hafnia systems may
help to elucidate problems concerning zirconia systems.

We have shown that the interactions of $ZrO_2$ and rare-earth
oxides begins at 1200°C. To obtain an equilibrium product at this
temperature it is necessary to maintain it at temperature for several
hours. The interaction of $HfO_2$ and rare-earth oxides begins at a
still higher temperature and proceeds more slowly.

We have also shown that, whereas the reaction of rare-earth
oxides and $ZrO_2$ results in the immediate development of an equilibrium product, the reaction of $HfO_2$ with rare-earth oxides is
sometimes complicated by the development of intermediate phase.[8]

## 2. RELATED SYSTEMS

### 2.1 $HfO_2$–$Sm_2O_3$

The X-ray investigation of mixtures of $HfO_2$ and $Sm_2O_3$ after
being fired at various temperatures shows that development of a

new phase (cubic solid solution) begins at 1450°C. Solid solution containing about 30–40 mol% of $Sm_2O_3$ develops as a primary reaction product which then interacts with the excess hafnium dioxide, producing a solid solution of composition corresponding to that of the equilibrium product.

It should be noted that the development of the final reaction product at temperatures of 1500°–1600°C proceeds so slowly that, even at a temperature of 1600°C for 3 h, equilibrium is not reached; the X-ray patterns show the lines of the initial components and of the two cubic solid solutions, one of which corresponds to the maximum possible saturation of $Sm_2O_3$ solid solution in $HfO_2$ and the other to the equilibrium phase composition. Only after heating at 1700°C for about 10 h was equilibrium achieved.

The need for such high temperatures for the synthesis of equilibrium products from oxide mixtures considerably complicates the whole procedure. Consequently particular importance was attached to synthesis by co-precipitation of the initial components in their amorphous state, when the systems investigated were based on $ZrO_2$ and, more particularly, $HfO_2$.

Solutions of rare-earth nitrates were prepared with zirconium nitrate and with hafnium nitrate by mixing the components in the proportions corresponding to the compound under investigation, and then slowly pouring, with constant stirring, into an aqueous solution of ammonia. The solution containing the residue was heated at 70°–80°C for 1 h; the residue was then filtered off, thoroughly washed, and dried. The behaviour of the co-precipitated mixtures was then investigated.

By means of thermogravimetric and differential thermal analysis we established that dehydration is normally complete by 250°–300°C. On further heating (400°–800°C), a sharp exothermic effect, corresponding to the solid solution crystallization, was observed. As the quantity of rare-earth oxide in the solid solution increases, the crystallization temperature rises. In addition, the crystallization temperature of $HfO_2$ systems exceeds by about 100°C the crystallization temperatures of the analogous systems of $ZrO_2$. Figure 1 shows thermographs of co-precipitated mixtures with $HfO_2$ and with $ZrO_2$

The X-ray investigation of the samples calcined above the crystallization temperature showed that, irrespective of the composition of the co-precipitated mixture, a single-phase cubic product developed.

## 2.2 $Nd_2O_3$–$ZrO_2$

It is of interest to determine whether the solid solution develops in the mixture prepared by co-precipitation at the moment of

16

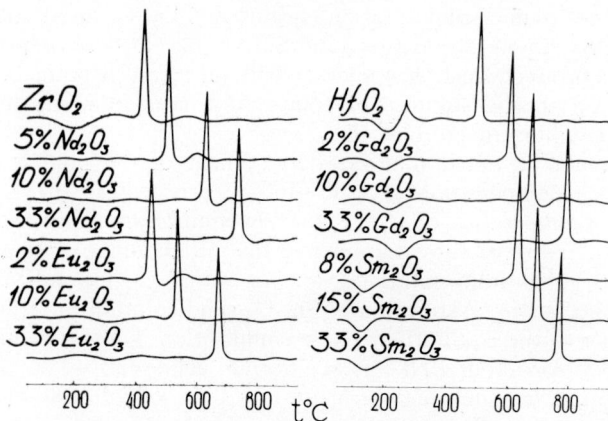

FIGURE 1.—The thermographs of the co-precipitated mixtures containing various additions of neodymium oxide and europium oxide to zirconium dioxide and of gadolinium oxide and samarium oxide to hafnium dioxide.

co-precipitation or at the crystallization temperature, by the interaction of a fine mixture of hydroxides or oxides. The thermogravimetric behaviour of co-precipitated mixtures of different compositions for the system $Nd_2O_3$–$ZrO_2$ at temperatures up to 900°C was investigated. The products resulting from various thermal treatments were also investigated by infra-red spectra. The results were compared with the behaviour of samples of $Zr(OH)_4$, $Nd(OH)_3$ and the mixture $Nd(OH)_3 + Zr(OH)_4$ at the same temperatures.

It was found that in co-precipitation a hydrated product containing about 20% of water develops. When dried in a thermostat at 150°C, most of the water is evaporated. The change in weight of the co-precipitated product stops at temperatures considerably lower than the crystallization temperatures of these compounds (600°–750°). The infra-red spectrum of the dehydrated amorphous co-precipitated product was similar to the spectrum of the $ZrO_2$ cubic solid solution stabilized by CaO (investigated in reference 9) and to the spectrum of the cubic form of $ZrO_2$.

Thus we have shown that solid solution does develop during co-precipitation.

LEFÈVRE[1] also observed the dependence of the crystallization temperature on the composition of the developing solid solution (for mixtures rich in zirconium dioxide) and on this basis was able to assume the formation of the solid solutions at the moment of co-precipitation.

He has also shown that, in a series of solid solutions of identical structure, the crystallization temperature (for the same content of rare-earth oxide additions) is lower as the size of the rare-earth ion approaches that of the zirconium ion.

As we have already shown,[3-6] the fluorite-type cubic solid solutions developed during the crystallization of co-precipitated mixtures are metastable in the zirconia-rich region, and are certain to separate into two solid solutions. But at low temperatures this separation is retarded and observed only at temperatures above 1100°C.

As an example let us examine the separation and structural transformation of a metastable solid solution containing 10% $Nd_2O_3$ and 90% $ZrO_2$.

Separation proceeds in two stages. As a result of the first stage, two solid solutions differing in composition and having the same type of crystalline structure develop. The composition of one of the solid solutions approaches the composition of $ZrO_2$; the composition of the other comes close to the composition of the equilibrium solid solution. The solid solutions developed as a result of such separation are also metastable, though they are comparatively stable, and treatment for a long period at high temperature is needed to transform them further. When the temperature rises, the solid solution the composition of which is close to that of $ZrO_2$ changes into a tetragonal solid solution (and in cooling, into a monoclinic one); simultaneous ordering of the structure of the solid solution rich in rare-earth oxide takes place. In this case, for the systems with light rare-earth oxides, including Nd, we observed a change into a structure of the pyrochlore type.

### 2.3 $ZrO_2$–$Eu_2O_3$

The investigation of the behaviour of co-precipitated mixtures in the system $ZrO_2$–$Eu_2O_3$ confirmed the two-stage separation of the metastable solid solutions obtained on the crystallization of co-precipitated mixtures, with the intermediate development of two solid solutions of cubic structure (also metastable) but differing in composition. We have also shown that the increase of rare-earth elements in a metastable solid solution retards separation. For example, heating samples containing 2, 3 and 5 mol% $Eu_2O_3$ at 1200°C 5 h does not lead to the development of equilibrium products, whereas the solid solution with 1 mol% $Eu_2O_3$ has already separated at 1000–1100°C.[5]

### 2.4 $ZrO_2$–$Er_2O_3$ $Yb_2O_3$

The metastable solid solution in systems containing heavy rare-earth oxides separates even more slowly. Samples containing 1 mol% $Er_2O_3$ or $Yb_2O_3$ separate after heating at 1200°C for 2 h; to obtain the equilibrium product, compositions containing 2–3 mol% of these oxides need to be heated for 2 h at 1500°C.

## 2.5 HfO₂–M₂O₃

In the crystallization of co-precipitated mixtures of rare-earth oxides and $HfO_2$ we also observed the development of metastable cubic solid solutions ($CaF_2$ type) which, when further heated, separate, transforming into the equilibrium phases.

Metastable cubic solid solutions based on $HfO_2$ have approximately the same stability as the corresponding solid solutions based on zirconium dioxide. For example, on heating cubic metastable solid solutions of samarium oxide in hafnium dioxide, separation was observed at 1200°C.

## 2.6 ZrO₂–M₂O₃ and HfO₂–M₂O₃

Previous work[10] showed that the decomposition of hafnium hydrate occurs at about 500°C and results in the development of monoclinic $HfO_2$. The decomoposition of zirconium hydrate results in the development of a metastable cubic form $ZrO_2$, the relative stability of which increases with the increase of impurities in it. This difference is also reflected in the process of separation of the metastable solid solutions based on $HfO_2$ and $ZrO_2$. Thus, zirconium solid solutions separate in two stages, whereas metastable solid solutions based on $HfO_2$ separate at once into monoclinic $HfO_2$ and the equilibrium cubic solid solution of the composition corresponding to the phase diagram.

By investigating a number of systems of $ZrO_2$–$M_2O_3$ and $HfO_2$–$M_2O_3$ (where M is a rare-earth element) we determined the phase relationships in the zirconia-rich or hafnia-rich region. The investigations were carried out in the temperature range 600–2050°C using both co-precipitation and mechanical mixing of the oxides. The general appearance of the phase diagrams obtained is shown in Figures 2 and 3. The phase diagrams with $La_2O_3$ have been constructed according to the data of references 11 and 12, the rest on the basis of the joint papers of the authors with DAVTYAN, ISUPOVA and SAVONOVA.[3,4,5,6,8,10]

The high-temperature investigations of the mechanical mixtures based on $ZrO_2$, containing from 1 to 33 mol% of rare-earth oxides, have shown that in all cases (except the $ZrO_2$–$La_2O_3$ system) at high temperature a single-phase product develops which is a solid solution, in which some zirconium ions are replaced by the trivalent rare-earth element ions. With a low content of rare-earth oxide (up to 3–5 mol%) the solid solution structure is tetragonal, transforming by increase of the rare-earth oxide content into the cubic

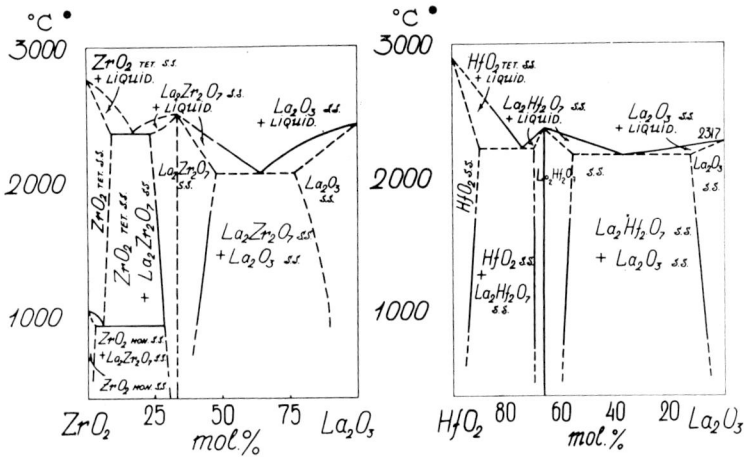

FIGURE 2

Phase diagrams of the systems $ZrO_2$–$La_2O_3$ and $HfO_2$–$La_2O_3$ according to the data of references 11 and 12.

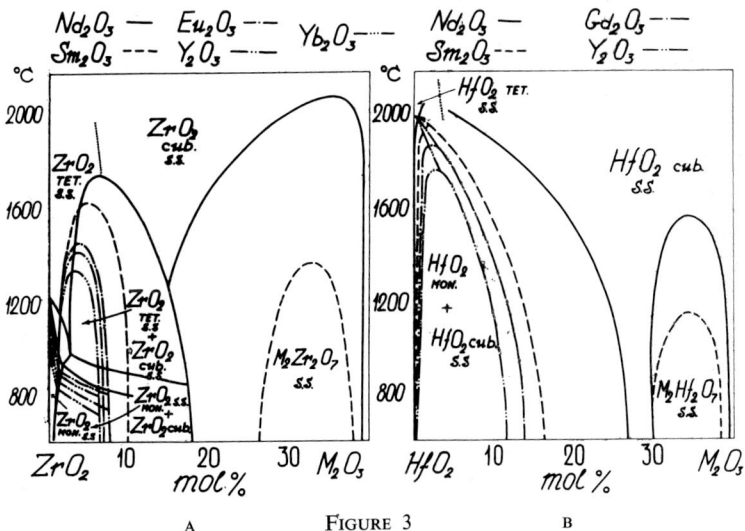

A          FIGURE 3          B

Phase diagrams of the systems:

A. $ZrO_2$–$Nd_2O_3$,[1] $ZrO_2$–$Sm_2O_3$,[2] $ZrO_2$–$Eu_2O_3$,[3] $ZrO_2$–$Y_2O_3$,[4] $ZrO_2$–$Yb_2O_3$.[5]
B. $HfO_2$–$Nd_2O_3$,[1] $HfO_2$–$Sm_2O_3$,[2] $HfO_2$–$Gd_2O_3$,[3] $HfO_2$–$Y_2O_3$.[4]

(CaF$_2$ type) solid solution. A further increase of the rare-earth oxide content transforms the solid solution (for the rare-earth elements with large ionic radius), into a cubic solid solution, but of the pyrochlore type.

## 3. COMPARISON OF DIFFERENT SYSTEMS

On slow cooling, or on firing at lower temperatures, the cubic (or tetragonal) solid solution obtained at high temperatures separates into a mixture of two solid solutions—the tetragonal (which on cooling becomes the monoclinic form) approaches ZrO$_2$ in composition and the cubic (the fluorite or pyrochlore type) being richer in rare-earth oxide. The smaller the ionic radius of the rare-earth element, the lower the temperature limiting the two-phase region (Figure 3). We have also shown that the decrease of the ionic radius of the rare-earth elements results in a decrease in the width of the two-phase region. Thus for complete stabilization of ZrO$_2$ with Nd$_2$O$_3$, an addition of 19 mol% is necessary, with europium oxide 7–8 mol%, and with erbium oxide about 5–6 mol%. The dilatometric curves of these compositions (Figure 4) do not show the hysteresis loop, which demonstrates the complete stabilization of ZrO$_2$. The solid solutions of such compounds are stable and do not separate when heated for a long time at high temperature.

The two-phase region in HfO$_2$ systems is larger than that in analogous ZrO$_2$ systems. Thus, to obtain in the HfO$_2$–Nd$_2$O$_3$ system a cubic solid solution stable at 1000°C, more than 25 mol% of Nd$_2$O$_3$ must be added, and in the HfO$_2$–Sm$_2$O$_3$ system more than 15 mol% must be added, whereas in analogous systems with ZrO$_2$ the addition of 19 mol% of Nd$_2$O$_3$ or 10 mol% of Sm$_2$O$_3$ is enough to obtain a stable single-phase cubic solid solution. In HfO$_2$ systems the maximum temperatures of the two-phase region are also considerably higher: 1950°C in the HfO$_2$–Nd$_2$O$_3$ system and 1800°C in the HfO$_2$–Y$_2$O$_3$ system, and correspondingly 1600°C in the ZrO$_2$–Nd$_2$O$_3$ system and 1350°C in the ZrO$_2$–Y$_2$O$_3$ system.

The rate of separation of high-temperature solid solutions in the ranges of temperature and composition of the two-phase region was studied in the ZrO$_2$–Nd$_2$O$_3$ and ZrO$_2$–Sm$_2$O$_3$ systems.

To this end, test samples of ZrO$_2$ to which had been added between 2 and 20 mol% of M$_2$O$_3$ were fired for 3 h at 1900°C and then rapidly cooled (the period of cooling to 800°–900°C amounted to not more than 10 min.). X-ray analysis has shown the development of the high-temperature solid solutions for all mixtures except ZrO$_2$+2% Nd$_2$O$_3$ and ZrO$_2$+2% Sm$_2$O$_3$ which also contained

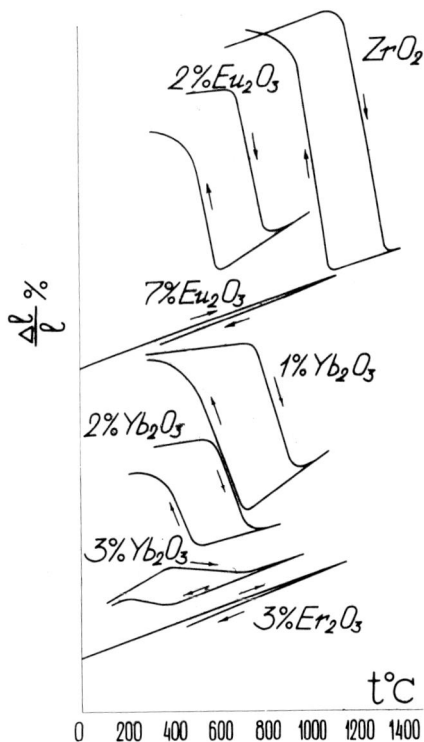

FIGURE 4.—Dilatometric curves of zirconium dioxide samples with various rare-earth oxide additions.

some of the monoclinic phase. Figure 5 shows the expansion curves of the samples both after firing at 1900°C and after firing for half an hour at 1300–1500°C (the two-phase region) with subsequent slow cooling. The 2% $Nd_2O_3 + 98$% $ZrO_2$ sample, which was solid and hard after firing at 1900°C, crumbled to fine powder during the first cycle of expansion measurements (20°–1100°–20°C).

The remaining samples withstood several heating-cycles without fracturing, though hysteresis loops appeared in the samples containing 2 mol% $Sm_2O_3$ and 4 mol% $Nd_2O_3$.

Thus we have shown that the rate of separation of a high-temperature solid solution depends considerably on the quantity of rare-earth oxide in it. In practice, $ZrO_2$ can be stabilized by smaller additions of oxides than those shown to be necessary on the equilibrium phase diagram, especially of the heavy rare-earth oxides,

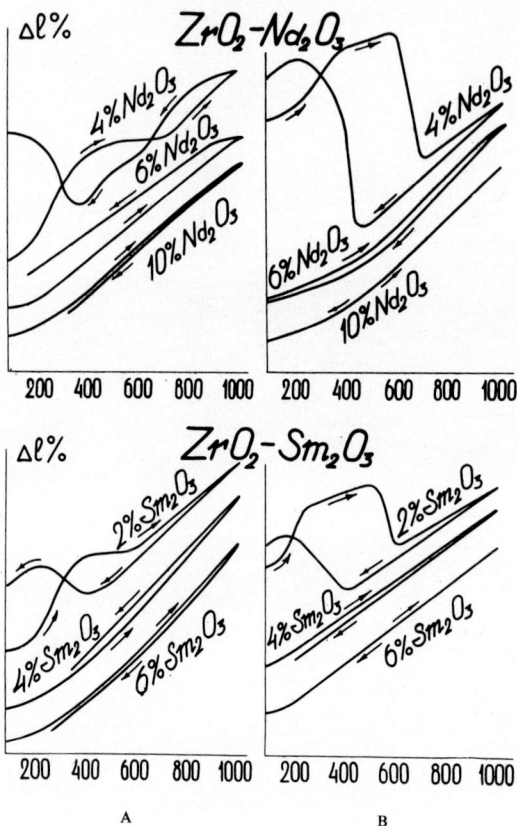

FIGURE 5.—Dilatometric curves of samples of zirconium dioxide with additions of neodymium and samarium oxide after: A. Firing at 1900°C and B. Subsequent firing at 1500°C.

the addition of 3–4 mol% of which prevents the destruction of samples based on $ZrO_2$.

Peculiar behaviour is exhibited by solid solutions based on $ZrO_2$ with additions of $Ce_2O_3$, $Pr_2O_3$ and $Tb_2O_3$. The solid solutions of $ZrO_2$ with $Ce_2O_3$, investigated in detail by LEONOV,[14] are prone to intense oxidation when heated in air. Both oxidation and reduction result in large volume changes which lead to fracturing and loosening of the material.

The similarity of the phase diagrams of $ZrO_2$ and $HfO_2$ systems is already clear from the data presented. The difference lies mainly in the existence of different temperature ranges for the various phases. Monoclinic $HfO_2$ is considerably more stable than monoclinic $ZrO_2$ (the temperature of the monoclinic–tetragonal transformation of $HfO_2$ is $1950°C$[13] and of $ZrO_2$ is $1250°C$); at the same time the pyrochlore phase is characteristic for the systems containing $ZrO_2$. This can be explained by the thermochemical data relating to these compounds.

The influence of additions of rare-earth oxides on the monoclinic–tetragonal transformation of zirconia was studied in detail. When the other oxides are added, solid solutions of the general formula $Zr^{4+}{}_{1-x} M^{3+}{}_x O^{2-}{}_{2-x/2} \square_{x/2}$ develop. The X-ray patterns show that the peaks of lines that are characteristic of $ZrO_2$ change their position, and the d.t.a. and dilatometer curves show the decrease in temperature of the monoclinic–tetragonal transformation.

The solubility of rare-earth oxides in the monoclinic form amounts to not more than 1–2 mol%, and is almost unchanged in the series of rare-earth elements; the solubility in the tetragonal form is about 4 mol% at $1200°C$, and increases as the temperature rises.

Table 1 illustrates the change in the temperature of the polymorphic transformation after the addition of rare-earth oxide. Data for additions of $Y_2O_3$ are also presented for comparison.

**Table 1**

**The Temperature of the Monoclinic–Tetragonal Transformation (°C) for the Zirconia Systems and the Hafnia Systems.**

| Systems | $M_2O_3$ (mole %) | $Nd_2O_3$ | $Sm_2O_3$ | $Gd_2O_3$ | $Eu_2O_3$ | $Er_2O_3$ | $Y_2O_3$ |
|---|---|---|---|---|---|---|---|
| $ZrO_2$ (temperature of transformation 1180–1250°)° | 1 | 900–1050 | 720–870 | — | 650–800 | 550–750 | 750–850 |
| | 2 | 850–1020 | — | — | 550–750 | 450–620 | 500–650 |
| | 3 | 830–1000 | 650–840 | — | — | 300–450 | 450–580 |
| | 5 | 750–970 | 620–820 | — | 500–700 | absent | 450–550 |
| | 10 | 570–920 | absent | — | absent | absent | absent |
| $HfO_2$ (temperature of transformation ∼1980°) | 3 | 1950 | 1920 | 1850 | — | — | 1800 |

The table also shows that the transformation temperature decreases as the quantity of rare-earth oxides in the samples increases, the fall in transformation temperature being observed not only with homogeneous solid solutions but also in the two-phase region. The smaller the ionic radius of the rare-earth cation, the greater the fall of the transformation temperature with similar oxide additions.

The volume change $(V/V_m)$ in the phase transformation amounts to 8–9% for $ZrO_2$, and to about 4% for the solid solution containing

1 mol% of $Eu_2O_3$. It should be noted that the volume changes occurring in the monoclinic–tetragonal transformation of solid solutions are not only less in absolute value, but they also occur over a wider range of temperature, particularly on the cooling curves (Figure 4).

## 4. SUMMARY

The influence of minor additions of rare-earth oxides on the polymorphism of hafnium dioxide was studied by the authors in collaboration with Isupova by means of the high-temperature X-ray unit designed by Boganov.[15] Light rare-earth oxides are practically insoluble in monoclinic $HfO_2$, whereas the solubility of heavy rare-earth elements does not seem to exceed 1–1·5% even at 1800–1900°C. This is reflected by the fact that the fall in temperature of the monoclinic–tetragonal transformation for mixtures of $HfO_2$ with light rare-earth oxides is not large (of the order of 50–100°C) and increases slightly with decrease of the ionic radius of the rare-earth element, reaching 150–200°C for heavy rare-earth elements and yttrium.

It has been shown that the best stabilizers of $ZrO_2$ in the cubic form are heavy rare-earth oxides and $Y_2O_3$. Using them only, the smallest oxide additions are needed to obtain stable solid solutions, and the solid solutions developed are stable in various gaseous media and are free from phase changes up to high temperatures (2000°C).

$HfO_2$ systems are very similar to the analogous $ZrO_2$ systems.

## REFERENCES

1. Lefèvre, I., *Ann Chim.*, **8**, 1–2, 119, 1963.
2. Collongues, K., Perez y Jorba, M., Lefèvre, T., *Bull. Soc. Chim. France*, 149, 1962.
3. Glushkova, V. B., and Sazonova, L. V., (Coll.) "Chemistry of High-temperature Materials". (Nauka M.-L: 1967).
4. Davtyan, I. A., Glushkova, V. B., and Koehler, E. K., *Izv. AN SSSR ser. Neorganicheskie materialy*, **1**, (5), 743, 1965.
5. Davtyan, I. A., Glushkova, V. B., and Koehler, E. K., *Izv. AN SSSR ser. Neorganicheskie materialy*, **2**, (5), 890, 1966.
6. Sazonova, L. V., Davtyan, I. A., and Glushkova, V. B., *Izv. AN SSSR ser. Neorganicheskie materialy*, **1**, (11), 1965, 1965.
7. Toropov, N. A., Barzakovsky, V. P., Lapin, V. V., and Kurtzeva, N. N., Handbook "Phase Diagrams of Silicate Systems", Pt 1, (Nauka, M.-L: 1965).
8. Isupova, E. N., Glushkova, V. B., and Koehler, E. K., *Izv. AN SSSR ser. Neorganicheskie materialy* **3**, 1967; in publication.
9. McDevitt, N. T., and Boun, W. L., *J. Amer. Ceram. Soc.*, **47**, 622, 1964.
10. Isupova, E. N., Glushkova, V. B., and Koehler, E. K., *Izv. AN SSSR ser. Neorganicheskie materialy* (in publication).

11. ROTH, R. S., *J. Res. Nat. Bur. Stand.* **56**, 1, 17, 1956.
12. KOMISSAROVA, L. N., VAN DEN-SHI, SPITSYN, V. I., and SIMAKOV, I. P., *Z. Neorganich. Khimii*, **9**, No. 3, 693, 1964.
13. BOGANOV, A. G., RUDENKO, V. S., and MAKAROV, L. P., *DAN SSSR*, **160**, 5, 1061, 1965.
14. LEONOV, A. I., KOEHLER, E. K., and ANDREEVA, A. B., *Izv. AN SSSR ser. Neorganicheskie materialy* **2**, (1), 137, 1966.
15. BOGANOV, A. G., MAKAROV, L. P., and RUDENKO, V. S., *DAN SSSR* **161**, 2, 332, 1965.

# 17.—Zirconia Solid Solutions with Rare-earth Oxides

By U. Spitsbergen and P. M. Houpt

*Central Laboratory TNO, Delft*

*ABSTRACT* E181/E21

*An arc-melting procedure was studied for the rapid heating of mixtures of zirconia and rare-earth oxides. The oxides of each element of the rare-earth series, except Pm, were considered. The products of melting were examined by X-ray powder diffraction and by X-ray fluorescence methods. The hardness of some of the products was measured.*

## Solutions solides dérivées de la zircone

*Un procédé utilisant un arc électrique fut étudié pour fondre rapidement des mélanges de zircone et d'oxydes des terres rares. Les oxydes de chaque élément de la série des terres rares, sauf Pm, furent considérés. Les produits de la fusion furent examinés au moyen de la diffraction des rayons X et de fluorescence des rayons X. Les duretés de quelques produits furent déterminées.*

## Feste Lösungen des Zirkonoxyds

*Ein Lichtbogen-Schmelzverfahren für die schnelle Erhitzung einer grösseren Anzahl von Mischungen aus Zirkondioxyd und Oxyden der seltenen Erden wurde erprobt. Alle Oxyde der seltenen Erden, mit Ausnahme von Pm, wurden geprüft. Die Schmelzprodukte wurden mit Hilfe der Röntgenbeugung und Röntgenfluoreszenz-spektrometrie untersucht und Härte einiger dieser Produkte ermittelt.*

## 1. INTRODUCTION

A better understanding is needed of the zirconium-rich regions in zirconia-based systems, especially with respect to low-temperature behaviour :

(1) Small amounts of monoclinic precipitates may have a de-stabilizing effect on the structure of the solid and hence on its properties;

(2) In the numerous investigations of systems based on zirconia that have already carried out, among which are the extensive studies

247

of COLLONGUES[1,2] and his school and of FEHRENBACHER[3,4] and co-workers, it was mainly the high-temperature ranges that were explored. These authors investigated combinations of zirconia and various rare-earth oxides and made use of the lanthanide contraction, which reveals the influence of the size of the substituted rare-earth ions.

The present contribution deals mainly with the preheating stage of heating cycles to which mixtures of zirconia and rare-earth oxides are subjected. The products of this preheating, which is an arc-melting procedure, and their characteristics are described below.

All the rare-earth elements except Pm were taken into consideration, and so a general approach was adopted:

(1) To facilitate in particular an understanding of the less accessible lower temperature ranges to be studied when complete heat treatments have been worked out;

(2) To make more selectively a subsequent study on multi-substitutions leading to solid solutions:

$$Zr_{(1-x)}M_{\frac{1}{2}x}M'_{\frac{1}{2}x} O_{(2-\frac{1}{2}x)}$$

Both with single- and with multi-substitutions, the abundant literature on the subject indicates that M and M' must be ions with a noble gas type of outer electronic shell (irrespective of other factors such as ionic size, valency, etc.).

## 2. METHOD OF PREPARATION

The present approach requires experimental methods that enable many samples to be prepared within a reasonable time. Since the reaction between two refractory oxides such as zirconia and a rare-earth oxide proceeds very slowly, and thus makes the experiments very time-consuming, a rapid prefiring treatment was desirable, and particularly one that avoided regrinding between two heat-treatments, since the samples proved to be hard.

For this stage an arc-melting procedure was adopted, as illustrated in Figure 1. Since the samples are small (3 mm diam.), and the procedure is very rapid (5 sec.), $4 \times 14$—i.e. $4 \times$ lanthanide series—melts can be run within $2\frac{1}{2}$ h.

The melts represented in Figure 2 show colours which are specific for the rare-earth addition. Provisional microscope examination showed that the solid drops consisted of a transparent, poly-crystalline, cracked bulk. The drops are slightly hollow. With Ce and Eu, blackening occurs, which however completely vanishes on heating to 1000°C in air in a resistance furnace. This blackening

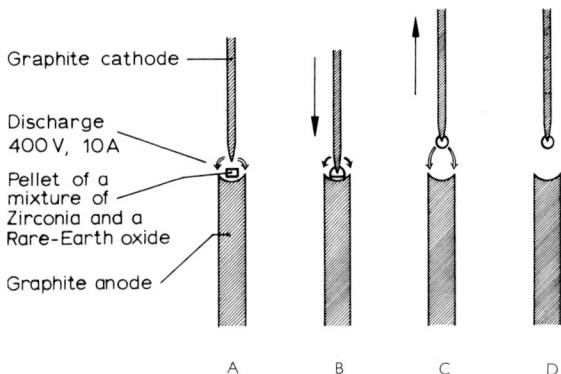

Graphite cathode

Discharge
400 V, 10 A

Pellet of a
mixture of
Zirconia and a
Rare-Earth oxide

Graphite anode

A          B          C          D

FIGURE 1.—Arc melting

A. Loosely pressed pellet of mixture of zirconia and a rare-earth oxide is placed on a graphite anode, the top of which is slightly hollowed. When current is switched on, an arc discharge is started by spark ignition between electrodes (these electrodes are spectrographically pure).
B. The cathode is lowered and dipped in partly molten sample.
C. The cathode is raised and current passes through the now conducting oxide melt for 5 seconds. This raising is necessary to keep the sample in a relatively cold spot ($T \simeq 2700°C$) within the distance between the electrodes; the hot area is the anode ($T \simeq 5000°C$) where the electron collisions take place.
D. The sample is cooled down to the ambient temperature on the raised cathode or can directly be quenched by pushing into water.

within each grain of the drop has local variations in intensity. It may be that phenomena occur identical to those that occur while a constant charge is passed through solid stabilized zirconia, i.e. BERANGER and LACOMBE'S [5] "zircone noire", in which oxygen deficiency was assumed to occur by migration of oxygen towards the anode. With Tb also the specific coloration varies in intensity

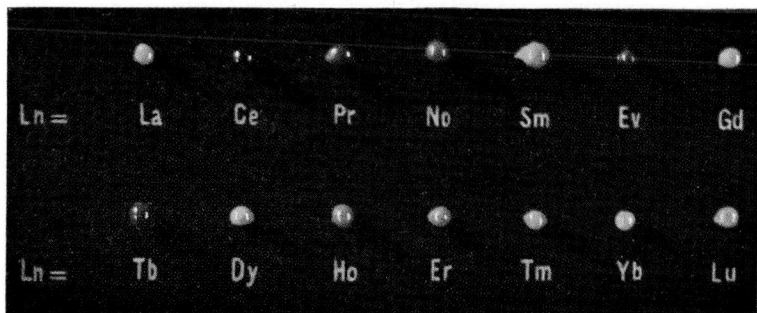

FIGURE 2.—Arc-melted mixtures of zirconia and 20 mole % $LnO_{1.5}$.

locally and, in addition, depends on the cooling rate. Different valencies of Tb are assumed to be present.

## 3. EXAMINATION OF THE SAMPLES

The small size of the samples and the rigorous circumstances during heating necessitates an analysis of the Zr/Ln ratio (Ln = lanthanide element). Deviations in the composition of the sample owing to the arc melting were established by measuring the intensity of certain selected X-ray wavelengths of the Zr and Ln elements before and after the melting. From these intensity ratios the values in Table 1 have been calculated.

Powder photographs were taken in a Guinier–de Wolff focusing camera, (R = 229·2 mm) with $CuK\alpha_{1,2}$ radiation. The resolving power of this type of camera allowed a close examination of the low $2\theta$ range.

In this range, possible super-structure lines are most likely to be detected. Figure 3 shows these powder diagrams* for additions to zirconia of 2·5, 5, 8 and 20 mole % $LnO_{1.5+\delta}$, nominally, with Ln = La, Ce, . . ., and Lu:

### (1) x = 0·025

The patterns are predominantly monoclinic. The tetragonal 111 reflection has considerable intensity only in the case of La. The relatively large size of the La ion inhibits its entering the monoclinic structure. The reaction might be incomplete; a prolonged heating or remelting seems necessary.

### (2) x = 0·05

The tetragonal pattern has a relatively greater intensity. The monoclinic pattern has undergone a change; some of its reflections are shifted,** which suggests that at this concentration some rare-earth cations are taken up into the monoclinic lattice.

### (3) x = 0·08

In the case of Dy, Ho, Er, and Tm, the powder photographs only show a tetragonal fluorite pattern. The mixtures containing the larger rare-earth ions La, Ce, and Pr seem to need a remelting, both the tetragonal and monoclinic patterns being diffuse. In the case of Yb and Lu monoclinic reflections were observed repeatedly. The question whether indeed within the lanthanide series a minimal

*Thanks are due to Drs. J. W. Visser, Institute of Applied Physics TNO-TH, for taking these powder photographs, and allowing us to publish them.

**The increase of the 200, 002 and 300 spacing and the converging of the 11$\bar{1}$ and 111 reflections indicate larger $a$ and $c$ parameters, and a $\beta$ value closer to 90°. (These Miller indices are not indicated in Figures 3 and 4.)

## Table 1

**Values of x of Mixtures $(1-x)ZrO_2$ and $xLnO_{1.5+\delta}$ after the Arc Melting, derived from X-ray Fluorescence Measurements**

| Ln | Nominal value of x | | | |
|---|---|---|---|---|
| | 0·025 | 0·05 | 0·08 | 0·20 |
| La | 0·036 | 0·047 | 0·088 | — |
| Ce | 0·025 | 0·051 | 0·069 | 0·201 |
| Pr | 0·024 | 0·056 | 0·08 | 0·240 |
| Nd | 0·023 | 0·057 | 0·07 | 0·175 |
| Sm | 0·025 | 0·049 | 0·079 | 0·176 |
| Eu | 0·021 | 0·041 | 0·069 | 0·220 |
| Gd | 0·020 | 0·048 | 0·073 | 0·206 |
| Tb | 0·025 | 0·048 | 0·08 | 0·202 |
| Dy | 0·022 | 0·048 | 0·081 | 0·198 |
| Ho | 0·027 | 0·052 | 0·074 | 0·206 |
| Er | 0·025 | 0·055 | 0·08 | 0·218 |
| Tm | 0·027 | 0·05 | 0·078 | 0·204 |
| Yb | 0·026 | 0·053 | 0·083 | 0·174 |
| Lu | 0·025 | 0·05 | 0·078 | 0·192 |

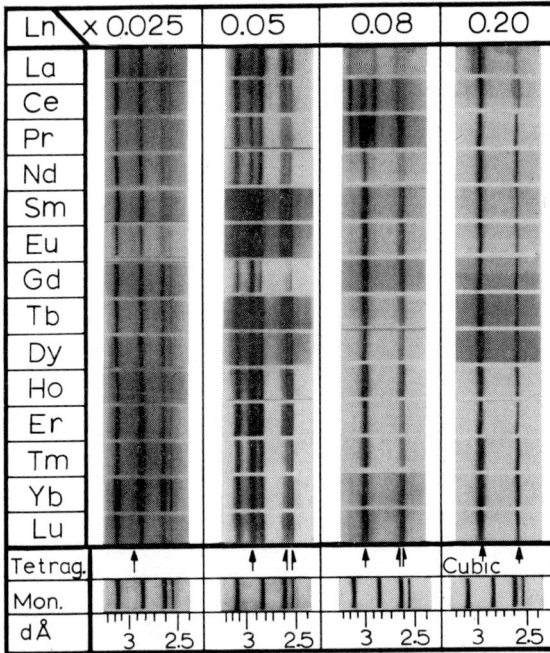

FIGURE 3.—Low $2\theta$ ranges of powder photographs of the products of arc melting. The arrows at the bottom of the figure indicate: the tetragonal 111 reflection at $x=0·025$; the tetragonal 111, 002 and 200 reflections at $x=0·05$ and $x=0·08$; and the cubic 111 and 200 reflections at $x=0·020$.

17

ionic size favours the stabilization of a more symmetric structure, in this case the tetragonal modification, remains to be answered.

*(4) x=0·20*

The solid solutions have the cubic fluorite structure. In the case of La and Ce again the reaction seems incomplete.

In Figure 4, powdered arc-melted samples containing 8 mole% $LnO_{1.5+\delta}$, as shown in Figure 3, are compared with those of samples which were prepared by the co-precipitation method [1] and were afterwards heated in a resistance furnace for 40 h at 2050°C. The latter samples had therefore undergone a completely different thermal treatment. Except for weak monoclinic reflections present in the powder diagrams of the arc melts containing Gd and Tb, both series of heatings have in common:

(1) An equal separation of the tetragonal 002 and 200 lines;
(2) With Yb and Lu monoclinic reflections occur.

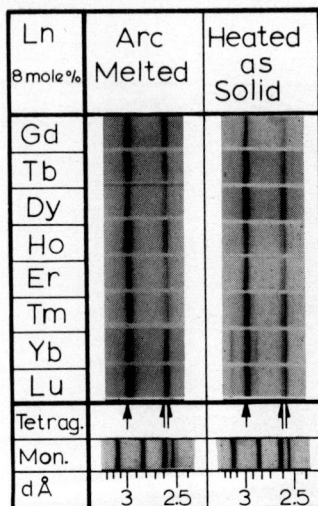

FIGURE 4.—Powder diffraction lines of mixtures of zirconia and 8 mole % $LnO_{1.5+\delta}$ after different heat treatments. The representation of the patterns is as in Figure 3.

It should be emphasized that the powder patterns discussed here belong to rapidly cooled, and not to purposely quenched, samples. These quenching techniques will be studied in future experiments.

In none of the products dealt with here could extra lines be ascribed to super-structures, not even in some of the samples containing 8 mole % $LnO_{1.5+\delta}$ which had been annealed at 1000°C during one week. The products with 20 mole % $LnO_{1.5+\delta}$ are not studied on their ageing behaviour.

However, at very low $2\theta$ angles, below the ranges shown in Figures 3 and 4, the strongest reflections of mullite and/or silica (agate) were detected, which were due to pick-up from the mortar used for grinding the products before their powder photographs were taken. This had induced us to have hardness measurements made.*

In Table 2 the Vickers microhardness, determined from 3 indentations for each sample, under a load of 50 g (HV 50 g), of zirconia mixtures containing Gd, Tb, . . ., Lu, respectively, are shown together with that of pure zirconia which was arc-melted under the same conditions.

**Table 2**

**Vickers Microhardness of Arc-melted Mixtures of**
**$(1-x)ZrO_2$ and $xLnO_{1.5+\delta}$**

| Ln | x | | | |
|----|---------|---------|---------|---------|
|    | 0·025 | 0·05 | 0·08 | 0·20 |
| Gd | 800  700  950 | 1050 1250 1300 | 1650 1900 1850 | 1500 1500 1500 |
| Tb | 800  650  650 | 1400 1050 1100 | 1800 1700 1900 | 1500 1450 1500 |
| Dy | 800  800  950 | 1400 1400 1400 | 1650 1700 1800 | 1450 1400 1450 |
| Ho | 900  900  800 | 1400 1400 1300 | 1650 1700 1500 | 1450 1500 1500 |
| Er | 900 1050  950 | 1300 1500 1300 | 1700 1800 1650 | 1500 1450 1450 |
| Tm | 1050 1050 1050 | 1300 1300 1250 | 1900 1650 1800 | 1500 1450 1500 |
| Yb | 950 1050  900 | 1400 1300 1250 | 1800 1650 1800 | 1450 1450 1500 |
| Lu | 900  950 1050 | 1400 1300 1300 | 1700 1550 1550 | 1450 1450 1450 |

$ZrO_2$ (pure) 739 701 666

From these values it is noted that:

(1) Even small quantities of the second oxide cause a considerable increase in hardness. The hardness of pure zirconia is comparable to that of quartz, both oxides have a hardness $\simeq 7$ on the Mohs scale, whereas all the melted mixed oxides have a hardness equal to or higher than that of topaz,[6] 8 Mohs $\simeq 1200$ Vickers.

(2) The hardness seems not to depend upon the kind of the rare-earth element.

(3) The tetragonal structure leads to greater hardness as compared to the cubic structure, as is seen comparing the 8 mole % mixtures in Table 2 with the 20 mole % mixtures.

*These measurements were performed at the Metal Research Institute TNO.

## 4. RESULTS

From small quantities of mixed oxides of zirconium and rare-earth elements arc-melted in air for 5 sec. and afterwards cooled to the ambient temperature, it was found that:

(1) A complete solid solution of $(1-x)ZrO_2$ and $xLnO_{1.5+\delta}$ was formed in the case of $Ln = Dy$, Ho, Er, and Tm, at $x = 8$ mole% $LnO_{1.5+\delta}$; and in the case of $Ln = Pr$, Nd, Sm . . . and Lu, at $x = 20$ mole %. In the case of La and Ce, longer and/or repeated melting seemed necessary;

(2) In the case of 8 mole % $YbO_{1.5}$ and $LuO_{1.5}$, repeatedly the monoclinic phase was observed, as opposed to Dy, Ho, Er, and Tm at the same concentration;

(3) Thus far no super-structures have been observed at a Ln concentration 8 mole %.

(4) The arc-melted zirconia solid solutions, in particular the tetragonal ones, exhibit considerable hardness compared with arc-melted pure zirconia.

### ACKNOWLEDGMENTS

The authors wish to express their gratitude to Professor Dr. Ir. P. M. de Wolff, Technological University of Delft, for valuable advice in interpreting the X-ray powder diagrams. Their thanks are also due to Drs. Ch. A. Kruissink, Central Laboratory TNO, Delft, for stimulating discussions. They are indebted to Miss Dicky Geuze, Central Laboratory TNO, Delft, who carried out many experiments.

### REFERENCES

1. LEFÈVRE, J., *An. Chimie Paris*, **8**, 117, 1963.
2. COLLONGUES, R., QUEYROUX, F., PEREZ Y JORBA, M., and GILLES, J., *Bull. Soc. Chim.* p. 1141, 1965.
3. FEHRENBACHER, L. L., JACOBSON, L. A., and LYNCH, C. T., 4th Rare Earth Research Conf., Phoenix, Arizona, April 1964.
4. FEHRENBACHER, L. L., 68th Annual Meeting the American Ceramic Society, April 1966, 14–B–66.
5. BÉRANGER, G., and LACOMBE, P., *Rev. Hautes Temp. Réfr.*, **3.**, 235, 1966.
6. RYSHKEWITCH, E., "Oxide Ceramics" (Academic Press: New York and London, 1960), pp. 158, 379.

# 18.—Solid–Liquid Reactions and Brick/Clinker Adhesion in Rotary Cement Kilns

By D. S. Buist and J. R. Gelsthorpe

General Refractories Ltd, Central Research
Laboratories, Sandy Lane, Worksop, Notts.

*ABSTRACT* C512/D78

An investigation has been carried out to determine some of the factors which influence the formation and retention of a clinker coating on the refractory lining of a rotary cement-kiln. A mineralogical study of the interface between refractories used in the burning-zones of cement kilns and cement clinkers of various chemical composition has shown the extent of these surface reactions. Experiments have also been carried out to compare the adhesive forces operating across the interface at temperatures similar to those found in cement-kiln burning-zones.

*Reactions solide–liquide et adhérence entre la brique et le clinker dans les fours à ciment rotatifs*

Une étude a été effectuée pour déterminer quelques uns des facteurs qui influencent la formation et la fixation d'une couche de clinker sur le garnissage réfractaire d'un four à ciment rotatif. Une étude minéralogique de l'interface entre les réfractaires utilisés dans les zones de cuisson des fours à ciment et les clinkers de ciment de différentes compositions chimiques a montré l'étendue de ces réactions de surface. Des expériences ont également été faites pour comparer les forces d'adhérence qui se développent au travers de l'interface à des températures semblables à celles que l'on trouve dans les zones de cuisson des fours à ciment.

*Fest-flüssig-Reaktionen und die Adhäsion zwischen Klinker und feuerfestem Material in Zementdrehrohröfen*

Untersucht wurde der Einfluß einiger Faktoren auf die Ausbildung und den Abbau von Klinkeransätzen auf Ofenfuttern in Zementdrehöfen. Mineralogische Untersuchungen der Zwischenschicht zwischen Feuerfest-Material der Brennzone und Zementklinkern verschiedener chemischer Zusammensetzung zeigten Oberflächenreaktionen. Weiter wurden Untersuchungen durchgeführt, um die Adhäsionskräfte, die *über die Zwischenschicht wirken, mit anderen vergleichen zu können bei Temperaturen, die denen der Brennzone in Zementdrehrohröfen entsprechen.*

# 1. INTRODUCTION

Portland cement is produced by heating proportioned mixtures of calcareous and clay materials to a temperature at which partial fusion occurs. In various parts of the world the physical nature of these raw materials may differ, e.g. the calcareous material may vary from a limestone to a chalk, but the chemical composition of the fired product, known as the cement clinker, usually falls within defined limits. A typical cement clinker analysis is shown in Table 1.

**Table 1**

|  | $SiO_2$ | $Al_2O_3$ | $Fe_2O_3$ | $CaO$ | $MgO$ | $Na_2O$ | $K_2O$ | $SO_3$ |
|---|---|---|---|---|---|---|---|---|
| Clinker analysis (%) | 21·9 | 6·0 | 2·5 | 66·8 | 1·2 | 0·2 | 0·6 | 0·5 |

Liquid content     26%

Silica ratio     $\dfrac{SiO_2}{Al_2O_3 + Fe_2O_3} = 2·57$

$Al_2O_3/Fe_2O_3$     2·4

Phase composition (%)     $C_3S$, 51·4; $C_2S$, 24·0; $C_3A$, 11·7; $C_4AF$, 7·6; free lime, 2·5.

Cement clinkers contain four main phases: tricalcium silicate ($C_3S$), dicalcium silicate ($\beta$-$C_2S$), tricalcium aluminate ($C_3A$) and a ferrite solid solution usually approximating to brownmillerite ($C_4AF$). A typical phase analysis is shown in Table 1 and may be estimated by the Bogue calculation.[1] The raw cement material is fired or clinkered between 1400° and 1550°C, depending on its composition and within this temperature range there is usually 20–30% liquid in the clinker. The liquid content may be calculated for a given temperature by the empirical formula:

$$Liquid = 3·13\, C_3A + 1·35\, C_4AF + MgO + alkalis \text{ (for 1450°C)}$$

A term widely used by cement chemists is the silica ratio:

$$\frac{SiO_2}{Al_2O_3 + Fe_2O_3}$$

which is widely used as a measure of the ease of combination of the raw materials, the value usually lying between 1·5 and 3·0. A low silica ratio indicates relatively easy combination of the raw materials and hence a low burning-temperature. A high value indicates difficult combination and a higher burning-temperature, e.g. 1550°C will be required to produce an equivalent product.

Most of the refractory problems encountered in rotary cement kilns occur in the burning zone, where the raw materials are clinkered and temperatures in the region of 1500°C are recorded. The life of refractories used in the burning zone of rotary Portland cement kilns depends on the formation and retention of a clinker coating on the refractory lining. The formation of this coating not only protects the brickwork from the severe chemical and abrasive action of the clinker as the kiln rotates, but also lowers the working temperature of the hot face of the refractory lining. The thickness of this coating is governed by the nature of the raw materials and the method of cement production, but is usually between 1 in. and 9 in.

The expansion of the cement industry, coupled with increased production in recent years, has led to the need for greater efficiency in firing techniques, e.g. the reduction of heat losses through the kiln shell, and as a result there has been an increase in refractory-lining temperatures. For this reason it has been necessary to use magnesite–chrome and fired dolomite refractories in the burning zone in place of the 70% alumina refractories used previously. Typical analyses of these refractories are shown in Table 2.

**Table 2**

|  | $SiO_2$ | $Al_2O_3$ | $Fe_2O_3$ | $TiO_2$ | $Cr_2O_3$ | $CaO$ | $MgO$ | $LoI$ |
|---|---|---|---|---|---|---|---|---|
| Alumina | 24·4 | 70·9 | 2·1 | 2·0 | — | 0·3 | 0·3 | — |
| Magnesite/chrome   (a) | 4·06 | 4·71 | 8·96 | 0·13 | 16·62 | 2·40 | 61·86 | 0·47 |
| (b) | 2·65 | 18·23 | 8·24 | 0·40 | 19·36 | 0·92 | 49·63 | 0·04 |
| (c) | 2·50 | 7·45 | 4·40 | 0·07 | 6·89 | 1·62 | 76·77 | 0·22 |
| Fired dolomite | 1·57 | 1·04 | 1·37 | 0·04 | 0·03 | 52·30 | 43·60 | — |

If a substantial clinker coating can be maintained, the life of these refractory linings may be 2 or 3 years, but if not, a lining may last for only a few weeks; the average lining life is usually about 18 months. The loss of a clinker coating and, as a result, direct attack on the refractory lining may be brought about by various factors, the major ones being:

(1) Changes in the raw materials or in the burning conditions.
(2) Mechanical distortions of the kiln.
(3) Mechanical breakdowns which cause loss of clinker coating due to spalling.

Since the manufacture of Portland cement is a dynamic process, it is difficult to reproduce the working conditions of the kiln on a laboratory scale and a study of the refractory behaviour under these conditions is not possible. In order to determine the factors which influence the formation and retention of a clinker coating under working conditions, a comprehensive survey[2] of cement production was made both in Great Britain and on the Continent of Europe, in which the manufacturing conditions at many works were studied with particular reference to raw materials, burning conditions, fuel and refractory wear. However, the formation of a clinker coating was usually found to depend on a combination of factors which under laboratory conditions could only be studied independently. A chemical and mineralogical study[3] of refractories taken from cement kilns revealed the general pattern of behaviour under working conditions; but as physical and chemical changes take place on cooling, a study at working temperatures would be more informative.

## 2. EXPERIMENTAL PROCEDURE

The present investigation was initially concerned with measuring the adhesive forces between refractories and cement clinkers at working temperatures under static laboratory conditions. This was followed by a mineralogical examination of the specimens, both after normal cooling in the furnace and also after quenching from temperature.

The measurement of the adhesive forces acting between brick and cement clinker was carried out with the apparatus shown in Figure 1, which consists of an electric furnace capable of attaining 1500°C, in which a 3-in. square slice of refractory brick is placed in contact with a 2-in. diameter cylinder of cement slurry, pressed at 1 t.s.i. A vertical load of 1000 g is applied to the cylinder and the furnace is then heated to the required test temperature, e.g. 1450°C. After a soaking at temperature for 10 h, the vertical load is removed and a horizontal load is applied to the clinker pellet. By this means, the force required to detach the clinker from the brick can be determined. This apparatus has been used to compare the adhesive properties of cement materials of differing composition with magnesite–chrome and with dolomite bricks. Results have been obtained covering the range 1000°C to 1450°C and can be correlated with microscopic evidence based on the examination of polished sections.

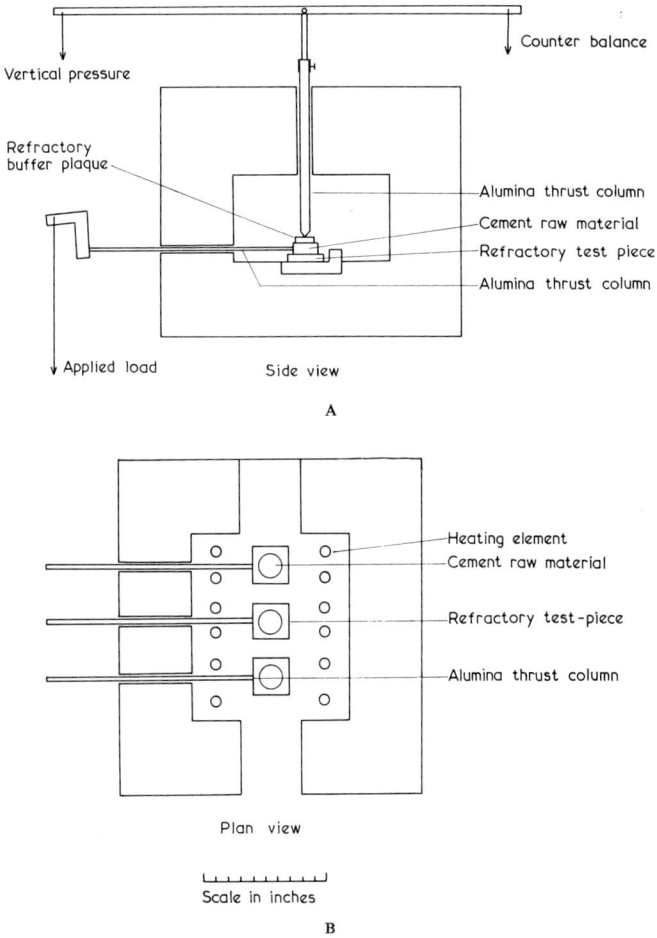

FIGURE 1.—Apparatus for measurement of brick/clinker adhesion.

## 3. CLINKER/BRICK ADHESION AT CEMENT-KILN HOT-ZONE TEMPERATURES (1450°C)

### 3.1 High-alumina Refractories

Most of the present investigation has been concerned with the reactions of cement raw materials with magnesite–chrome and dolomite refractories, but a brief investigation has been made into the reactions of alumina refractories, since they are still used in processes for which low burning-temperatures are required.

Studies involving 70% alumina refractories have been restricted because, on cooling, the cement clinker breaks away from the refractory. This process, known in the cement industry as "dusting", is due to the inversion of $\beta$-$C_2S$ to $\gamma$-$C_2S$ on cooling and is accompanied by a 10% expansion. However, a sample of this alumina refractory in contact with cement materials has been quenched after heating to 1450°C and a macroscopic examination showed a dark interfacial reaction zone extending 2–3 mm below the interface.

A mineralogical study indicated that the reaction between the two materials had been quite considerable, with the formation of large quantities of liquid in the region of the interface. The most evident feature was the breakdown of mullite by a solution of lime in the cement-clinker liquids. This gave rise to several well-defined reaction zones corresponding to a decrease in the lime concentration gradient as the liquid penetrated into the brick structure (Figures 2 and 3). The interface is defined by a layer of $C_2S$ set in a matrix of $C_3A$ and $C_4AF$, and was followed by a zone of anorthite and gehlenite in which crystals of corundum and small quantities of calcium hexaluminate had formed. Studies of samples tested at 1300°C and at

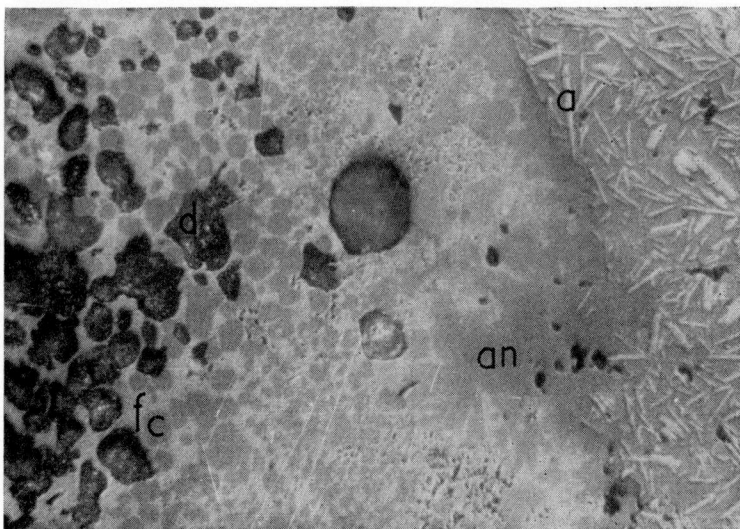

FIGURE 2.—Reaction zone of 70% alumina refractory with clinker ($\times 300$). **a,** $Al_2O_3$; **an,** $CAS_2$; **f,** $C_4AF$; **d,** $C_2S$; **c,** $C_3A$.

1350°C indicated that the reaction between the materials had been less and that the breakdown of mullite was negligible, thereby confirming their value in processes using low burning-temperatures. Further work in this field is described by HUGGETT [4] and by BRISBANE and SEGNIT. [5]

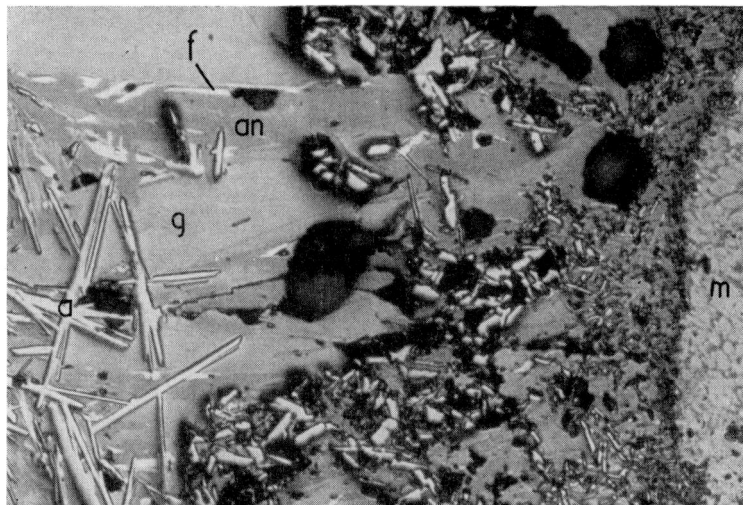

FIGURE 3.—Continuation of Figure 2 showing penetration of calcium-base liquids into the refractory ($\times 240$). a, $Al_2O_3$; an, $CAS_2$; g, $C_2AS$; m, $A_3S_2$; f, $C_4AF$.

Experiments to determine the adhesive forces between 70% alumina refractories and cement clinker showed that, between 1000° and 1450°C, these forces were superior to those of basic refractories, owing to the formation in reactions involving alumina refractories of a highly viscous aluminosilicate liquid.

### 3.2 Magnesite–chrome Refractories

Magnesite–chrome refractories show much greater resistance to chemical attack by cement raw materials than do alumina refractories. An intensive study of the reactions involved has led to a possible explanation for the development of the strong forces which exist at the interface between magnesite–chrome refractories and cement clinkers. As one might expect, the mechanism of the adhesion at initial working temperatures, i.e. in the region of 1450°C, is based on the assumption that the liquid phases of the cement material penetrate into the pores of the refractory. A visible reaction zone is formed and microscopic studies both on naturally-cooled and on quenched specimens showed the following features whenever adhesion had occurred:

(1) A layer of $C_2S$ extensively wetted by $C_4AF$ was present in the clinker immediately adjacent to the interface.

(2) The liquid penetrating into the brick attacked severely the chrome spinel grains which, on cooling, recrystallized out in a modified form.

(3) Calcium penetration further into the brick resulted in the formation of merwinite ($C_3MS_2$) and monticellite (CMS).

With quenched samples the basal layer of $C_2S$ usually extends about 400 $\mu$m into the clinker. The transition from the typical clinker structure to the $C_2S$ layer is shown in Figure 4 and is quite sudden. There is evidence of solid/solid bonding with $C_3S$ and $C_2S$ in physical contact, which indicates that the $C_3S$, the dominant phase in the clinker, reacts with the silicate matrix of the brick to form $C_2S$. The basal layer may be sub-divided as follows:

(1) A zone of columnar $C_2S$ crystals, up to 250 $\mu$m long, often direct-bonded to the $C_3S$ in the original clinker. Presumably these elongated $C_2S$ crystals have been formed as a result of the attack on the clinker by the silicate matrix of the brick. Their shape is due to a lime concentration gradient set up along the crystal, the end adjacent to the brick being slightly lime-deficient; uniaxial grain growth of the $C_2S$ would then develop.

(2) The zone adjacent to the refractory consists of a more rounded form of $C_2S$, similar to that found in industrial clinkers. These crystals are present in an area of high liquid content which suggests that they might have been formed by the reaction of the $C_3A$–$C_4AF$ liquid in the clinker with the magnesium-silicate-based matrix of the brick.

The major factor in producing adhesion in magnesite–chrome bricks therefore appears to have been the penetration of the liquid phases ($C_3A$–$C_4AF$) of the cement materials into the refractory. These liquid phases then reacted with the magnesium silicates of the refractory matrix to form a calcium-based liquid system which severely attacked the grains of mixed spinel, the composition of which would be mainly $FeO(Cr_2O_3.Al_2O_3)$ (Figures 5 and 6). Variable but significant proportions of $C_2S$ have been formed within the chrome grains, together with a modified spinel, the composition of which would be $MgO.Cr_2O_3$.

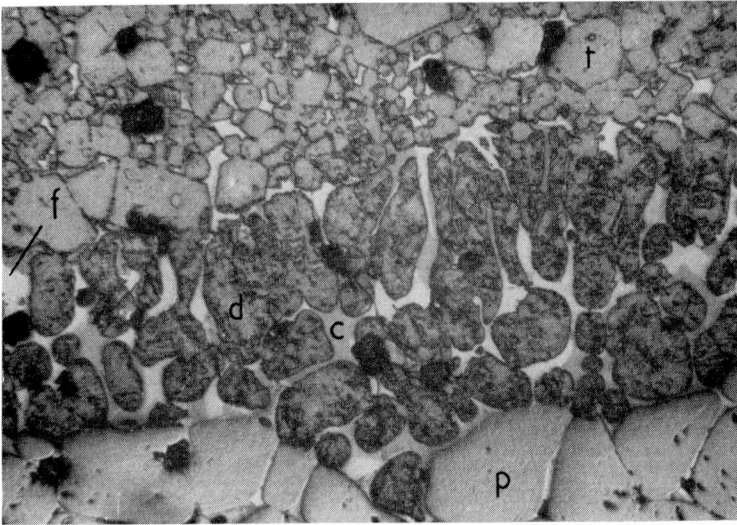

FIGURE 4.—Reaction zone of magnesite–chrome refractory with cement clinker ($\times 300$). f, $C_4AF$; d, $C_2S$; c, $C_3A$; p, MgO; t, $C_3S$.

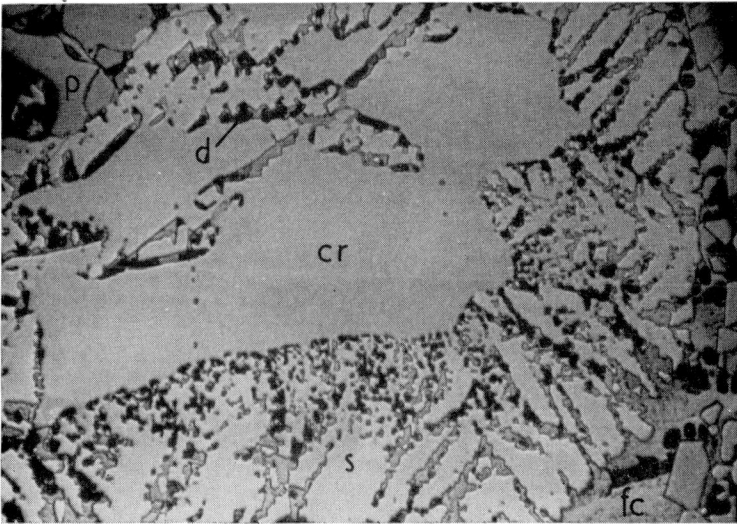

FIGURE 5.—Initial decomposition of chrome grains by clinker liquids ($\times 300$). cr. $FeO(Cr_2O_3 . Al_2O_3)$; d, $C_2S$; s, MgO. $Cr_2O_3$; p, MgO; fc, $C_4AF$-$C_3A$ ss.

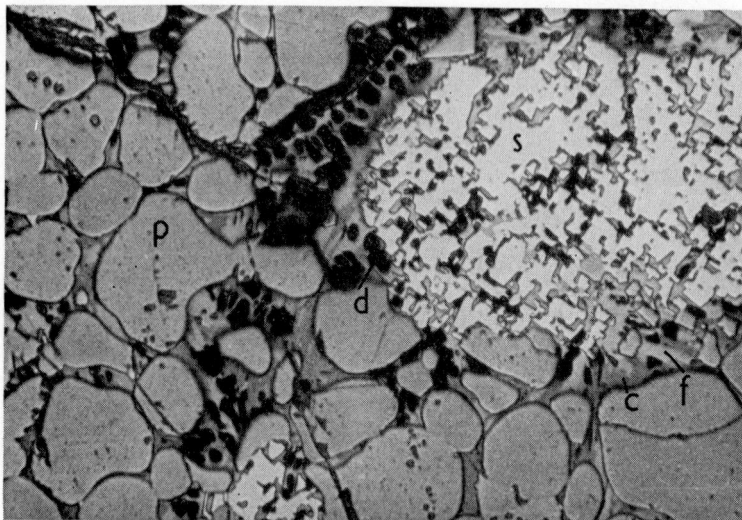

FIGURE 6.—Advanced stage of chrome grain decomposition ($\times 300$). s,
MgO.$Cr_2O_3$; p, MgO; f, $C_4AF$; c, $C_3A$; d, $C_2S$.

It was of interest that adhesion on cooling only occurred when the
chrome grains near the interface entered into the reaction, no doubt
owing to small quantities of chromic oxide that had migrated into
the clinker section of the interface and stabilized the basal layer of
$\beta$-$C_2S$, preventing it from inverting into the $\gamma$ form.[6] The green
coloration of the clinker section was further evidence of the migration
of chrome.

The grains of magnesia in the refractory had been attacked less
and evidence of such reaction was observed only occasionally at the
magnesia grain boundaries (Figure 7). The solubility of MgO in
cement clinker liquids has been found experimentally to be only
5–6% at 1450°C[7] and results in the formation of a monticellite-
based liquid phase. Apart from this liquid phase taking part in the
reaction with the chrome grains, there was further evidence of it at a
greater distance from the interface: a layer of merwinite followed by
a zone of monticellite was present up to 7 mm from the interface,
the extent of the chrome grain attack being about 3 mm.

The reactions described above have been observed, but to a greater
extent, in polished sections of magnesite–chrome refractories taken
from cement-kiln linings after service. These refractories usually
show a highly porous zone behind the cement clinker interface, where
the silica content of the brick is found to be greatly decreased owing

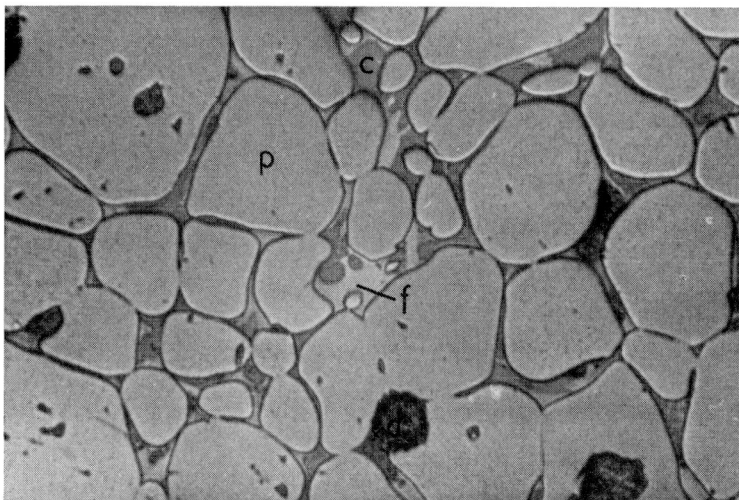

FIGURE 7.—Impregnation of clinker liquids into magnesite–chrome ($\times 300$). p, MgO; d, $C_2S$; c, $C_3A$; f, $C_4AF$.

to migrations into the cement clinker to form $C_2S$ and into the refractory in the form of a monticellite-based liquid. This highly porous zone is followed by a dense zone consisting of modified mixed spinel—$C_2S$, CMS and $C_3MS_2$. Consequently the interface between these two zones is the region where spalling occurs if the refractory is exposed to thermal or mechanical shock.

### 3.3 Dolomite Refractories

Because dolomite is a recent addition to the range of burning-zone refractories, few technical investigations have been made on its adhesive properties with cement materials.

Our initial experiments were concerned with comparing the adhesive properties of dolomite with magnesite–chrome refractories, followed by microscopic observations of the brick/clinker interface. The samples examined had been fired at 1450°C and cooled rapidly. Unfortunately, very few test-pieces showing extensive adhesion of clinker have been obtained, owing to the effect of "dusting" on cooling. However, on closer examination, many samples have disclosed the presence of a thin film of adhering clinker. Microscopic investigations have shown that this layer, extending to 150 $\mu$m thick, consisted of euhedra of $C_3S$ extensively "wetted" by a $C_3A$–$C_4AF$ liquid. Where this zone was thickest, the presence of irregular $C_2S$ plates at the surface indicated "dusting". The adhesion appeared to be due in part to solid/solid bonding of $C_3S$ to a thin continuous layer of rounded CaO plates, making up the brick interface (Figure 8). However, in some cases $C_3AF–C_4AF$ liquid had intervened between the two phases. The "dusting" of the clinker section of the sample is difficult to explain, since $C_3S$ would be expected to be the more stable phase in the presence of lime. However, it appears that lime in the dolomite catalyses the decomposition of $C_3S$ in the clinker into $C_2S$ and lime. This reaction usually takes place between 1300° and 1000°C. On further cooling, the $C_2S$ inverts at 675°C to the $\gamma$ form and "dusting" results. The layer of $C_3S$ at the interface appears to have recrystallized from a lime-rich liquid, since there has been considerable growth of the lime grains in this region. The stability of this $C_3S$ must be due to high concentrations of aluminous liquids at the interface.

Observations on the remainder of the brick showed that the original matrix ($C_2S$, $C_3MS_2$, and $C_2F$) had been extensively displaced by $C_4AF–C_3A$ liquid. Calcium silicate in a solution of the latter composition had entered the brick and the $C_3S$ in this area had been formed by recrystallization from the lime-rich liquid (Figure 9). The many observations revealed that the recrystallized $C_3S$ is most prominent in liquids high in alumina but this will depend mainly on the composition of the raw materials.

### 3.4 Basic Refractories below 1450°C

The reactions described above have involved the adhesion of cement clinker at an arbitrary working temperature of 1450°C. Further experiments have been carried out in order to determine the variation of the interfacial adhesive forces with temperature (Figure 10). As one might expect, as the quantity of "liquid" at the interface increases with rise in temperature, the degree of adhesion decreases but it was surprising that strong adhesive forces were operating across the interface between 1000°C and 1300°C. Since it is generally accepted that the liquid phases in the cement clinkers do not become apparent until about 1260°C, the bonding must be due to the reaction between the constituents of the cement slurry and the matrix of the refractory. A mineralogical study of samples quenched from temperatures in this range indicated a possible explanation for these results.

At 1000°C the calcination of the slurry is just completed and the only other reaction taking place in the slurry is the initial stage of the combination of lime with the aluminium silicates. A study of a magnesite–chrome refractory showed that the reaction was confined to areas where chrome grains were exposed to the surface. It appeared that these areas carried a surface film of CMS and $M_2S$ and that this

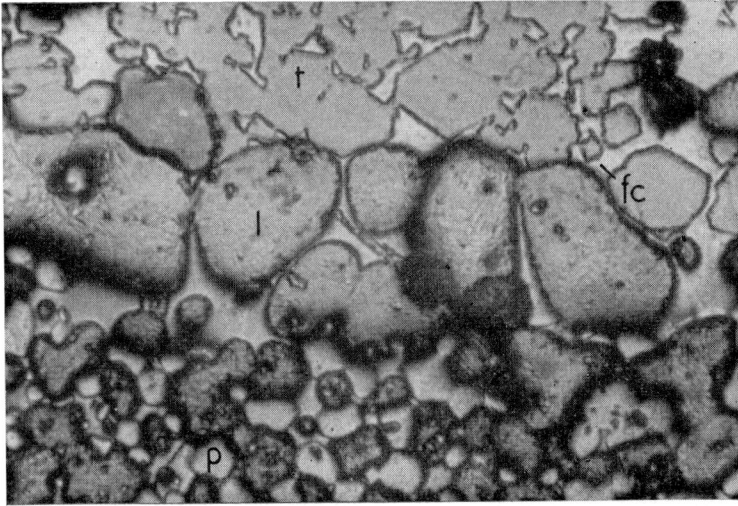

FIGURE 8.—Reaction zone of dolomite refractory with clinker ($\times 240$). t, $C_3S$; fc, $C_4AF$-$C_3A$ ss; 1, CaO, p, MgO.

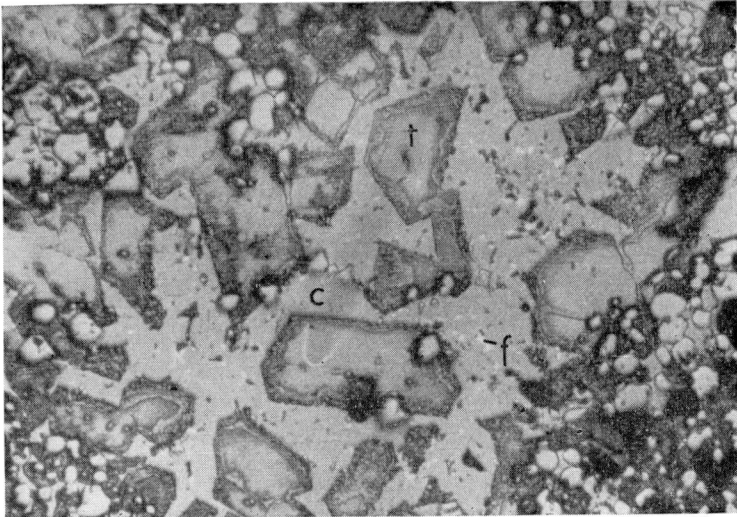

FIGURE 9.—Recrystallization of $C_3S$ in dolomite refractory ($\times 240$). c, $C_3A$; t, $C_3S$; f, $C_4AF$.

18

film had reacted with the cement materials to form at the interface a highly viscous phase consisting of a liquid in the CMS–$C_3A$ system.

At 1100°C the cement materials had reacted with the interstitial silicates of the brick to form a CMS-based liquid which had penetrated up to 100 $\mu$m into the brick and had dissolved portions of the periclase crystals. Clinker formation at this temperature was far from complete and small $C_2S$ platelets, set in brownmillerite, were the only recognizable clinker phases.

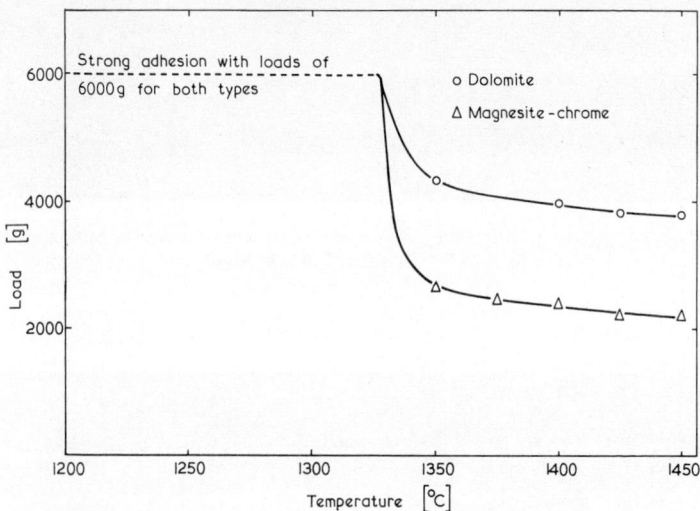

FIGURE 10.—Brick/clinker adhesive forces for basic refractories.

By 1200°C, surface development of a CMS-based liquid was extensive and considerable solution of the exposed periclase was observed. The presence of larger concentrations of calcium, probably $C_2S$ in solution, was evident by the attack on the chrome grains, which had taken place to a depth of about 2 mm (Figure 11).

Penetration of brownmillerite and tricalcium aluminate into the brick was first observed at 1300°C and extended about 250 $\mu$m from the interface. The usual features of the chrome grain disintegration were observed and these disintegrated grains were followed by layers of interstitial $C_2S$ and $C_3MS_2$ about 300 $\mu$m thick.

In the samples treated between 1000°C and 1300°C it was evident that, owing to the pickup of lime, iron oxide or alumina by the interstitial brick silicates, a CMS-rich liquid of high viscosity had formed at the interface. Some solution of periclase into the CMS-based liquid was noticeable at 1100°C and it increased with temperature. However, the attack on the chrome grains by the CMS-based liquid was evident at 1000°C and their corroded margins were surrounded by a film of $C_3A$, $C_3MS_2$, $C_4AF$ and reprecipitated periclase globules.

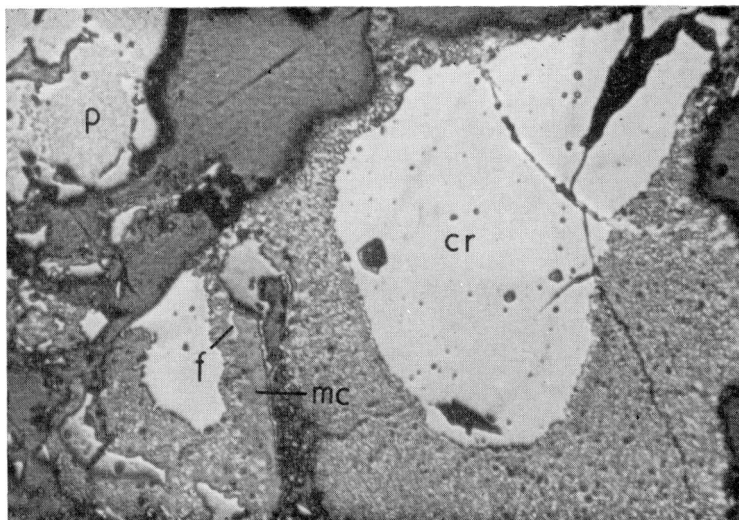

FIGURE 11.—Formation of highly viscous monticellite-based layer around chrome grain at 1200°C. ($\times$300), **cr**, FeO($Cr_2O_3$. $Al_2O_3$); **mc**, CMS; **p**, MgO; **f**, $C_4AF$.

The incipient $C_4AF$ "wetting" and consequent penetration into the brick at 1300°C is believed to represent a change in the type of "wetting" mechanism. Once the $C_4AF$ and $C_3A$ liquids from the cement clinker exceeded a critical depth, the clinker adhesion decreased owing to the reduction in the surface energy associated with the transition from "adhesional wetting" below 1300°C to "immersional wetting" developed above that temperature.[8] With "adhesional wetting", the refractory grains are partially surrounded by small quantities of a highly viscous liquid, but the penetration of $C_4AF$ and $C_3A$ liquids gives rise to a condition of "immersional wetting" in which the grains are surrounded by larger quantities of low-viscosity liquid.

Although mineralogical studies have not been carried out on dolomite refractories at these temperatures it would be fair to assume that the strong adhesive properties are due to "adhesional wetting", as in magnesite–chrome refractories.

### 3.5 Comparison of the Adhesive Properties of Magnesite–Chrome and of Dolomite Refractories

Figure 10 shows that, at all temperatures, fired dolomite refractories show better adhesive properties than magnesite-chrome which is thought to be due to:

(1) The presence of $C_3S$–CaO solid/solid bonding in dolomite refractories.

(2) The high viscosity of the liquid phase at the interface, which appears to be greatest in dolomite samples in which the highly viscous aluminous liquids appear to predominate.

(3) With magnesite–chrome refractories the penetration of the cement materials causes attack not only with the matrix but with the chrome grains of the brick, and as a result more "liquid" is formed than in dolomite bricks, where the majority of the reaction is confined to the matrix.

Further experiments were carried out to compare the adhesion of dolomite and of magnesite–chrome refractories with raw cement materials of various compositions. Tests were carried out at 1450°C and showed that the brick/clinker adhesion for magnesite–chrome was always of the same order of magnitude, irrespective of the composition of the brick or the clinker. With dolomite refractories, however, the adhesive forces appeared to depend to some extent on the composition of the cement materials, especially with cement clinkers of a high alumina/iron ratio, when stronger adhesion was observed.

These results may be explained by comparing the composition limits of the liquid produced when cement materials react with magnesite–chrome and with dolomite refractories. With magnesite–chrome refractories the liquid composition is confined within narrow limits owing to the controlled release of iron and alumina into the impregnating clinker liquids when the chrome grains in the brick decompose. In dolomite refractories, where the iron and alumina components are present in small amounts and are confined to the matrix, there is a much wider range in liquid composition. This is because the constitution of the liquid in the impregnated brick is primarily influenced by the alumina/iron ratio of the clinker. Therefore, in general terms, the breakdown of the chrome grains substantially influences the composition of the impregnating clinker liquids in magnesite–chrome refractories, but with dolomite refractories the matrix is modified by the liquids from the clinker.

## 4. PHASE-EQUILIBRIA RELATIONSHIPS IN MAGNESITE–CHROME AND DOLOMITE REFRACTORIES IN CONTACT WITH CEMENT CLINKER

Very detailed investigations have been made into the phase relationships of the constituents of Portland cement itself, but there is a paucity of information concerning the reactions developed between cement clinker and basic bricks (the question of clinker and high-alumina bricks in terms of phase equilibria has been adequately discussed by HUGGETT[4] and BRISBANE and SEGNIT[5]). One of the few more or less relevant references is that of SOLACOLU,[9] who has dealt with the phase relationships encountered in magnesites and dolomites. In his Figure 1 (p. 141), he indicates the eight paragenetic tetrahedra making up the quaternary system $MgO–CaO–Al_2O_3–SiO_2$; these fields cover magnesite and dolomite compositions, but unfortunately his diagram does not show the effect of $Fe_2O_3$, which is of particular importance with reference to magnesite–chrome bricks. The superimposition of $Fe_2O_3$ on his field III (MgO-MA-$C_3MS_2$-$C_2S$) would adequately account for the fact that only $C_2S$ is stable within the brick and in the clinker at the interface. The presence of this $Fe_2O_3$ modification of field III provides us with a concept of the reaction zone developed within the brick in contact with the clinker, the only significant difference being that alumina is not present as MA, but lies in the $C_4AF–C_3S$ ss. field.

The reactions between clinker and dolomite are covered by Solacolu's field III ($MgO–CaO–C_3A–C_3S$), which shows the principal stable phases in the "impregnated" dolomite in contact with clinker, except that a variable quantity of $C_4AF$ is also present.

The stable phases in clinker-impregnated regions of basic bricks observed under the microscope, as noted during our investigations were as follows on normally-cooled and on quenched specimens:

| Magnesite–chrome | Dolomite |
|---|---|
| $C_4AF$ | $C_4AF$ |
| $C_3A$ ss | $C_3A$ ss |
| $C_2S$ | $C_3S$ |
| | CaO |

## 5. DISCUSSION

The formation and retention of a clinker coating is dependent on the development of a stable, continuous reaction zone in the region of the interface.

The initial formation of the coating is governed by the production of a highly viscous liquid at the interface, which in turn depends on the degree of reaction between the refractories and the cement clinker. If the refractory and cement clinker are considered as separate systems, then at burning-zone temperatures the liquid phases are confined almost completely to the clinker and, when the two systems come into contact, the initial reaction is governed predominantly by the liquid phases of the cement material. It is for this reason that the initial development of a clinker coating depends, in the main, on the composition of the raw materials and not on the particular constitution of the burning-zone refractory. In general, this confirms the fact that materials having a low silica ratio form a coating more readily than those with a higher value.

In some processes the initial development of a coating is catalysed by covering the burning-zone lining with a layer of molten coal ash from the burner before the raw feed is admitted to the kiln. It has already been shown[10] that, under burning-zone conditions, the reaction of coal ash with cement clinkers terminates in the conversion of $C_3S$ to $C_2S$ and similar modifications may take place when coal ash reacts with the refractory lining. A full investigation of these reactions is in progress at our laboratory.

Once the clinker coating has been formed its retention is essential and our investigation shows the following factors to be important in this respect:

(1) The stabilization of any $\beta$-$C_2S$ in the region of the interface.

(2) The presence of conditions favouring the stability of $C_3S$.

(3) The existence of some direct bonding between brick and clinker constituents.

With high-alumina refractories the retention of a coating is not difficult because of the high degree of reaction. However, the problem of "dusting" is always evident and the refractories on cooling lose their clinker coating.

The major factor in the retention of the clinker coating on magnesite–chrome refractories is the stability of the interfacial layer of $\beta$-$C_2S$ by the migration of chromic oxide from the brick into the clinker. However, the presence of a porous zone, low in silica, often found about 2 in. behind the interface in magnesite–chrome refractories taken from cement kilns is sometimes responsible for the loss of the clinker coating. This porous zone does not necessarily have a detrimental effect, since the formation of a stable clinker coating throughout the burning-zone may hold the lining together. A significant advantage of magnesite–chrome refractories appears to be the wide range of raw materials and refractory compositions over which strong brick/clinker adhesion is observed, confirming their applicability to various cement-making processes.

With dolomite the brick/clinker adhesive forces are noticeably stronger at burning-zone temperatures than with magnesite–chrome refractories but, owing to the instability of a layer of $\beta$-$C_2S$ just above the interface in the clinker, the maintenance of the majority of the coating appears to be difficult. The retention of at least part of the coating is due to the presence of large quantities of $C_3S$ in the interface region. Both phase-diagram studies and microscopic investigations suggest that the stability of $C_3S$ at the interface of dolomite refractories is favoured by a $C_3A$ environment and is substantiated by the stronger adhesion of materials having a high alumina content. Therefore it would appear that the maintenance of strong interfacial forces is favoured by raw materials having a high alumina/iron ratio, despite the successful use of dolomite in many processes.

It must be emphasized that this investigation has been carried out into a system which is, in practice, subject to many variable conditions which cannot be studied together, e.g. the fall in the interfacial temperature as the thickness of the coating increases, the effects of kiln atmosphere and the effects of coal ash. However, an insight has been given into the conditions which might favour the use of certain burning-zone refractories in a particular process and the extent and nature of the adhesive forces across the interface.

## ACKNOWLEDGMENTS

The authors thank the Board of Directors of General Refractories Limited for permission to publish this paper, and Mr D. R. Shepherd for his assistance with microscope investigations.

## REFERENCES

1. BOGUE, R. H., *Indust. Eng. Chem.*, **1**, 192, 1929.
2. GELSTHORPE, J. R., and WHITELEY, P. G. Unpublished work, Central Research Laboratories, General Refractories Limited, Worksop.
3. BUIST, D. S., and SHEPHERD, D. R. Unpublished work, Central Research Laboratories, General Refractories Limited, Worksop.
4. HUGGETT, L. G., *Trans. Brit. Ceram. Soc.*, **56**, 87, 1957.
5. BRISBANE, S. M., and SEGNIT, E. R., *Trans. Brit. Ceram. Soc.*, **56**, 237, 1957.
6. NEWMAN, E. S., and WELLS, L. S., *J. Res. Nat. Bur. Stand.*, **36**, 137, 1946.
7. BOGUE, R. H., "The Chemistry of Portland Cement", 2nd Ed., 1955.
8. OSTERHOF, H. J., and BARTELL, F. E., *J. Phys. Chem.*, **34**, 1399, 1930.
9. SOLACOLU, S., *Ber. Deutsch. Keram. Ges.*, **34**, 141, 1957.
10. HEILMANN, T., "Chemistry of Cement", Proc. of Fourth International Symposium, p. 87, 1960.

# 19.—Phase Equilibria in the System CaO.SiO$_2$–2CaO.SiO$_2$–ZrO$_2$

By M. H. QURESHI and N. H. BRETT

*Department of Ceramics with Refractories Technology*
*The University of Sheffield*

**ABSTRACT**                    A423(CaO.SiO$_2$/2CaO.SiO$_2$/ZrO$_2$)

*The work reported is part of a programme reappraising the reported data on the ternary system CaO–SiO$_2$–ZrO$_2$. The system has been studied by the usual techniques of quenching in air, followed by ceramographic and X-ray examination. The phase boundaries on the liquidus surface of the system CS–C$_2$S–ZrO$_2$ have been determined and are found to differ from those previously reported. A new compound, of approximate composition 3CaO.ZrO$_2$.2SiO$_2$, appears on the liquidus surface of the system. This compound melts incongruently at 1635°C ± 5°C to form 2CaO.SiO$_2$, ZrO$_2$–CaO solid solution and liquid. X-ray crystallographic data for this compound are given. A narrow primary field of 3CaO.2SiO$_2$ extends into the ternary system. The remaining primary phases are CaO.SiO$_2$, 2CaO.SiO$_2$ and ZrO$_2$–CaO solid solution. The compositions and melting points of four ternary invariant points are reported.*

### Equilibre de phase dans le système CaO.SiO$_2$–2CaO.SiO$_2$–ZrO$_2$

*Ce travail constitue une partie du programme relatif au ré-examen des données fournies pour le système ternaire CaO–SiO$_2$–ZrO$_2$. Ce dernier est étudié par les méthodes usuelles de trempe dans l'air suivies d'examen céramographique et roentgénographique. Les phases limites sur la surface du liquidus dans le système CS–C$_2$S–ZrO$_2$ ont été déterminées; elles diffèrent de celles décrites précédemment. Un nouveau composé, de composition approximative: 3CaO.ZrO$_2$. 2SiO$_2$ apparaît sur la surface du liquidus dans le système. Il fond incongrûment à 1635°C + 5°C en donnant 2CaO.SiO$_2$, une solution solide de ZrO$_2$–CaO et un liquide. Les données cristallographiques aux rayons X de ce nouveau composé sont fournies. Un domaine. primaire étroit de 3CaO.2SiO$_2$ s'étend dans le système ternaire Les phases primaires restantes sont CaO.SiO$_2$, 2CaO SiO$_2$ et une solution solide de ZrO$_2$–CaO. Les compositions et les points de fusion des 4 points ternaires invariants sont présentées.*

## Phasengleichgewichte im System $CaO.SiO_2$–$2CaO.SiO_2$–$ZrO_2$

*Die hier mitgeteilte Arbeit ist Teil eines Programms, die bisher angegebenen Daten des ternären Systems $CaO$–$SiO_2$–$ZrO_2$ erneut abzuschätzen. Das System wurde mit der üblichen Abschreckmethode in Luft untersucht mit anschließenden keramographischen und röntgenographischen Bestimmungen. Die Phasengrenzen der Liquidus-fläche des Systems $CS$–$C_2S$–$ZrO_2$ wurden bestimmt, und es wurde gefunden, daß sie sich von den zuvor bekannten unterscheiden. Eine neue Verbindung der ungefähren Zusammensetzung $3\ CaO.ZrO_2$. $2SiO_2$ erscheint auf der Liquidusfläche des Systems. Diese schmilzt bei $1635 \pm 5°C$ inkongruent und bildet dabei $2CaO.SiO_2$, $ZrO_2$–$CaO$ als feste Lösung bzw. Flüssigkeit. Kristallographische Röntgendaten für diese Verbindung werden angegeben. Ein schmales Primärfeld von $3CaO.2SiO_2$ erstreckt sich in das ternäre Feld. Die verbleibenden primären Phasen sind $CaO.SiO_2$, $2CaO.SiO_2$ und $ZrO_2$–$CaO$ als feste Lösung. Die Zusammensetzung und Schmelzpunkte von vier invarianten ternären Punkten werden angegeben.*

## 1. INTRODUCTION

The high refractoriness and chemical inertness of zirconia has led to its use in fusion-cast blocks for glass-tank linings. The well known Z.A.C. composition approximates to 12% silica, 34% zirconia and 58% alumina, the crystalline phases being corundum and zirconia in a glassy matrix. The polymorphism of zirconia precludes its use at higher concentrations, since the tetragonal⇌monoclinic inversion about 1000°C is accompanied by a large volume change. Stabilization of zirconia in a cubic form can be achieved by the addition of calcia and, according to DUWEZ, ODELL and BROWN,[1] the cubic solid solutions formed are stable from the liquidus temperature down to room temperature. BUSBY[2] reports that the corrosion resistance to a soda–lime glass at 400°C of slip-cast lime-stabilized zirconia is superior to that of slip-cast zircon ($ZrSiO_4$) but inferior to that of fusion-cast zirconia–corundum. The lime–silica–zirconia system is interesting, since there appears at first sight possible compositions which would yield a crystalline phase of lime-stabilized zirconia in a high-melting siliceous glass matrix. Moreover, if the melting points of such compositions are in the region of 2000°C, fusion-casting techniques could be used to produce a dense corrosion-resistant body.

## 2. LITERATURE

### 2.1 The Binary Systems $CaO$–$SiO_2$, $CaO$–$ZrO_2$ and $ZrO_2$–$SiO_2$

Phase equilibria in the system $CaO$–$SiO_2$ have been extensively investigated. The most recent equilibrium diagram for the system is that described by PHILLIPS and MUAN.[3] In the CaO.SiO$_2$–2CaO.SiO$_2$ portion of this diagram, α-CaSiO$_3$ and α-Ca$_2$SiO$_4$ melt congruently at 1544°C and at 2130°C respectively. Rankinite or Ca$_3$Si$_2$O$_7$ melts incongruently at 1464°C and forms a eutectic with CaSiO$_3$ at 1460°C. The eutectic and peritectic points are separated by a narrow primary phase field of Ca$_3$Si$_2$O$_7$. CaSiO$_3$ inverts from the α to the β form at 1125°C[4] and BREDIG[5] reports that Ca$_2$SiO$_4$ has at least four distinct crystalline phases. The inversions of Ca$_2$SiO$_4$ have been discussed at length by ROY.[6]

RUFF, EBERT and STEPHEN[7] and CURTIS[8] investigated the liquidus region of the CaO–ZrO$_2$ system and established the phase boundaries of the CaZrO$_3$ phase. Later, DUWEZ et al.[1] re-examined the system, especially at low temperatures, where incomplete information was available. They found that the addition of 15–28 mol% lime to zirconia forms solid solutions which are stable up to the liquidus. This feature has important technological implications.

The reported data on the ZrO$_2$–SiO$_2$ system are often conflicting and therefore uncertain, owing in part to the inertness of zirconia and to the difficulty of reaching equilibrium conditions in the presence of viscous siliceous liquids. The dissociation temperature of ZrSiO$_4$ has been variously reported from as low as 1500°C to above 2000°C. As the result of a careful study, CURTIS and SOWMAN[9] concluded that the dissociation rate is slow at 1538°C and increases rapidly above 1760°C. COCCO and SCHROMEK[10] have since reported that ZrSiO$_4$ is decomposed only at temperatures above 1720 ± 20°C. In the revised SiO$_2$–ZrO$_2$ phase diagram of GELLER and LANG[11] ZrSiO$_4$ is shown to melt in a peritectic manner at 1775°C to ZrO$_2$ s.s. and liquid. A eutectic of ZrSiO$_4$ and SiO$_2$ is formed at 1675°C and at the composition 88% by SiO$_2$ weight.

This differs somewhat from the phase diagram proposed by TOROPOV and GALAKHOV,[12] who studied the liquidus region of the system. Their diagram shows a two-liquid region and a eutectic of ZrO$_2$ s.s. and SiO$_2$ (96 $^w/_o$ SiO$_2$) at 1675°C.

## 2.2 The Ternary System CaO–SiO$_2$–ZrO$_2$

Information available on this system is scarce and somewhat sketchy. MATSUMATO, SAWAMOTO and KOIDE[13] made a preliminary study which involved electro-casting ternary compositions on the joins CaO–ZrO$_2$, 2CaO.SiO$_2$–ZrO$_2$ and 2CaO.SiO$_2$–CaO.ZrO$_2$. From an X-ray and microstructural examination of the cast blocks the authors presented a tentative melting diagram which is shown in Figure 1. The main features of this diagram are the presence of six primary-phase fields of the compounds present in the binary systems

—$ZrO_2$, $SiO_2$, $CaO$, $CaSiO_3 . Ca_2SiO_4$ and $CaZrO_3$,[3] together with three ternary eutectic points and one ternary peritectic point. It is interesting to note the author's conclusion that there are no ternary compositions which satisfy the conditions for a refractory.

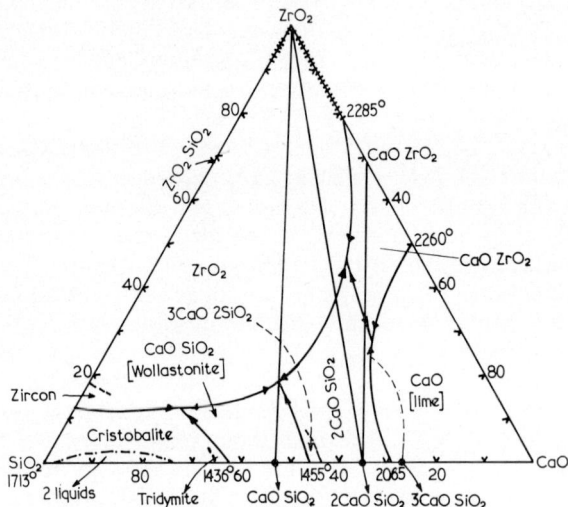

FIGURE 1.—Liquidus projection of the system CaO–SiO₂–ZrO₂ according to MATSUMATO *et al.*[13]

## 3. EXPERIMENTAL PROCEDURE

The starting materials were zirconia powder of 3N purity, Analar calcium carbonate dried in an oven at 110°C, and Belgian sand of high purity. The powders were ground separately to pass through a 300-mesh B.S. screen and the desired proportions were weighed out and thoroughly mixed in a mechanical agate mortar. Batches, each weighing 2–3 g, prepared in this way, were moistened with 4% starch solution and pressed at 5–7 t.s.i. to give pellets of 6 mm diam. and weighing 0·3–0·4 g.

The pellets were contained in small platinum foil crucibles which were suspended by a platinum wire in the hot zone of a molybdenum-wound furnace. A Pt/Pt–13% Rh thermocouple was attached to the wire with its tip in contact with the crucible. Specimens were fired in air at temperatures up to 1750°C for times ranging up to 18 h, then air-quenched by being quickly withdrawn from the furnace. Both firing-temperature and composition markedly affect the time taken to reach equilibrium and at solidus temperatures it was necessary to crush and refire the pellets several times to be sure that this was achieved. The crystalline phases present at the firing-temperature could be recognized in the polished sections from the fact that they occurred as relatively coarse crystals which coarsened with time. In contrast, material that had been liquid occurred as finely divided crystallites or glass.

Quenched specimens were removed from the platinum foil, mounted in Araldite epoxy resin and roughly ground before being polished on successively finer silicon carbide papers and finished on a nylon cloth impregnated with $\gamma$-Al$_2$O$_3$. Polished sections were examined by means of a reflected-light microscope to determine the crystalline phases present. The appearance of the crystalline phases was usually distinctive: for example, dicalcium silicate occurred as rounded crystals with some twinning above 1450°C, CaSiO$_3$ or pseudo-wollastonite appeared as grey tabular crystals, ZrO$_2$ s.s. was rounded and highly reflecting, and monoclinic zirconia was in the form of irregularly shaped crystals though still highly reflective.

Where necessary, specimens were examined by X-ray techniques. Powder diffraction photographs were obtained on a Debye-Scherrer type camera (diameter 11·46 cm) and diffractometer traces were obtained on a Philips 1010 diffractometer both employing nickel-filtered CuK$\alpha$ radiation, wavelengths $\alpha_1 = 1\cdot5405$Å, $\alpha_2 = 1\cdot5443$Å.

Electron microprobe analysis was carried out on an A.E.I. S.E.M. 2 instrument. The minimum K$\alpha$ X-ray wavelengths for Ca, Si and Zr radiations were 6·72, 7·13 and 0·80Å respectively. With the first two long-wavelength radiations, absorption corrections would be large and so linear relationships between X-ray intensity and weight content of the elements could not be expected. With Zr, however, the short wavelength should lead to low absorption corrections and it was found that X-ray intensity was proportional to the weight(%) of Zr in three standards—ZrO$_2$, ZrSiO$_4$ and CaO–SiO$_2$–ZrO$_2$ glass containing 10% ZrO$_2$.

Infra-red absorption spectra were obtained using a Parkin-Elmer 337 spectrometer employing the KBr disc technique.

## 4. RESULTS AND DISCUSSION

### 4.1 General

In establishing phase relationships in the CaO.SiO$_2$–2CaO.SiO$_2$–ZrO$_2$ system, compositions along 12 sections were examined, all radiating from the ZrO$_2$ apex and intersecting the CS–C$_2$S join. In five of these sections the form of the liquidus and the composition of the boundaries between the primary phase fields were determined only. The remaining sections received more-detailed investigation; their phase diagrams are presented in Figures 2–9, together with the

liquidus projection of the ternary system derived from them. Figure
10 shows the isotherms on the liquidus surface and the composition
triangles in the system. Heavy lines in these sections represent boundaries determined by phase identification, dashed lines are boundaries
predicted by considering the cooling of melts. In all sections the high
liquidus temperatures precluded the investigation of compositions
containing more than 40 $^w/_o$ $ZrO_2$. The liquidus projection of the
ternary system, derived from the binary joins, is shown in Figure 10

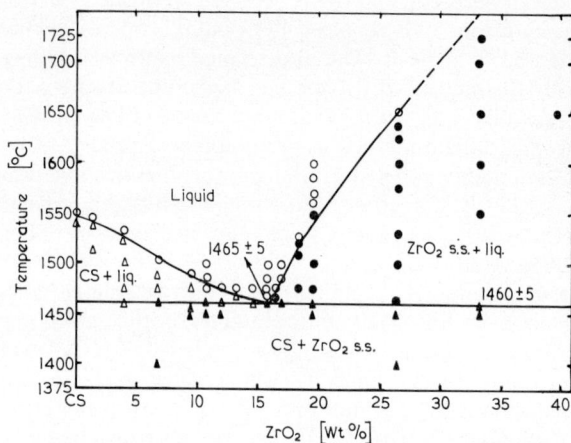

FIGURE 2.—Phase diagram of the section CS–$ZrO_2$ of the pseudo-ternary system
CS–$C_2S$–$ZrO_2$.

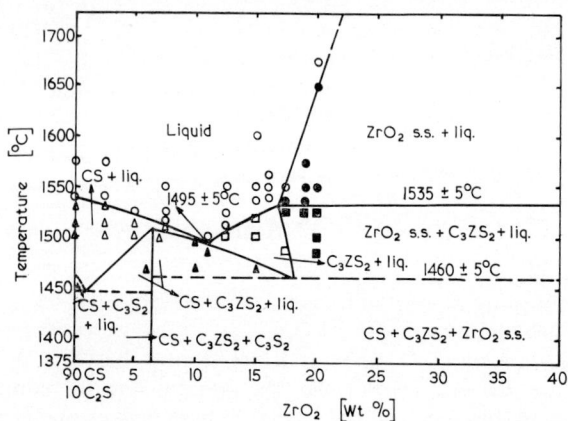

FIGURE 3.—Phase diagram of the section 90CS.10$C_2S$–$ZrO_2$ of the pseudo-
ternary system CS–$C_2S$–$ZrO_2$.

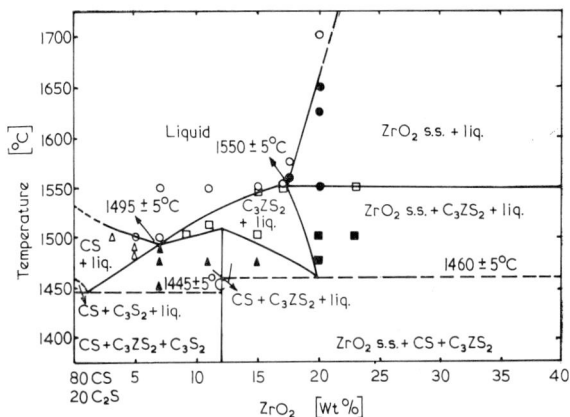

FIGURE 4.—Phase diagram of the section 80CS.20C$_2$S–ZrO$_2$ of the pseudo-ternary system CS–C$_2$S–ZrO$_2$.

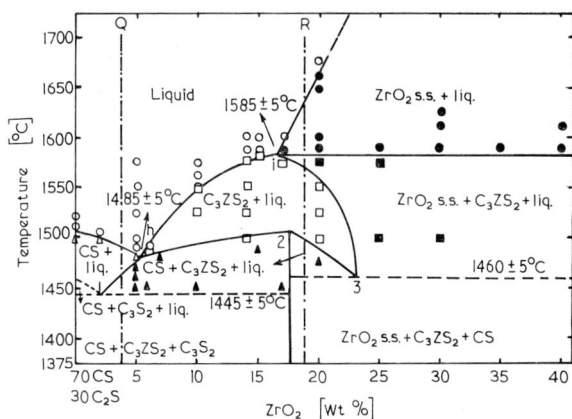

FIGURE 5.—Phase diagram of the section 70CS.30C$_2$S–ZrO$_2$ of the pseudo-ternary system CS–C$_2$S–ZrO$_2$.

and the microstructures of phases present are shown in Figures 11–13. The composition and temperatures of four ternary invariant points are given in Table 1 and are shown as points **a**, **q**, **t**, and **w** in Figure 9. The primary-phase field of Ca$_3$Si$_2$O$_7$ is very small and **t** and **w** are difficult to locate accurately. Many compositions were examined in this area and the boundaries were fixed by identifying the primary phases of the adjacent fields. The appearance of Ca$_3$Si$_2$O$_7$ was quite distinctive but microstructures usually contained

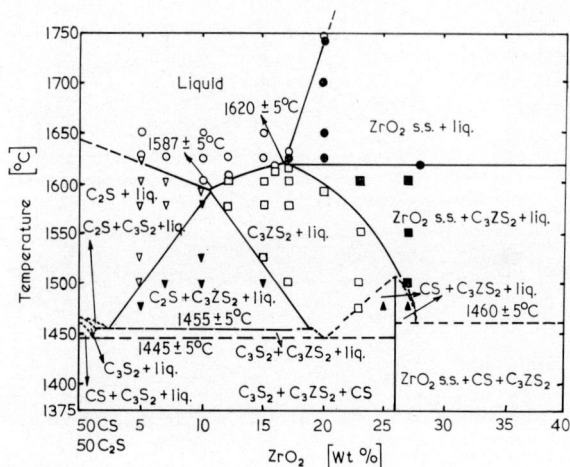

FIGURE 6.—Phase diagram of the section 50CS.50C$_2$S–ZrO$_2$ of the pseudoternary system CS–C$_2$S–ZrO$_2$.

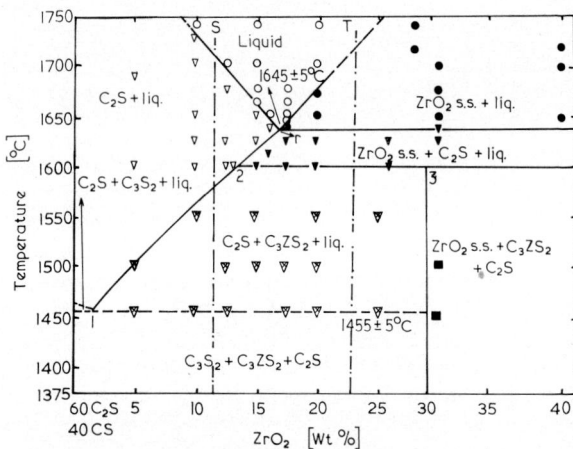

FIGURE 7.—Phase diagram of the section 40CS.60C$_2$S–ZrO$_2$ of the pseudoternary system CS–C$_2$S–ZrO$_2$.

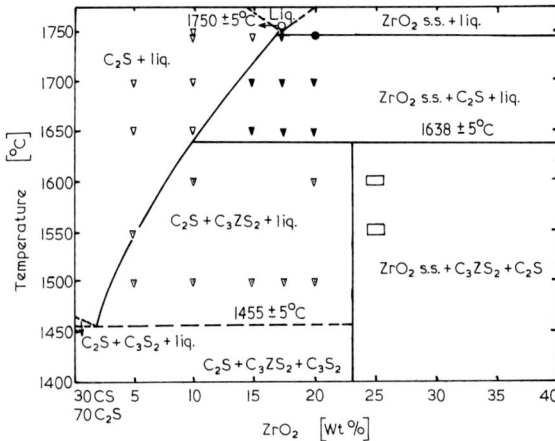

FIGURE 8.—Phase diagram of the section 30CS.70C$_2$S–ZrO$_2$ of the pseudo-ternary system CS–C$_2$S–ZrO$_2$.

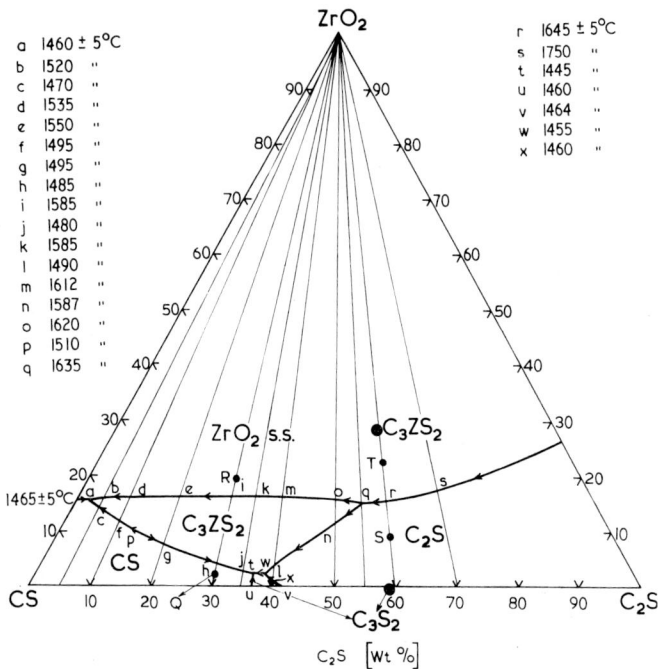

FIGURE 9.—Liquidus surface of the pseudo-ternary system CS–C$_2$S–ZrO$_2$ constructed from the phase diagrams of binary sections within the system.

19

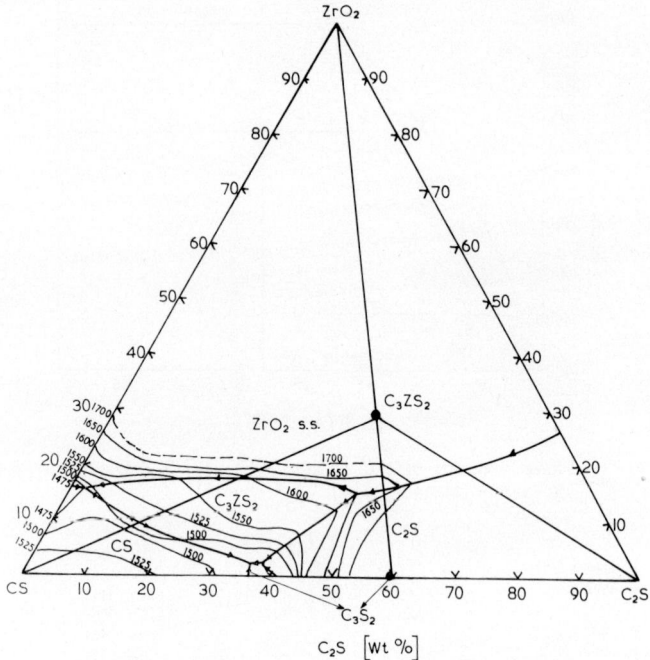

FIGURE 10.—Liquidus surface of the pseudo-ternary system CS–C₂S–ZrO₂ showing isotherms (light lines) and compatibility triangles within the system.

small amounts of $CaSiO_3$ or $Ca_2SiO_4$. Comparison of the X-ray data obtained for specimens containing $Ca_3Si_2O_7$ with the data reported by HELLER and TAYLOR[14] confirmed the presence of this compound.

## 4.2 The Join CS–ZrO₂

MATSUMOTO et al.,[13] in their diagram of the ternary system $CaO–SiO_2–ZrO_2$, show the join $CS–ZrO_2$ intersecting a ternary peritectic composition which implies that the system is not a binary. Figure 2 shows that, whereas the system cannot be described as a binary, only two crystalline phases are present—$CaSiO_3$ and $ZrO_2$ s.s. These phases form a eutectic which melts at 1460°C and the composition 16 w/o CS, 84 w/o $ZrO_2$. The $ZrO_2$ s.s. phase was rounded in appearance and highly reflecting (Figure 13). Measurements of lattice parameter gave values of 5·128 to 5·160Å in different specimens, which correspond to a replacement of more than 20 molar % $ZrO_2$ by CaO.[1]

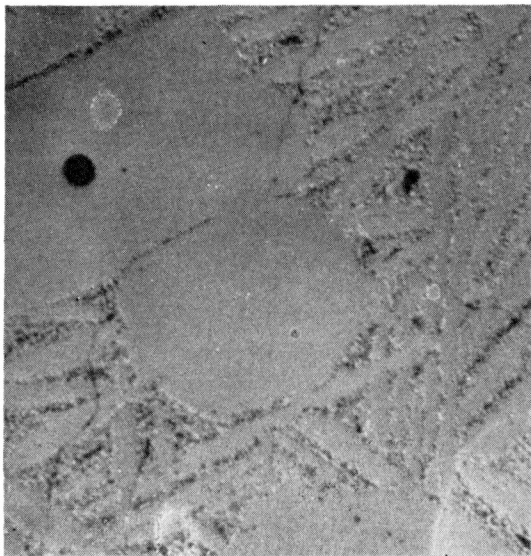

FIGURE 11.—Photomicrograph (×360) of a mixture containing 97% (80CS.20C₂S) and 3% ZrO₂ fired at 1500°C for 3 h, showing large grey grains of CS in a glassy matrix.

### 4.3 The Compound $Ca_3ZrSi_2O_9$

As CS is progressively replaced by $C_2S$, there appears on the liquidus surface a new phase which has a composition that approximates to $3CaO.ZrO_2.2SiO_2$ (or $Ca_3ZrSi_2O_9$) and is therefore on or close to the join between $Ca_3Si_2O_7$ and $ZrO_2$. The composition of the ternary compound was deduced from an electron microprobe analysis of crystals and melt of specimens in its primary phase field. Table 2 shows counting data for the glass sample and the corresponding weight (%) of Ca, Si and Zr. From these data, the element counts for the crystal sample have been converted to weight (%) by direct proportion, the values for Ca and Si being uncorrected. The last column in the table shows the calculated contents of elements of the compound $Ca_3ZrSi_2O_9$. The agreement is excellent for Zr and sufficiently good for Ca and Si to warrant the assumption that the crystal has this composition. A photomicrograph of a crystal of $Ca_3ZrSi_2O_9$, with the Zr trace simultaneously recorded as the specimen is scanned, is shown in Figure 14.

Confirmation of the composition of the ternary compound was obtained from the position of the maximum in the eutectic boundary between the primary-phase fields of CS and $C_3ZS_2$ (Figure 9, point **P**). By Van-Alkamades principle a temperature maximum occurs

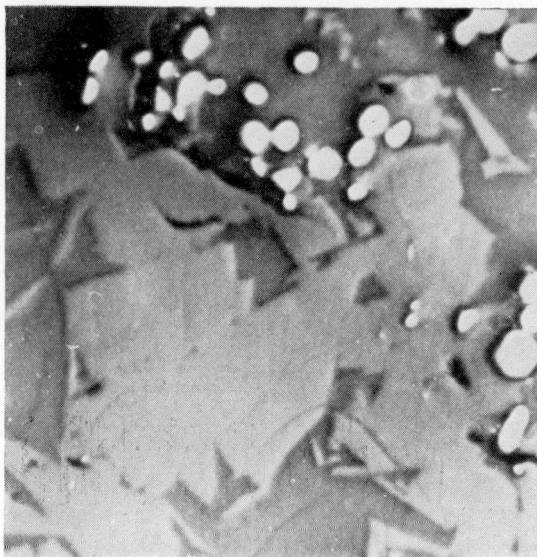

FIGURE 12.—Photomicrograph ($\times 256$) of a mixture containing 75% (70CS.30C$_2$S) and 25% ZrO$_2$ fired at 1575°C for 4 h, showing rounded grains of ZrO$_2$ s.s., angular grains of C$_3$ZS$_2$ and a matrix which was liquid at firing temperature.

on the boundary between two primary-phase fields, where they are crossed by the line joining the compositions of those primary phases. Thus a line from CaSiO$_3$ through **P** should intersect the composition of the compound, and this was found to occur.

N.B. A maximum is not observed at **i**, where the C$_3$ZS$_2$–CS join crosses curve **aq**, since the phases at this boundary are C$_3$ZS$_2$ and ZrO$_2$ s.s. Infra-red spectra for Ca$_3$ZrSi$_2$O$_9$ were compared with those obtained from $\beta$-C$_2$S and C$_3$S$_2$ and are presented in Figure 15 (curves *a*, *b*, and *c* respectively). In curve *a* the peak at wave number 1060 cm$^{-1}$ is distinctive and could be due either to the silicate glass **a** associated with the crystals of compound, or to the presence of large silicate ions in the crystal. The sharp peak at 635 cm$^{-1}$ could be explained as a straightening of the Si–O–Si bond in the (Si$_2$O$_7$)$^{6-}$ ion. (It will be noted that this peak is split in C$_3$S$_2$ (curve *c*) owing to the non-linearity of the Si–O–Si bond.) Alternatively, the peak may indicate the presence of large (SiO$_3{}^{2-}$)$_n$ chains. Taken together, these two peaks suggest the presence of a cross-linked chain structure.

X-ray data for Ca$_3$ZrSi$_2$O$_9$ are presented in Table 3. The observed d spacings do not correspond to any compound known to the authors. The structure is of low symmetry and has not been deduced from the powder data.

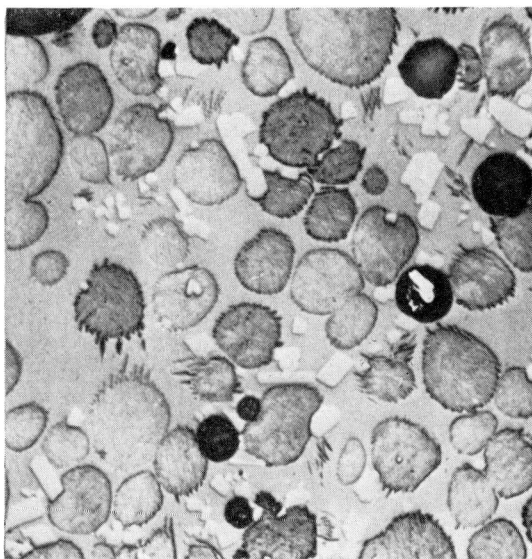

FIGURE 13.—Photomicrograph (×256) of a mixture containing 95%
(40CS.60C₂S) and 5% ZrO₂ fired at 1500°C for 5 h, showing angular
grains (light) of C₃ZS₂ and C₂S crystals (dark) in a glassy matrix.

Table 1

**Compositions and Melting Points of Invariant Points
in the CS–C₂S–ZrO₂ system**

| Invariant point (*Figure 9*) | CaO (ʷ/ₒ) | SiO₂ (ʷ/ₒ) | ZrO₂ (ʷ/ₒ) | Temperature (°C) | Type |
|---|---|---|---|---|---|
| q | 48·6 | 5·9 | 15·5 | 1635 ± 5 | Peritectic |
| a | 41·1 | 3·4 | 15·5 | 1460 ± 5 | Eutectic |
| t | 53·5 | 44·0 | 2·5 | 1455 ± 5 | Peritectic |
| w | 53·0 | 44·5 | 2·5 | 1445 ± 5 | Eutectic |

## 4.4 Crystallization Paths in the Ternary System

Crystallization paths in the ternary system will be discussed by
reference to the freezing of compositions in two sections only, since
those are sufficient to illustrate the variety of conditions which prevail.

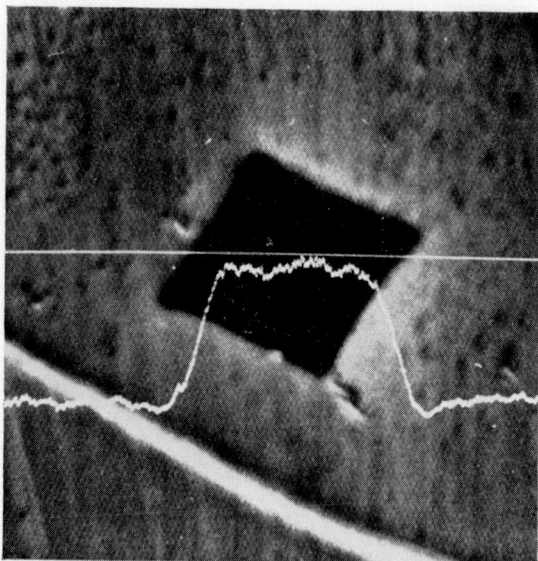

FIGURE 14.—Photomicrograph (× 1000) showing a crystal of $Ca_3ZrSi_2O_9$ in a glass matrix, with the $ZrK\alpha$ X-ray trace recorded.

**Table 2**
**Electron Probe Counting Data for $Ca_3ZrSi_2O_9$**

| Element | Glass | | Crystal | | $Ca_3ZrSi_2O_9$ |
|---------|-------|-------|---------|-------|------------------|
|         | (p.p.s.) | ($^w/_o$) | (p.p.s.) | ($^w/_o$) | Calculated ($^w/_o$) |
| Ca | 1850 | 35·4 | 1400 | 26·8* | 29·2 |
| Si | 450 | 18·9 | 350 | 14·7* | 13·6 |
| Zr | 1020 | 7·4 | 3040 | 22·1 | 22·2 |

\* Uncorrected values

### 4.41 The Join $70CS.30C_2S$–$ZrO_2$

On cooling composition **Q** (Figures 5, 9) from temperatures above the liquidus, CS is the first phase to crystallize. The composition of the melt moves in the direction of a line from CS through **Q** produced, reaching the eutectic boundary **a–t** between **h** and **j**. A eutectic of $Ca_3ZrSi_2O_9$ ($C_3ZS_2$) and CS will then separate out while the liquid composition moves towards **t**. At **t**, ternary eutectic crystallization of CS, $C_3ZS_2$ and $C_3S_2$ occurs until the liquid is consumed.

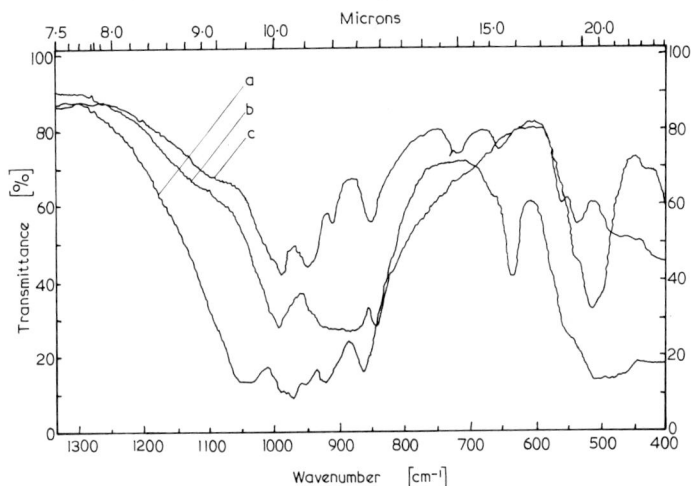

FIGURE 15.—Infra-red absorption data for the compounds Ca$_3$ZrSi$_2$O$_9$ (curve a) $\beta$-Ca$_2$SiO$_4$ (curve b) and Ca$_3$Si$_2$O$_7$ (curve c), showing differences in the peaks attributed to silicate ions.

**Table 3**
**X-ray Data for the Compound Ca$_3$ZrSi$_2$O$_9$**

| $d(Å)$ | $I/I_0$ | $d(Å)$ | $I/I_0$ |
|---|---|---|---|
| 7·31 | 50 | 2·43 | 19 |
| 6·38 | 4 | 2·28 | 10 |
| 5·22 | 16 | 2·25 | 6 |
| 5·09 | 5 | 2·05 | 14 |
| 4·58 | 13 | 2·04 | 20 |
| 4·29 | 16 | 2·02 | 17 |
| 4·18 | 14 | 1·998 | 10 |
| 3·87 | 13 | 1·981 | 5 |
| 3·46 | 8 | 1·978 | 5 |
| 3·22 | 66 | 1·897 | 6 |
| 3·03 | 85 | 1·842 | 25 |
| 2·99 | 84 | 1·812 | 18 |
| 2·87 | 85 | 1·794 | 18 |
| 2·85 | 100 | 1·782 | 12 |
| 2·75 | 10 | 1·777 | 10 |
| 2·70 | 6 | 1·699 | 18 |
| 2·49 | 18 | 1·675 | 10 |
| 2·47 | 10 | 1·431 | 11 |

Compositions to the left of point **l** in Figure 5 precipitate C$_3$S$_2$ when the liquid reaches the eutectic boundary **tu**. To the right of **h**, C$_3$ZS$_2$ is the first phase to crystallize.

Next, consider the cooling of a melt of composition **R**. ZrO$_2$ s.s crystallizes first, the liquid reaches the boundary **aq** at **i** (Figure 9) and proceeds along **ie**, precipitating $C_3ZS_2$ and redissolving $ZrO_2$ s.s. When the melt composition lies on the line from $C_3ZS_2$ through **R** produced, all the $ZrO_2$ s.s. has gone. The melt then moves across the phase field of $C_3ZS_2$ until it reaches the boundary **a–t**. For composition **R**, this occurs to the left of **P**. The liquid then moves along the eutectic boundary in the direction **fa**, with CS and $C_3ZS_2$ crystallizing together. At **a**, freezing is completed with the crystallization of the ternary eutectic.

Compositions to the left of point 2 (Figure 5) proceed in a similar manner but move down the curve **Pt** and hence precipitate $C_3S_2$ eventually. Compositions lying to the right of point 3 do not pass through the primary phase field of $C_3ZS_2$.

### 4.42 The Join $40CS.60C_2S–ZrO_2$

On cooling melt **S** (Figures 7, 9), $C_2S$ crystallizes initially until the boundary **qw** is reached. As the composition of the melt moves along **qw**, $C_3ZS_2$ separates and $C_2S$ dissolves. At **w** crystallization is completed, giving $C_3ZS_2$ and $C_3ZS_2$ and, since the section lies slightly to the right of the join $C_3S_2–C_3ZS_2$, a small amount of $C_2S$. (Actually completion of the peritectic reaction at **w** was never observed in any of the sections and appreciable residual $C_2S$ always remained in the specimens.) For compositions to the left of point **l** (Figure 7) there is a small field of $C_3S_2 + C_2S +$ liquid, since the freezing path intersects the curve **wv** (Figure 9).

Composition **T**, on cooling from the melt, yields a primary phase of $ZrO_2$ s.s. When the liquid composition reaches **r** (Figure 9), $C_2S$ appears and along **rq** this phase separates whilst $ZrO_2$ s.s. is redissolved. At **q**, re-solution of $C_2S$ is completed. The liquid composition then follows **qw**, $C_3ZS_2$ and $C_2S$ crystallizing together until **w** is reached, at which point freezing finishes. Compositions to the right of point 3 will complete their freezing at the ternary invariant point **q**, $ZrO_2$ s.s., $C_3ZS_2$ and $C_2S$ being the final phases.

## 5. SUMMARY AND CONCLUSIONS

The ternary system $CS–C_2S–ZrO_2$ has been re-examined and the phase diagram has been found to differ appreciably from the diagram previously reported. The present studies reveal the existence of a new compound of approximate composition $Ca_3ZrSi_2O_9$ which melts incongruently at $1635 \pm 5°C$ to form $ZrO_2$ s.s., $C_2S$ and liquid.

The diagram is of technological interest from two points of view. Firstly, the corrosion resistance of existing zirconia bodies to lime–silica glasses can be better understood, since information is provided on possible high-temperature phases which may be formed. Secondly, the form of the liquidus surface suggests that the solubility of zirconia in CaO–SiO$_2$ melts is low. Thus compositions can be predicted where the zirconia exists as a solid solution phase in a high-melting siliceous matrix. The temperatures involved are such that refractories based on these compositions are worth considering. More information is, however, required on the CS–SiO$_2$–ZrO$_2$ area of the diagram and is being sought.

## ACKNOWLEDGMENTS

The authors are grateful for the study leave and financial support made to one of them (H. M. Qureshi) by the Pakistan Government, which enabled the work to be carried out. They thank Dr W. F. Ford (Ceramics Department) for the electron microprobe results, Dr Mitchell (Metallurgy Department) for the infra-red spectra data and Professor White (Ceramics Department) for his valuable discussions and comments on the work.

## REFERENCES

1. DUWEZ, P., ODELL, F., and BROWN, H. R., *J. Amer. Ceram. Soc.*, **35**, (5), 107, 1952.
2. BUSBY, T. S., "Tank Blocks for Glass Furnaces" (Society of Glass Technology, 1966).
3. PHILLIPS, B., and MUAN, A., *J. Amer. Ceram. Soc.*, **42**, (9), 414, 1959.
4. OSBORN, E. F., and SCHAIRER, J. F., *Amer. J. Sci.*, **239**, 715, 1941.
5. BREDIG, M. A., *J. Amer. Ceram. Soc.*, **33**, (6), 188, 1950.
6. ROY, D. M., *J. Amer. Ceram. Soc.*, **41**, (8), 293, 1958.
7. RUFF, O., EBERT, F., and STEPHAN, E., *Ceram. Abstracts*, **8**, (10), 749, 1929.
8. CURTIS, C. E., *J. Amer. Ceram. Soc.*, **30**, (6), 180, 1947.
9. CURTIS, C. E., and SOWMAN, H. G., *J. Amer. Ceram. Soc.*, **36**, (6), 190, 1953.
10. COCCO, A., and SCHROMEK, N., *La Ceramica*, **12**, (8), 45, 1957.
11. GELLER, R. F., and LANG, S. M., "Phase Diagrams for Ceramists", Figure 361. (The American Ceramic Society, 1964.)
12. TOROPOV, N. A., and GALAKHOV, F. YA., *Izvest. Akad. Nauk. S.S.S.R.*, p. 160, 1956.
13. MATSUMATO, K., SAWAMOTO, T., and KOIDE, S., *Asahi. Garasu. Kenkyu. Hokoku.*, **4**, (2), 8, 1954.
14. HELLER, L., and TAYLOR, H. F. W., "Crystallographic Data for the Calcium Silicates" (Her Majesty's Stationery Office, 1956).

# 20.—Contact Reactions between Forsterite and Tricalcium Silicate and Aluminate

By W. L. DE KEYSER and R. DERIE

Université Libre de Bruxelles, Institut de Chimie Industrielle

**ABSTRACT** E141/L21

*In reactions at high temperature between magnesite refractories and the constituents of Portland Cement, the action of compounds containing a high proportion of CaO on the forsterite and monticellite phases of magnesite refractories is important in the formation of a protective layer on the brick. These reactions were studied by maintaining in close contact pellets (1) of forsterite and of tricalcium silicate and (2) of forsterite and of tricalcium aluminate, and subsequently heating them at 1380°C and at 1290°C, these temperatures being chosen to avoid the formation of liquid. The location of the area of contact was indicated by platinum reference markers. In each case the progress of the reaction was followed by taking successive sections parallel to the area of contact and subjecting them to X-ray fluorescence and X-ray crystallographic analysis; thin and polished sections cut perpendicular to the area of contact were examined optically. In both systems the profile showed that the reaction was due to part of the CaO from the silicate or aluminate diffusing into the forsterite and progressively replacing magnesia; all the techniques employed confirmed that only CaO diffuses either as CaO or as $Ca^{2+}$ and $O^{2-}$, but the exact mechanism is not fully understood. The reaction zones consisted of: for $C_3S/M_2S$ in contact for 81 h at 1380°C, $C_3S$, $\gamma$-$C_2S$, $\gamma$-$C_2S+M$, $C_3MS_2+M$, $CMS+M$, $M_2S$; and $C_3A/M_2S$ in contact for 90 h at 1290°C, $C_3A$, $C_{12}A_7$, "$C_6A_4MS$"$+M$, $C_3MS_2+M$, $M_2S$. Similar diffusion of CaO was observed during the study of the reaction between calcium oxide and zirconium silicate in contact.*

*Réactions de contact entre la forsterite et le silicate et l'aluminate tricaeciques*

*Lors des réactions aux hautes températures entre les réfractaires de magnésie et les constituants du ciment Portland, l'action exercée par les composés à haute teneur en CaO sur la forstérite et la monticellite, phases constitutives des réfractaires de magnésie, est importante pour la formation d'une couche protectrice sur les blocs. Ces réactions sont étudiées en maintenant en contact étroit des*

293

*pastilles (1) de forstérite et de silicate tricalcique et (2) de forstérite et d'aluminate tricalcique et en les portant ensuite à 1380°C et à 1290°C, ces températures ayant été choisies pour éviter la formation de liquide. L'emplacement de la zone de contact est indiqué par des marqueurs de référence en platine. Dans chaque cas, la progression de la réaction est suivie sur des sections successives, parallèles à la zone de contact qu'on soumet à des analyses par fluorescence X et à des examens cristallographiques aux rayons X. Des lames minces, découpées perpendiculairement à la zone de contact sont soumises à un examen optique. Le profil obtenu dans ces deux systèmes montre que la réaction est due à ce qu'une partie de CaO du silicate ou de l'aluminate diffuse dans la forstérite et remplace progressivement la magnésie. Toutes les méthodes employées confirment que seul CaO diffuse soit sous forme de CaO soit de $Ca^{2+}$ et $O^{2-}$, mais jusqu'à présent le mécanisme de diffusion n'est pas encore entièrement élucidé. Les zones de réaction sont formées: pour $C_3S/M_2S$ en contact pendant 81 h à 1380°C de $C_3S$, $C_2S\gamma$, $C_2S\gamma + M$, $C_3MS_2 + M$, $CSM + M$, $M_2S$; et pour $C_3A/M_2S$ en contact pendant 90 h à 1290°C de $C_3A$, $C_{12}A_7$, "$C_6A_4MS$" $+ M$, $C_3MS_2 + M$, $M_2S$. Une diffusion similaire de CaO a été observée en étudiant les réactions entre l'oxyde de calcium et le silicate de zirconium mis en contact.*

## Kontaktreaktionen zwischen Forsterit und Trikalziumsilikat und-aluminat

*Bei den Hochtemperaturreaktionen zwischen Magnesitsteinen und den Bestandteilen des Portlandzementes ist die Reaktion von Verbindungen, die einen hohen Anteil an CaO entthalten, mit den Forsterit und Monticellitphasen der Magnesitsteine durch die Bildung einer Schutzschicht auf dem Stein von Bedeutung. Diese Reaktionen wurden an Pillen untersucht, die in engem Kontakt miteinander gehalten wurden und aus (1); Forsterit und Trikalziumsilikat und (2); Forsterit und Trikalziumaluminat bestanden. Sie wurden auf 1380°C bzw. auf 1290°C erhitzt. Diese Temperaturen wurden gewählt, um die Bildung von Flüssigkeit zu vermeiden. Die Stelle der Kontaktfläche wurde durch Platinbezugsmarken gekennzeichnet. In jedem Fall wurde der Fortschritt der Reaktion durch aufeinanderfolgende Schnitte parallel zur Kontaktfläsche verfolgt; sie wurden durch Röntgenfluoreszenz und durch kristallographische Röntgenanalyse untersucht. In beiden Systemen zeigte das Profil, daß die Reaktion auf das teilweise vom Silikat bzw. Aluminat in den Forsterit diffundierende CaO zurückzuführen ist, wo es nach und nach das MgO ersetzt. Alle benutzten Techniken bestätigen, daß CaO entweder als CaO oder als $Ca^{2+}$ und $O^{2-}$ diffundiert allein, der exakte Mechanismus aber noch*

*nicht ganz verstanden wird. Die Reaktionszonen bestehen aus: für*
$C_3S/M_2S$, *81 h bei 1380°C im Kontakt* $= C_3S$, $\gamma$-$C_2S$, $\gamma$-$C_2S + M$,
$C_3MS_2 + M$, $CMS + M$, $M_2S$ *und für* $C_3A/M_2S$, *90 h bei 1290°C im*
*Kontakt* $= C_3A$, $C_{12}A_7$, *"$C_6A_4MS$" $+ M$*, $C_3MS_2 + M$, $M_2S$. *Eine*
*ähnliche CaO-Diffusion wurde beim Studium der Kontaktreaktion*
*zwischen Kalziumoxid und Zirkonium Silikat beobachtet.*

## 1. INTRODUCTION

This work forms a new contribution to our study of the attack
of magnesian refractories by Portland cements during firing.[1]
Recent work[2,3] has shown the importance of reactions between the
forsterite of magnesian refractories and the constituents of Portland
cement clinker in the formation of the protective coating. We have
carried out a general study of the contact reactions between these
substances at high temperatures.

In the present work, pellets of forsterite and tricalcium aluminate
and of forsterite and tricalcium silicate were maintained in contact
at high temperatures; the various compounds formed and the concentration profiles in the various phases were determined by X-ray
diffraction and fluorescence. These concentration profiles were
checked by electron micro-probe analysis, and X-ray pictures of the
various phases were taken in the zones of interest.

## 2. EXPERIMENTAL TECHNIQUE

### 2.1 Preparation of the Starting Materials

The raw materials were UCB 1430 analytical purity magnesium
oxide, "Aérosil" high-purity amorphous silica, UCB 1223 analytical
purity calcium carbonate, and Merck 1093 analytical purity aluminium hydroxide.

### 2.11 Preparation of Forsterite

The mixture $2MgO + 1SiO_2$ is heated for two periods of 8 h at
1550°C, with intermediate grinding.

### 2.12 Preparation of Tricalcium Silicate [4]

The mixture, $3CaCO_3 + 1SiO_2$, previously decarbonated, is
melted with an oxy-acetylene burner; the resulting product is ground,
passed through a sieve of 16,900 mesh/cm², compressed into briquettes and annealed for 3 h at 1600°C.

### 2.13 Preparation of Tricalcium Aluminate

The mixture, $3CaCO_3 + 1Al(OH)_3$, compressed into briquettes,
is heated gradually to 1400°C and maintained for 6 h at this temperature.

All these products were ground until they passed through a sieve of 16,900 mesh/cm$^2$; their purity was checked by X-ray crystallography and microscopy.

## 2.2 Preparation of the Samples

We have used the marble-crucible technique[5] in order to maintain a good contact between the pellets at high temperatures; the crucible piston was held down by a weight of about 0·5 kg. Platinum markers (30 $\mu$m diam. wires) were placed between the pellets. Figure 1A

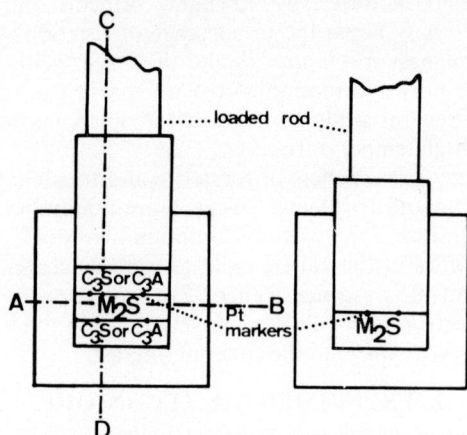

FIGURE 1.—Experimental set-up

shows schematically the experimental arrangement used. After the reaction, the crucible is impregnated under vacuum with Araldite resin, which takes about 10 h to set at 120°–130°C. It is then cut with a diamond saw and grindstone; the optical and electron microprobe examinations are carried out on cross-sections parallel to the CD plane, while for the X-ray crystallographic and X-ray fluorescence examinations the sample is cut along the AB plane and abraded progressively towards one of the contact boundaries in such a way that the surfaces thus formed are always parallel to the contact planes.

## 2.3 Analytical Technique

The X-ray crystallographic examinations were carried out with a Norelco Geiger counter spectrometer fitted with a copper anti-cathode. The electron probe microanalyser used in this investigation was the JEOL JXA–3 A model. The X-ray fluorescence examinations were made using a Philips PW 1050 fluorescence spectrometer fitted with a chromium anticathode operating at a current of 24 mA at 38 kV. We used gypsum (2d = 15·185 Å) as an analysing crystal, for the 1st order K$\alpha$ lines of the elements Ca, Si, Al and Mg; the counter used was a gas flux type and the whole of the spectrometer was evacuated to below 0·1 mm Hg.

## 3. EXPERIMENTAL RESULTS

### 3.1 C$_3$S/M$_2$S Contact: 81 h Heating at 1380°C

Figure 2 shows the variation in intensity of the fluorescence

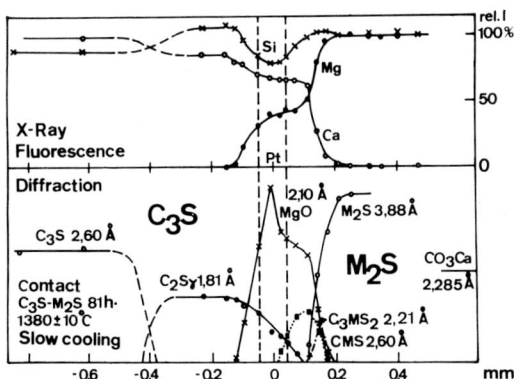

FIGURE 2.—C$_3$S/M$_2$S contact. 81 h at 1380°C. Results of X-ray diffraction and X-ray fluorescence analysis.

lines of Ca, Si and Mg, and the variation in intensity of the characteristic diffraction lines of the various substances identified, as a function of the distance from the platinum marker.

The diffraction lines chosen are compared with the diffraction line d = 2·285 Å of calcite; they are specific for the compounds identified, except for the C$_3$MS$_2$ line at 2·21 Å which coincides with weak lines of $\gamma$-C$_2$S and CMS.

The following reaction zones are observed between C$_3$S and M$_2$S:

Up to around the point − 0·4      C$_3$S
From the point − 0·4 to the point − 0·1    $\gamma$-C$_2$S (I)
From the point − 0·1 to the point + 0·05   $\gamma$-C$_2$S + M (II)
From the point 0·05 to the point 0·15   C$_3$MS$_2$ + M (III)
From the point 0·15 to the point 0·20   CMS + M (IV)
From the point 0·20 onwards     M$_2$S

According to the diffraction results the various zones seem to overlap each other; this is due both to the penetration of the X-rays (around 100 $\mu$m for an attenuation of 90%) and to the fact that the interface is not rigorously plane after heating. In fact, an optical transmission inspection enables one to determine with certainty the presence of four well-defined zones: 250 $\mu$m for zone I, 40 $\mu$m for zone II, 45 $\mu$m for zone III, 10 $\mu$m for zone IV; the increase in volume of $C_2S$ on cooling also caused cracks to appear between the zones (I) and (II).

We can therefore assume the following reaction scheme to hold: $C_3S$ decomposes into $C_2S$ and C
CaO thus formed diffuses into the forsterite where it progressively displaces the magnesia by the reactions

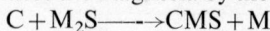

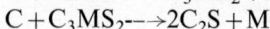

$$C + M_2S \longrightarrow CMS + M$$
$$C + 2CMS \longrightarrow C_3MS_2 + M$$
$$C + C_3MS_2 \longrightarrow 2C_2S + M$$

This scheme is somewhat simplified since, according to the C–M–S phase diagram:[6]

There exists a solid solution of the forsterite type, extending from $M_2S$ to $C_{0.26}M_{1.74}S$, which explains the fact that fluorescence shows calcium to be present in small concentrations very far into the forsterite phase.

There exists a solid solution of the monticellite type, extending from $C_{0.64}M_{1.36}S$ to CMS.

Part of the $C_2S$ results therefore from the total replacement of the magnesia in the forsterite by lime, the rest from the decomposition of the $C_3S$.

We have shown, by analysing the reaction zones with an electron micro-probe over a polished surface cut perpendicularly to the contact planes (Figure 3), that the magnesium definitely does not cross the boundary of the platinum markers: each point corresponds, for an element, to the mean number of counts per second determined over a "slice" of $20 \times 360$ $\mu$m, parallel to the contact plane (Figure 4). The definition is better than by the abrasion method, parallelism being assured to better than 5 $\mu$m over the scanning width. One observes in fact that the Mg concentration falls to zero beyond the point $-20$ $\mu$m. Moreover the concentration profiles of the three elements in the zone 0–200 $\mu$m are rather puzzling. This effect will be discussed later.

Figure 5 shows the X-ray photomicrographs for Mg, Ca and Si in the zones $C_3MS_2 + M$, $CMS + M$ and $M_2S$.

### 3.2  $C_3A/M_2S$ contact: 90 h Heating at 1290°C

This reaction had to be studied at a much lower temperature as beyond 1330°C a liquid phase is formed which floods the whole crucible.

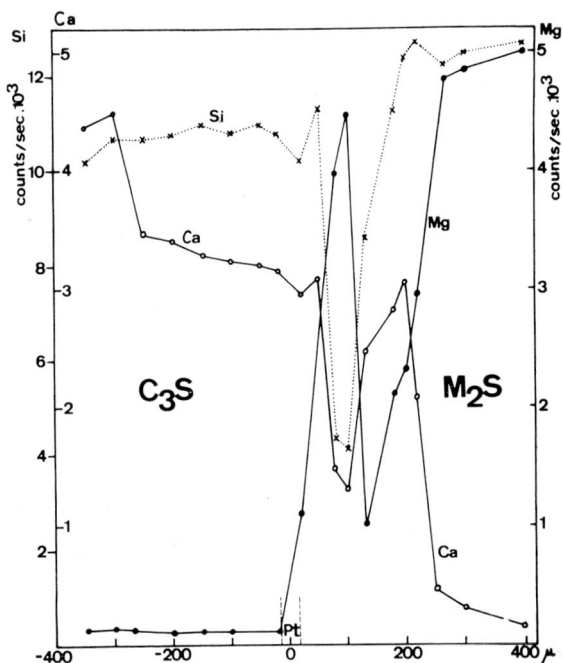

FIGURE 3.—C₃S/M₂S contact. 81 h at 1380°C. Results of electron microprobe analysis.

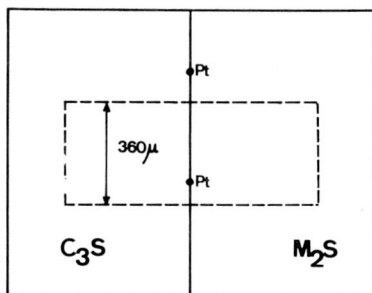

FIGURE 4.—"Slice" parallel to contact plane

20

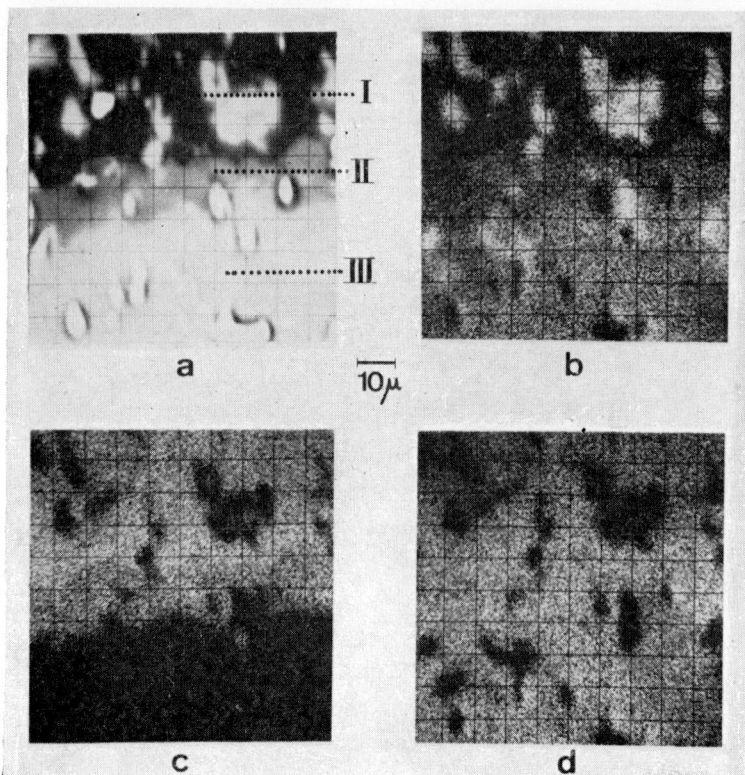

FIGURE 5.—$C_3S/M_2S$ contact. 81 h at 1380°C. Electron (a) and X-ray photomicrographs of: Mg, (b); Ca, (c); and Si, (d), of the zones $C_3MS_2+M$, $CMS+M$ and $M_2S$.

Figure 6 shows the variation in intensity of the fluorescence lines of Ca, Si, Al and Mg, and of the characteristic diffraction lines. The following reaction zones are observed, from $C_3A$ to $M_2S$:

Up to around the point $-0\cdot15$        $C_3A$
From the point $-0\cdot15$ to the point $-0\cdot05$ $C_{12}A_7$
    (by fluorescence: increase in Al, decrease in Ca).
From the point $-0\cdot05$ to the point 0    "$C_6A_4MS$" $+M$
From the point 0 to the point $0\cdot05$      $C_3MS_2+M$
From the point $0\cdot05$ onwards          $M_2S$

The presence of a layer of monticellite, which gives a weak line at $2\cdot60$ Å, is not confirmed by an optical inspection. Figure 7 shows a transmission photomicrograph of the reaction zones.

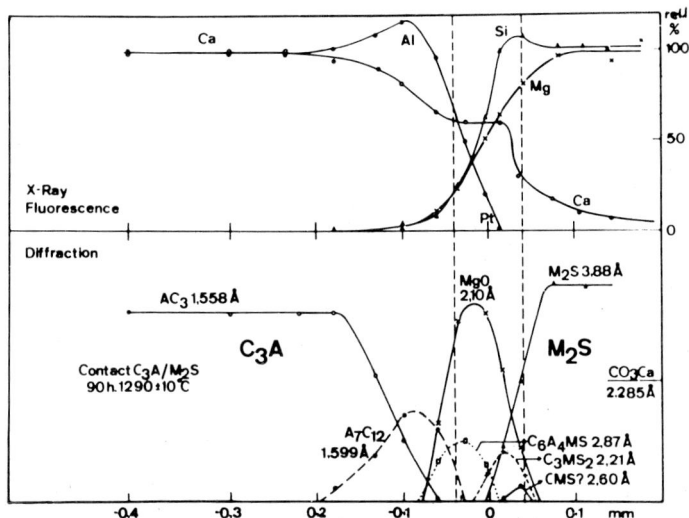

FIGURE 6.—$C_3A/M_2S$ contact. 90 h at 1290°C. Results of X-ray diffraction and X-ray fluorescence analysis.

The various layers are perfectly distinguished. The compound which we identify as being $C_6A_4MS$, for which the diffraction pattern and the weak birefringence correspond perfectly to MIDGLEY's data,[7] could be, according to WELCH,[8] a solid solution of $C_7A_5M$ and $C_2AS$. An acceptable formula in this case would be $C_9A_6MS$. We have carried out syntheses of compounds corresponding to the formulas $C_6A_4MS$ and $C_9A_6MS$ by firing oxide mixtures $2 \times 24$ h at 1340°C, with intermediate grinding. Although the diffraction patterns are analogous, the second compound is the only one which appears to be homogeneous on microscopic examination of a polished section.

We also observe here diffusion of Ca deep into the forsterite phase.

We may assume the following reaction scheme to hold:

The $C_3A$ decomposes into $C_{12}A_7$ and $C$;
CaO thus formed diffuses into the forsterite, where it displaces magnesia by the reaction:

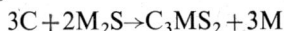

$$3C + 2M_2S \rightarrow C_3MS_2 + 3M$$

The quaternary compound is formed by a complex reaction at the interface, probably by a slight diffusion of alumina.

As in the previous contact, the most important diffusion, if not the only one, is that of CaO. This fact is confirmed by an examination of the concentration profiles determined with an electron microprobe over a polished surface cut perpendicularly to the contact plane (Figure 8). We observe once more the abnormal concentration profiles of Ca and Mg.

Figure 9 shows the electron (a) and the X-ray photomicrographs (b, c, d, e) for Mg, Al, Si and Ca of a region of this polished surface in the neighbourhood of the platinum wire.

We have also studied the $C_3A/M_2S$ reaction by heating to 1290°C pellets made up of grains of sintered forsterite (diam. between 0·1 and 0·3 mm) in a bedding of a fine powder (diam. < 40 $\mu$m) of $C_3A$. After heating, we examined polished surfaces of these pellets with the electron microprobe. We also determined concentration profiles over these pellets: the results are identical with those described above.

## 4. DISCUSSION OF THE RESULTS

The systems which we have studied show a certain similarity in that when in contact with forsterite, both tricalcium silicate and tricalcium aluminate are converted into a compound which contains less CaO and the CaO liberated diffuses into the forsterite, progressively displacing the magnesia.

These systems may seem relatively complex for studying the fundamental diffusion mechanism. However, in a previous work[3] we studied the contact reaction between lime and forsterite also by the marble crucible method, in which case the crucible itself acted as the reagent (Figure 1B).

The results are summarized in Figure 10: it can be seen that the reaction process is of the same type, in that only the CaO diffuses. Even by fluorescence, puzzling concentration profiles are found for Mg and Ca, of the same type as those determined with the microprobe over the contacts considered above. This leads us to discuss two main points.

### 4.1 The Direction of Diffusion

It is beyond doubt that only CaO crosses the interface. This follows both from the concentration profiles with respect to the platinum wire, and from the actual composition of the reaction zones.

It is quite likely that this diffusion takes place at the grain boundaries, but we observed that the previous sintering of the forsterite pellets, during 18 h at 1500°C, does not change the nature or the thickness of the reaction zones.

A similar mechanism of diffusion of CaO was observed by DE KEYSER et al.[10] during the study of the contact reactions between calcium oxide and zircon.

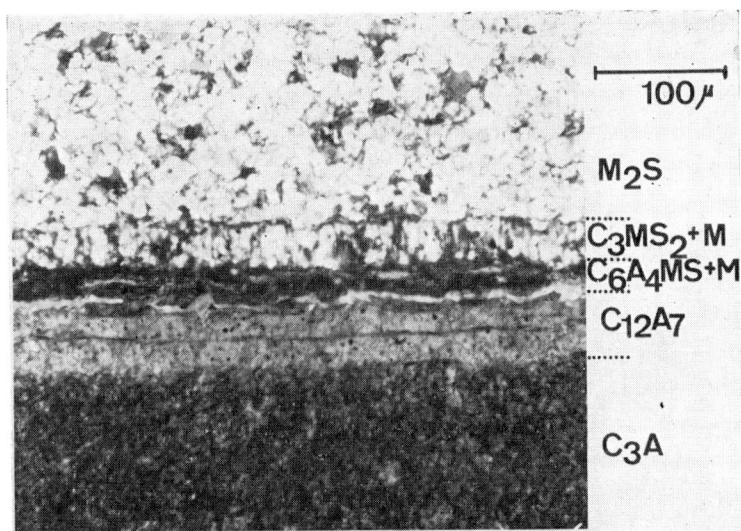

FIGURE 7.—C$_3$A/M$_2$S contact. 90 h at 1290°C. Transmission photomicrograph of the reaction zones (crossed nicols).

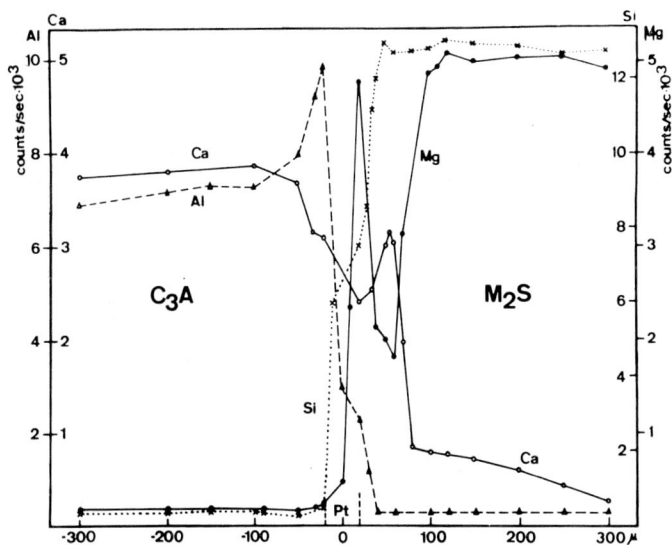

FIGURE 8.—C$_3$A/M$_2$S contact. 90 h at 1290°C. Results of electron microprobe analysis.

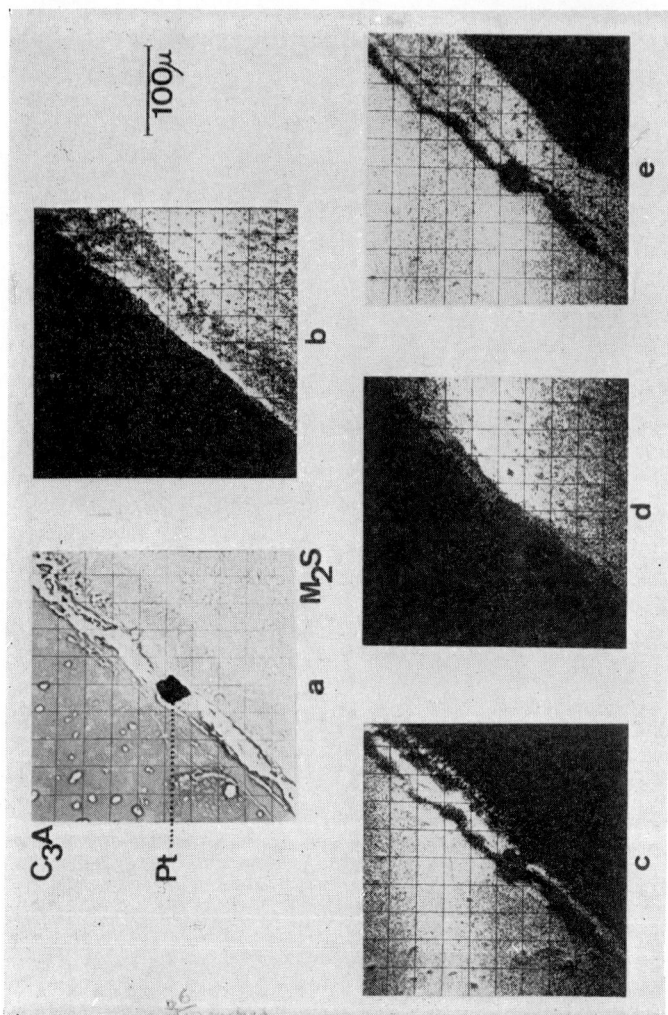

FIGURE 9.—C$_3$A/M$_2$S contact. 90 h at 1290°C. Electron (a); and X-ray photomicrographs of: Mg, (b); Al, (c); Si, (d); and Ca, (e), of a zone close to the platinum marker.

It was also possible for us to show the Kirkendall effect resulting from this unilateral diffusion, in the case of a contact $C/M_2S$ for 430 h at 1370°C. We prevented diffusion on a limited surface, by inserting a platinum foil 30 $\mu$m thick between the pellets; Figure 11 shows the shrinkage of calcium oxide and the swelling of the forsterite pellet, with respect to the plane defined by the platinum foil.

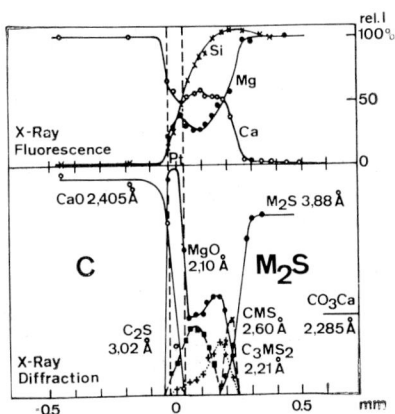

FIGURE 10.—$C/M_2S$ contact. 90 h at 1400°C. Results of X-ray diffraction and X-ray fluorescence analysis.

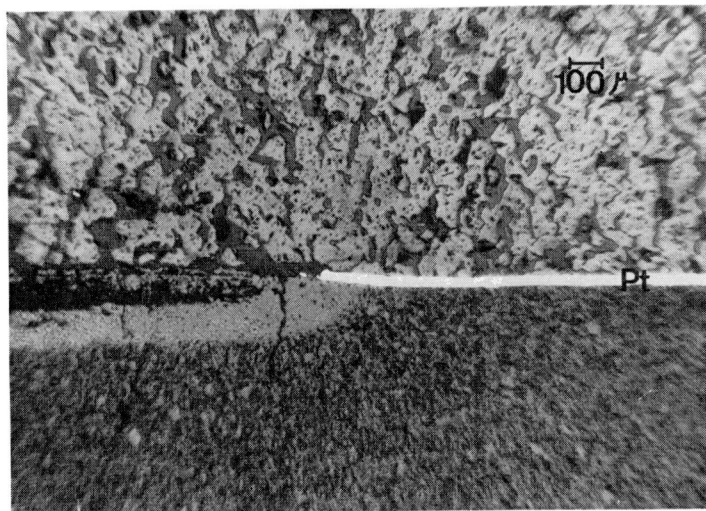

FIGURE 11.—$C/M_2S$ contact. 430 h at 1370°C. Movement of the interface with respect to the platinum foil. Above, CaO; below, $M_2S$.

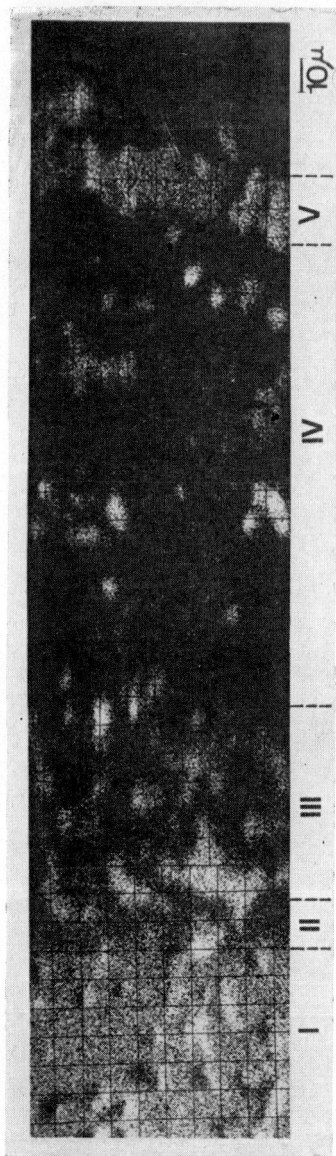

FIGURE 12.—C$_3$S/M$_2$S contact: 81 h at 1380°C. X-ray photomicrograph of Mg in the successive reaction layers. I, M$_2$S. II, CMS+M. III, C$_3$MS$_2$+M. IV, C$_2$S+M. V, MgO "barrier".

## 4.2  The Concentration Profiles

We have observed, during examination of the reaction zones by means of electron microprobe analysis, that a curious phenomenon occurs with magnesia: the MgO liberated by the reaction of forsterite with calcium oxide accumulates partly close to the interface. Figure 12 shows an electron microprobe X-ray picture of Mg in the whole reaction zone of the $C_3S/M_2S$ contact for 81 h at 1380°C. We can observe a real MgO "barrier". This phenomenon occurs in all the contacts investigated; further work is in progress.

### REFERENCES

1. DE KEYSER, W. L., and DERIE, R., Actes IX Congrès Int. Cér. Bruxelles 1964, pp. 105–114.
2. MAJDIC, A., and SCHWIETE, H. E., *Zement, Kalk, Gips*, (2), 45, 1962.
3. BUDNIKOV, P. P., SHUBIN, V. I., and LEPESHOVA, V. I., *Zh. Prikl. Khim.*, **38**, (6), 1193, 1965.
4. DE KEYSER, W. L., *Bull. Soc. Chim. Belges*, **62**, 235, 1953.
5. DE KEYSER, W. L., and CYPRÈS, R., Trabajos de la 3e Reunion Internacional sobre Reactividad de los Solidos, Madrid 1959, pp. 335–50.
6. LEVIN, E. M., ROBBINS, C. R., and McMURDIE, H. F., "Phase Diagrams for Ceramists", (American Ceramic Society, 1964), p. 210.
7. MIDGLEY, H. G., *Mag. Concrete Res.*, **17**, (25), 17, 1957.
8. WELCH, *Nature*, **191**, (4788), 559, 1961.
9. DERIE, R. Unpublished work.
10. DE KEYSER, W. L., WOLLAST, R., HANSEN, P., and NAESSENS, G., 5th International Symposium on the Reactivity of Solids, (Munich 1964. Schwab ed.: Elsevier 1965) pp. 658–666.

# 21.—Divalent and Trivalent Rare-earth Silicates and their Properties

By I. A. BONDAR, N. A. TOROPOV and L. G. SCHERBAKOVA

*Institute of Silicate Chemistry, Leningrad, U.S.S.R.*

**ABSTRACT** A5352

*The silicates of divalent rare-earth elements have been synthesized. It has been claimed that europium and ytterbium silicates develop more easily than samarium silicates, which can be explained on the basis of the data on the reduction potentials and the structure of the electronic shells of $Sm^{2+}$, $Eu^{2+}$ and $Yb^{2+}$. Correlation of X-ray, microscope and infra-red results indicated that the silicates of divalent europium and samarium are isostructural with the strontium silicates; there is also structural similarity between the silicates of $Yb^{2+}$ and $Ca^{2+}$. Phase equilibria in the $SmO-SiO_2$ and $EuO-SiO_2$ systems showed that the samarium compounds are of two types: an ortho- ($2SmO.SiO_2$) and a metasilicate ($SmO.SiO_2$). For europium and ytterbium four types of compound have been identified: oxyortho- ($3LnO.SiO_2$), ortho- ($2LnO.SiO_2$), diortho- ($3LnO.2SiO_2$) and meta- silicate ($LnO.SiO_2$). The silicates of trivalent rare-earth elements comprise three types of compounds which fall into three structural sub-groups. Single crystals of a number of silicates have been synthesized from solution-melts and by means of the Verneuil apparatus and their properties have been studied.*

### Silicates de terres rares d'éléments bivalents et trivalents: leurs propriétés

*Les silicates d'éléments de terres rares bivalents sont synthétisés. Il a été montré que les silicates d'europium et d'ytterbium se forment plus facilement que les silicates de samarium ce qui peut être expliqué sur la base des données que l'on possède sur les potentiels de réduction et la structure des couches électroniques de $Sm^{2+}$, $Eu^{2+}$ et $Yb^{2+}$. La mise en corrélation des résultats obtenus aux rayons X, par microscopie et aux infra-rouges indique que les silicates de l'europium et du samarium bivalents sont analogues, pour la structure, aux silicates de strontium. Il y a également une similarité de structure entre les silicates de $Yb^{2+}$ et $Ca^{2+}$. Les équilibres de phases dans les systèmes $SmO-SiO_2$ et $EuO-SiO_2$ montrent que les composés du samarium sont* *de deux types: un orthosilicate (2 SmO.SiO$_2$) et un métasilicate (SmO.SiO$_2$). Pour l'europium et l'ytterbium quatre types de composés sont identifiés: oxyortho- (3LnO.SiO$_2$), ortho- (2LnO.SiO$_2$), diortho- (3LnO.2 SiO$_2$) et métasilicate (LnO.SiO$_2$). Les silicates des éléments de terres rares trivalents comprennent trois types de composés qui se répartissent en trois sous-groupes au point de vue de la structure. Des monocristaux d'un certain nombre de silicates ont été synthétisés à partir de phases fondues et au moyen de l'appareil de Verneuil. Leurs propriétés sont étudiées.*

**Silikate mit zwei- und dreiwertigen Seltenen Erden und ihre Eigenschaften**

*Die Silikate zweiwertiger Seltener Erden wurden synthetisiert. Es wurde festgestellt, daß Europium- und Ytterbiumsilikate sich leichter herstellen lassen als Samariumsilikate; dies kann auf Grund der Reduktionspotentiale und der Struktur der Elektronenschalen von Sm$^{2+}$, Eu$^{2+}$, Yb$^{2+}$ erklärt werden. Die Zusammenfassung der Ergebnisse aus Röntgen-, mikroskopischen und Infrarotuntersuchungen zeigte, daß die Silikate des zweiwertigen Europiums und Samariums dieselbe Struk tur wie das Strontiumsilikat haben. Zwischen den Silikaten von Yb$^{2+}$ und Ca$^{2+}$ besteht ebenfalls eine strukturelle ähnlichkeit. Die Phasengleichgewichte in den Systemen SmO-SiO$_2$ und EuO-SiO$_2$ zeigten, daß die Samariumverbindungen zwei Typen angehören: einem Ortho- (2SmO.SiO$_2$) und einem Metasilikat (SmO.SiO$_2$). Für Europium und Ytterbium wurden vier Verbindungstypen identifiziert: Oxyortho- (3 LnO.SiO$_2$), Ortho-(2 LnO.SiO$_2$), Diortho- (3 LnO. 2 SiO$_2$) und Metasilikat (LnO.SiO$_2$). Die Silikate der dreiwertigen Seltenen Erden umfassen drei Verbindungstypen, die in drei strukturelle Untergruppen zerfallen. Von einer Anzahl von Silikaten wurden Einkristalle nach dem Verneuil-Verfahren hergestellt. Ihre Eingenschaften wurden untersucht.*

The synthesis of divalent and trivalent rare-earth silicates has opened a new page in inorganic chemistry. The practical interest of such synthetic products is associated with the production and development of ceramics, glass–ceramics and other materials which are being used in various fields of the national economy and show promise.

Many investigators from various countries are working on the problems of developing new materials based on rare-earth metal compounds of various valencies and are studying their properties. Most of the properties of elements are determined by the structure of the outer electron shells. Specific properties are determined by changes in the structure of the inner shells. Thus the trivalency that is characteristic of the whole rare-earth element group is based on the excited states $5d^1 6s^2$ or $5d^2 6s^1$ which occur as a result of the transition of an electron from 4f level to the 5d level. Another of the valencies of these elements depends on differences in the structure of the inner shells.

Rare-earth elements in simple and complex compounds are rather interesting for carrying out physico-chemical and crystallo-chemical investigations, as they enable a small change in the lanthanum ionic radius to determine the characteristics and the properties of the compounds developed, the isomorphous substitutions, melting points and so on. Being a special group of transitional metals, they exhibit specific properties in several valency states, peculiarities in their interaction with other compounds, interesting magnetic, electrical and spectral properties, refractoriness, high strength, etc. The ferromagnetism of some compounds of divalent elements—europium monoxide (EuO) and its orthosilicate ($Eu_2SiO_4$) in particular—is also a rather important property. These properties are exhibited to a considerable extent by divalent and trivalent silicates and are being widely investigated by the scientists of the Soviet, Dutch and American schools.

We have obtained divalent silicates (samarium, europium and ytterbium) and have investigated phase equilibria in the systems $SmO–SiO_2$ and $EuO–SiO_2$.[1,2,3] Of the divalent rare-earth silicates, europium silicates of the composition 3:1, 2:1 and 1:2[4] are mentioned in the literature and the magnetic properties of glasses containing divalent europium[5] are also referred to. For our synthesis we used a special furnace with molybdenum and tungsten heaters and a reducing atmosphere. But complete reduction of $Eu_2O_3$, and more particularly of $Sm_2O_3$ and $Yb_2O_3$, to the divalent state has not been possible; the superposition of divalent and trivalent oxides has been observed. Only with the mixture of silica and $Ln_2O_3$ has the reactivity with hydrogen led to almost complete reduction of $Ln_2O_3$ and to the development of divalent silicates. A similar phenomenon has also been recently noted by the French investigators studying divalent europium aluminates.[6]

Identification of the phases developed by means of crystallo-optical, X-ray, chemical and infra-red techniques has shown that samarium compounds are of two types: ortho- ($2SmO . SiO_2$) and metasilicate ($SmO . SiO_2$).

For europium and ytterbium we have obtained four types of compound: oxyortho- ($3LnO . SiO_2$), ortho- ($2LnO . SiO_2$), diortho- ($3LnO . 2SiO_2$) and metasilicate ($LnO . SiO_2$). It has been claimed that europium and ytterbium silicates are easier to form and that the reaction goes nearer to completion than with samarium silicates. This can be explained on the basis of the reduction potentials and of the structure of the electron shells of $Sm^{2+}$, $Eu^{2+}$ and $Yb^{2+}$. Europium has the lowest reduction potential (0·428 eV), samarium has the highest ($\sim 0·682$ eV) and ytterbium is intermediate (0·578 eV).

The three stable electron shells typical of $La^{3+}$, $Gd^{3+}$ and $Lu^{3+}$ respectively account for the existence of the ions having configurations similar to those of the elements near to them in the Periodic Table, each of the three and its neighbouring ion being an isoelectronic pair. Thus divalent samarium has the stable configuration of $Eu^{3+}$; divalent europium has the structure of $Gd^{3+}$ and forms an isoelectronic pair with it; $Yb^{2+}$ is isoelectronic with $Lu^{3+}$. As $Gd^{3+}$ and $Lu^{3+}$ have a more stable structure of electronic shells than $Eu^{3+}$, their isoelectronic ions ($Eu^{2+}$, $Yb^{2+}$) produce more stable compounds, than $Sm^{2+}$, which is isoelectronic with $Eu^{3+}$.

The investigations have enabled definite conclusions to be drawn about the isostructural characteristics of divalent lanthanum silicates and alkali-earth elements (strontium and calcium). Strontium silicates are isostructural with europium silicates and even more closely isostructural with samarium silicates; close structural similarity has been observed between calcium and ytterbium silicates. This similarity enables the possibility to be assumed of isomorphous substitutions in cation sub-lattices in the lanthanum–alkali-earth element series without lattice distortion.

The investigation has shown that the similarity of divalent rare-earth and alkali-earth silicates is reflected not only in the isostructural nature of the corresponding compounds but also in the general geometric agreement of individual regions of the phase diagrams in systems of the general type $MeO–SiO_2(Me = Sr^{2+}, Ca^{2+}, Sm^{2+}, Eu^{2+})$. This is due to a number of common component properties, which determine both structural phase relationships in the systems and their thermodynamic correspondence. Such similarity is to a certain degree connected with the cation dimensions. The ionic radii of $Sm^{2+}$ and $Eu^{2+}$ (according to Zachariasen 1·11 and 1·09Å) are close to that of $Sr^{2+}$(1·13 Å); $Yb^{2+}$(0·93 Å) is commensurate with $Ca^2$ (0·98 Å). At the same time definite differences between alkali-earth and rare-earth silicates should be noted. Thus whereas in the system $SrO–SiO_2$ there are three types of compound: $3SrO.SiO_2$, $2SrO.SiO_2$ and $SrO.SiO_2$, in the system $SmO–SiO_2$ only two compounds develop; we have not isolated samarium silicate of the composition $3SmO.SiO_2$. This can be explained both by the specificity of our experiment and by the limited stability of the "basic" silicates, which is determined by the structural peculiarities of these phases. Europium silicates are similar to strontium silicates, but at the same time they produce a compound of a diorthosilicate ($Eu_3Si_2O_7$) type, which is similar to rankinite ($Ca_3Si_2O_7$) in the system $CaO$–$SiO_2$ and which is missing from the system $SrO$–$SiO_2$. We can assume that europium silicates hold an intermediate position between the strontium and the calcium silicates, the so-called "basic" silicates being similar to strontium silicates and the "acidic" silicates to calcium silicates. The peculiarity of the rare-earth silicates is also reflected in their crystallo-optical properties (Table 1).

**Table 1**

**Some Crystallo-optical Characteristics of the Silicates of Samarium, Europium, Ytterbium, Strontium and Calcium**

| Compound | Ng | Nm | Np | Birefringence | Optical axes | Optical sign |
|---|---|---|---|---|---|---|
| 3SrO.SiO₂* | | 1·795 | 1·785 | 0·010 | uniaxial | |
| 3CaO.SiO₂ | | 1·723 | 1·717 | 0·006 | „ | (−) |
| 3EuO.SiO₂ | | 1·870 | 1·855 | 0·015 | „ | (−) |
| 3YbO.SiO₂ | | 1·815 | 1·805 | 0·010 | | |
| 2SrO.SiO₂ | 1·756 | | 1·727 | 0·029 | biaxial | (+) |
| β-2CaO.SiO₂ | 1·735 | | 1·717 | 0·018 | „ | (+) |
| 2SmO.SiO₂ | 1·865 | | 1·850 | 0·015 | „ | (+) |
| 2EuO.SiO₂ | 1·860 | | 1·840 | 0·020 | „ | (+) |
| 2YbO.SiO₂ | 1·800 | | 1·780 | 0·020 | | |
| 3CaO.2SiO₂ | 1·650 | | 1·641 | 0·009 | biaxial | (+) |
| 3EuO.2SiO₂ | 1·835 | | 1·820 | 0·015 | | |
| 3YbO.2SiO₂ | 1·780 | | 1·760 | 0·020 | | |
| SrO.SiO₂ | | 1·637 | 1·599 | 0·038 | uniaxial | (+) |
| CaO.SiO₂ | | 1·654 | 1·610 | 0·044 | | (+) |
| SmO.SiO₂ | | 1·760 | 1·740 | 0·020 | | (+) |
| EuO.SiO₂ | | 1·805 | 1·775 | 0·030 | uniaxial | (−) |
| YbO.SiO₂ | | 1·770 | 1·745 | 0·025 | | |

*According to the data of R. W. Nurse, *J. Appl. Chem.*, **2**, 244, 1952. For the other alkali-earth silicates the data are taken from "Physicochemical Systems of the Silicate Technology" by D. S. Belyankin, V. V. Lapin, N. A. Toropov, (Promstroyizdat, 1954).

As Table 1 shows, the silicates of $Sm^{2+}$, $Eu^{2+}$, and $Yb^{2+}$ have higher refractive indices than the corresponding silicates of $Sr^{2+}$ and $Ca^{2+}$. High birefringence is characteristic of europium and ytterbium diorthosilicates, in contrast to $Ca_3Si_2O_7$.

The phase equilibria in the systems $SmO$–$SiO_2$ and $EuO$–$SiO_2$ are demonstrated in Figures 1 and 2 respectively. Figure 1 shows that, for the system $SmO$–$SiO_2$, both samarium silicates melt congruently and are stable from the melting temperatures to 1500°C. A region of

immiscibility in the liquidus has been discovered in this system. We
have shown that for the system $EuO-SiO_2$ (Figure 2) two compounds
($Eu_2SiO_4$ and $EuSiO_3$) melt congruently and are stable over a wide
range of temperature; two other compounds ($Eu_3SiO_5$ and $Eu_3Si_2O_7$)
melt incongruently and are stable within a definite temperature range.
On the phase diagram the immiscibility region occupies a narrow
band.

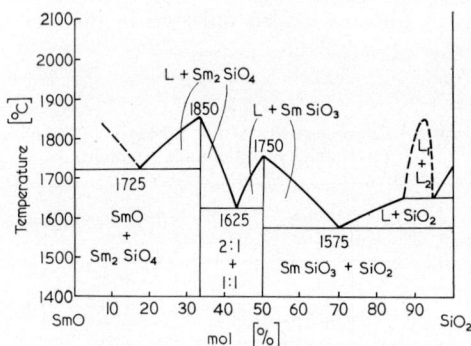

FIGURE 1.—Phase equilibria in the $SmO-SiO_2$ system.

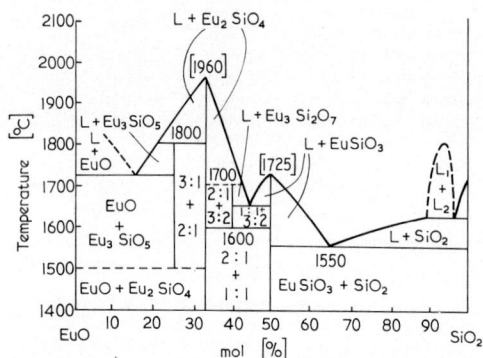

FIGURE 2.—Phase equilibria in the $EuO-SiO_2$ system.

MORRISH and SHAFER[7] show a phase diagram for the system
$EuO-SiO_2$ which is rather similar to the one we have derived. The
difference lies in the fact that Morrish and Shafer have not observed
the diorthosilicate ($Eu_3Si_2O_7$) and the immiscibility region. Photomicrographs illustrate structural inter-relations in the systems
$SmO-SiO_2$ and $EuO-SiO_2$. Thus Figure 3 demonstrates the primary

separation of $Eu_2SiO_4$ crystals as orthogonal plates and prisms. Figure 4 shows the microstructure that corresponds to incomplete crystallization: the light, round, dendrites represent the primary separations of $Eu_2SiO_4$, the large crystals in quantity represent $Eu_3Si_2O_7$, and the main field is presented by the peritectic alloy of ortho- and diorthosilicate of europium. The liquation in the system $SmO–SiO_2$ is demonstrated on Figure 5.

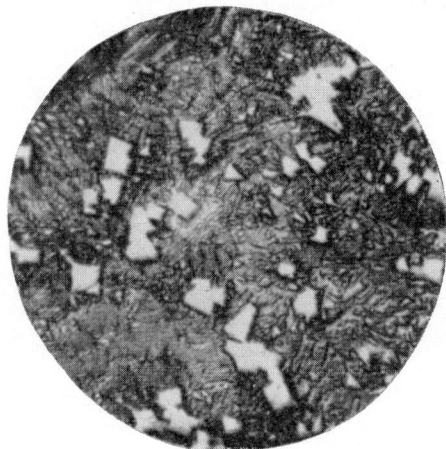

FIGURE 3.—Primary isolations of $Eu_2SiO_4$ as orthogonal plates and prisms.

As we have already stated, three types of chemical compound, oxyortho-, ortho- and diorthosilicates, are characteristic of the trivalent rare-earth silicates. The stability of oxyortho- and diorthosilicates ($Ln_2O_3 . SiO_2$ and $Ln_2O_3 . 2SiO_2$) increases with increase of the number of the lanthanide.

According to the latest data, the compound of the type $7Ln_2O_3 . 9SiO_2$, which has a defect structure of the apatitte type, is also stable over a wide temperature range in the case of large cations (La–Nd); with the heavier elements, the field of stability narrows, and for the cations Y, Er, Yb, the 7:9 compound exists only at comparatively low temperatures (of the order of 1000°–1100°C). For the latter cations, the stabilization of the orthosilicate $2Ln_2O_3 . 3SiO_2$, the composition of which approaches the 7:9 compound, has been observed at higher temperatures.

The $7Ln_2O_3 . 9SiO_2$ types of compound develop easily in the presence of potassium and fluorine ions; we have discovered this while growing single crystals from the solution in KF melt.

In addition, the investigations enable us to assume the possibility

21

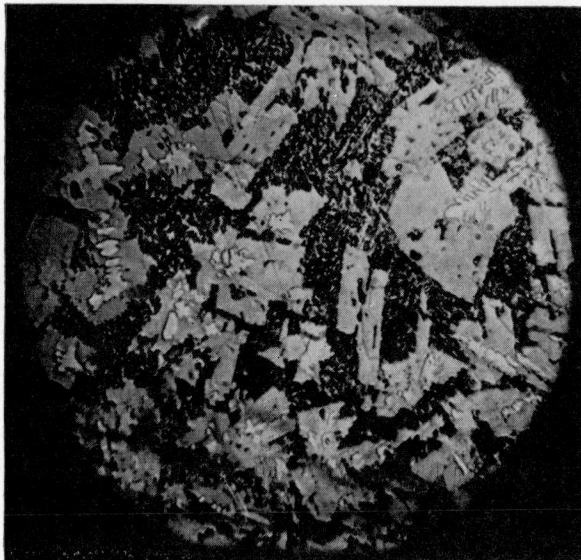

FIGURE 4.—The microstructure of the alloy satisfying the uncompleted crystallization:
Light dendrites of $Eu_2SiO_4$
Large crystals of $Eu_3SiO_7$
A mixture of europium ortho- and diorthosilicate.

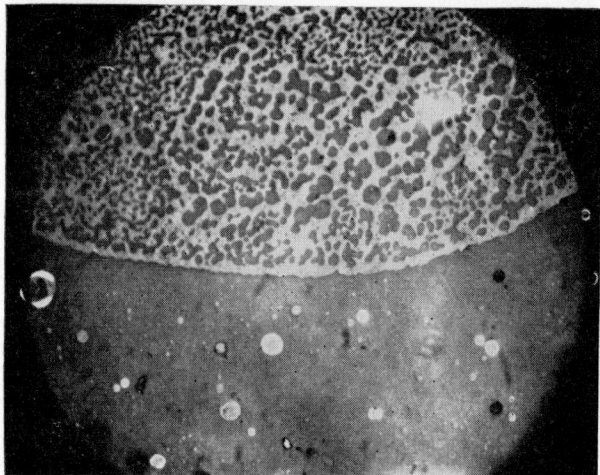

FIGURE 5.—Liquation in lke $SmO–SiO_2$ system.

of compounds of a metasilicate ($Ln_2O_3 . 3SiO_2$) type developing at 1300°C and below.

The compounds of divalent and trivalent rare-earth silicates are refractory materials (the melting temperatures range from 1750° to 2000°C), and are resistant to the action of a number of corrosive media; they have high refractive indices (from 1·79 to 2·0), high densities (4·5–7·0) and high microhardnesses.

We have investigated some optical properties (absorption and luminescence spectra) of lanthanide silicate single crystals and aluminium-silicate glasses. Absorption and radiation spectra are very important for studying the internal structure of the crystal. At the transition between the ion energy levels, corresponding to the properly protected internal f shells, characteristic spectra with the pronounced discrete structure are obtained. This structure is closely connected with the internal structure of the environment in which the ion is found. By changing the environment of the rare-earth ions we can observe the displacement of the line positions in the bands and of the bands as a whole, the effect of splitting of the electronic levels by the internal electric fields and a number of other phenomena.

The absorption spectra correspond to definite energy transitions; for the samples investigated they comply with the forbidden transitions between the f shell levels of the free trivalent ions, but they are possible for the ions in the field of a crystal or a glass. The absorption spectra of single crystals of $Nd_2O_3 . SiO_2$ and of the $Nd_2O_3–Al_2O_3–SiO_2$ glass in the wave-length range 400 to 750 m$\mu$ consist of a number of

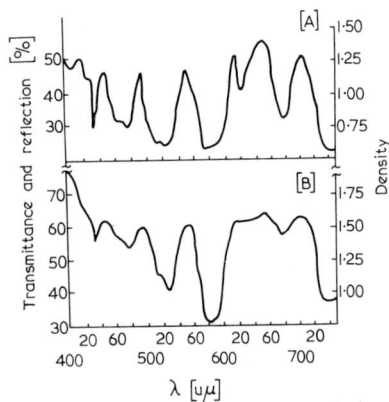

FIGURE 6.—Absorption spectra:
A. Single crystals of $Nd_2O_3 . SiO_2$.
B. $Nd_2O_3–Al_2O_3–SiO_2$ glass.

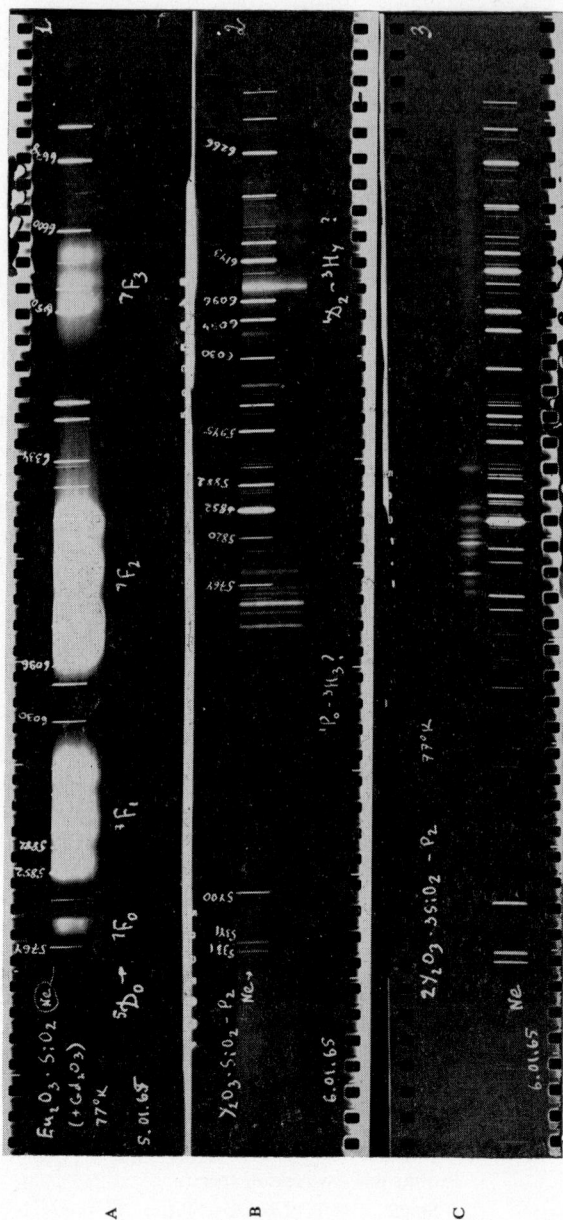

FIGURE 7.—Radiation spectra:

A. $Eu_2O_3.SiO_2(+Gd)$    B. $Y_2O_3.SiO_2(+Pr)$    C. $2Y_2O_3.3SiO_2(+Pr)$

bands (see Figures 6A and B). Their peak positions are in a good agreement with the electronic level diagram for $Nd^{3+}$. The different distribution of the intensities of separate absorption bands of trivalent neodymium in single crystals and in glass should be noted. Whereas with single crystals all the bands are of about the same intensity, the neodymium glass has a number of dominating bands. This phenomenon can be explained by the difference in the nearest environment of $Nd^{3+}$ in these media which in turn stipulates different degrees of resolution of separate transitions.

The radiation spectra for europium, samarium, praseodymium and yttrium silicates have a complex structure. Thus the radiation spectrum of $Eu_2O_3.SiO_2$ ($+Gd$) consists of one comparatively narrow line (the transition $^5D_0 \rightarrow ^7F_0$) and three complex diffusion bands corresponding to the transitions. $^5D_0 \rightarrow ^7F_{1,2,3}$ (Figure 7A). The presence of the zero transition in the spectrum indicates that the centre symmetry is lower than the cubic.

The luminescence spectra of the single crystals $Y_2O_3.SiO_2$ and $2Y_2O_3.3SiO_2$ activated by praseodymium consist of a number of rather narrow bands belonging to the transitions in the $Pr^{3+}$ ion (Figure 7 B, C). As the photographs show, they are characteristic of both types of crystal. Such differences in the spectra indicate the essential difference in the environment of $Pr^{3+}$ ions in these compounds, which confirms pyrochemical and microscope data on the existence of two compounds (1:1 and 2:3) in the system $Y_2O_3$–$SiO_2$.

Although for complete identification of the crystalline field symmetry more delicate methods of analysis are necessary than those being used at present, the use of luminescence analysis in its present form can give essential additional information about the structure of systems of $Ln_2O_3$–$SiO_2$ type.

## REFERENCES

1. BONDAR, I. A., TOROPOV, N. A., and KOROLEVA, L. N., *Izv.AN SSSR, Inorganic Materials* **1**, (4), 561, 1965.
2. TENISHEVA, T. F., and LAZAREV, A. N., *Izv. AN SSSR, Inorganic Materials,* **1**, (4), 569, 1965.
3. BONDAR, I. A., TOROPOV, N. A., and KOROLEVA, L. N., *Izv. AN SSSR, Inorganic Materials.* In the press.
4. SHAFER, M. W., McGUIRE, T. R., and SUITE, J. C., *Phys. Rev., Lett.,* **11**, (6), 251, 1963.
5. SHAFER, M. W., and SUITE, J. C., *J.Amer. Ceram. Soc.,* **49**, (5), 261, 1966.
6. ACHARD, JEAN-CLAUDE, and ALBERT, L., *Compt. Rend, Ser. C.,* **262**, (13), 1066, 1966.
7. MORRISH, A. H., and SHAFER, M. W., *J.Appl. Phys.,* **36**, (3), 1145, 1965.

# 22.—Mullite Changes in Fireclay Bricks at Different Temperatures

By K. Konopicky, G. Routschka and I. Patzak

*Forschungsinstitut der Feuerfest-Industrie, Bonn, West Germany*

***ABSTRACT*** E12/A5354

*As temperature increases, $Fe_2O_3$ and $TiO_2$ in the mullite crystals in firebricks decrease and the mullite becomes richer in $Al_2O_3$; the $SiO_2$ content of the mullite remains unchanged; as temperature decreases, the situation reverses. At and above $1400°C$ the equilibrium between mullite and the surrounding liquid phase is established almost at once, whereas at lower temperatures, e.g. below $1300°C$, the time factor becomes important. At a given temperature the content of $Fe_2O_3$ in mullite is determined by the total $Fe_2O_3$ content in the brick; this influence decreases as temperature increases. The changes in the $TiO_2$ content of the mullite with temperature are much less than in the case of $Fe_2O_3$. Iron and titania in mullite alter the peak intensities in X-ray diffraction patterns; the intensities of the peaks can be correlated with the impurity content.*

***Modifications de la mullite dans les briques d'argile réfractaire à différentes températures***

*Quand la température s'accroît, la quantité de $Fe_2O_3$ et $TiO_2$ présente dans les cristaux de mullite des briques d'argile réfractaire diminue et la mullite s'enrichit en $Al_2O_3$; la teneur en $SiO_2$ de la mullite reste inchangée; quand la température décroît, la situation se renverse. A $1400°C$ et au-dessus, l'équilibre entre la mullite et la phase liquide qui l'entoure est établi presque'à l'instant tandis qu'aux températures plus basses, par exemple au-dessous de $1300°C$, le facteur temps devient important. A une température donnée, la teneur en $Fe_2O_3$ de la mullite est déterminée par la totalité de $Fe_2O_3$ présent dans la brique; cette influence diminue à mesure que la température augmente. Les modifications, avec la température, de la teneur en $TiO_2$ de la mullite sont bien moindres que dans le cas de $Fe_2O_3$. La présence du fer et de l'oxyde de titane dans la mullite modifie les intensités des pics dans les diagrammes de diffraction des rayons X; celles-ci peuvent être mises en corrélation avec la teneur en impuretés.*

## Mullitumwandlungen in Fireclay-Steinen bei verschiedenen Temperaturen

*Mit steigender Temperatur nimmt der $Fe_2O_3$- und $TiO_2$-Gehalt von Mullitkristallen in feuerfesten Steinen ab und der Mullit wird $Al_2O_3$-reicher, während der $SiO_2$-Gehalt unverändert bleibt. Sinkt die Temperatur, kehrt sich das Ganze um. Über und bei 1400°C besteht ein Gleichgewicht zwischen Mullit und umgebender flüssiger Phase, während bei niedrigeren Temperaturen, z.b. unter 1300°C, der Zeitfaktor eine wesentliche Rolle spielt. Bei einer bestimmten Temperatur wird der $Fe_2O_3$-Gehalt des Mullits durch den Gesamtgehalt an $Fe_2O_3$ im Stein bestimmt. Dieser Einfluß nimmt mit zunehmender Temperatur ab. Die Änderung des $TiO_2$-Gehaltes des Mullits mit der Temperatur ist viel geringer als bei $Fe_2O_3$. Eisen und Titan im Mullit verändern die Intensitäten der Röntgenlinien. Diese Intensitäten können mit dem Fremdstoffgehalt in Beziehung gebracht werden.*

Occasionally in the literature the "self-purification" of mullite with increasing firing temperatures[1,2] has been reported but the investigations have been exclusively on slowly cooled samples of fireclay bricks or fired clays.[3] We heated small pieces of fireclay bricks at different temperatues and then quenched them in water. At and above 1400°C the equilibrium between mullite and surrounding liquid phase is established almost at once, whereas at lower temperatures, e.g. below 1300°C, the time factor becomes important. To complete the investigation the samples quenched from 1500°C were tempered at low temperature for 4 h, after which they were quenched in water; other samples were cooled slowly from 1500°C to 1200°C, at 0·2°C/min. All the tests were carried out in an oxidizing atmosphere (air). The fired samples were crushed to 60$\mu$m and the mullite was isolated by means of 10% HF and analysed.

Figure 1 shows an evident decrease in the content of $Fe_2O_3$ of mullite as the firing temperature increases which—even with the high-fired bricks as delivered (A) was comparatively high. It is surprising that, when cooled slowly from 1500°C to 1200°C, the high iron content of the mullites—present in the bricks as received—is restored.

The dashed line represents the results obtained by quenching the 1500°C sample in water after soaking for 4 h at various temperatures. Annealing at 1400°C the sample that had been quenched from 1500°C and firing the brick, as delivered, at 1400°C results in the same content of $Fe_2O_3$ in the mullite. It appears that, when the quenched samples are annealed at temperatures below 1300°C,

diffusion is so slow at these temperatures that annealing for 4 h
is not sufficient for equilibrium to be attained. At a given temperature the content of $Fe_2O_3$ in mullite is determined by the total $Fe_2O_3$
content in the brick; this influence decreases as temperature increases.

To determine the $TiO_2$ content of the mullite of fireclay bricks,
isolated by means of $10\%$ HF, is difficult because there is usually
still some free $TiO_2$ (rutile). In the literature the free $TiO_2$ is

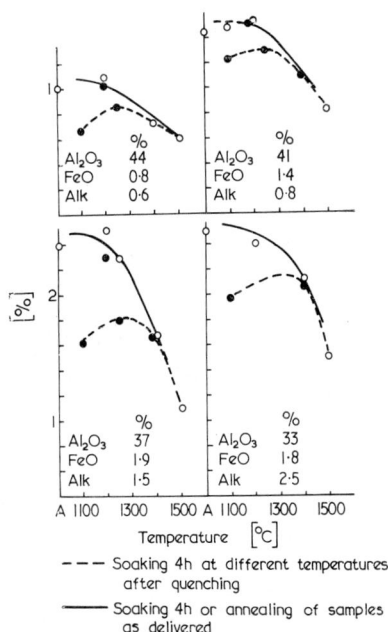

FIGURE 1.—Percentage of $Fe_2O_3$ in mullite as a function of temperature
(fireclay bricks).

reported as being between 1 and $1\cdot5\%$. When mullite is isolated by
means of $10\%$ HF, the rutile remains undissolved and is determined
as $TiO_2$ in the mullite. With a few exceptions our test bricks still
contained enough rutile to be detected by X-rays but, after being
fired at 1400°C and above, it had disappeared.

The $TiO_2$ content of mullite does not decrease as quickly with
firing temperature as does the iron content (Figure 2), but again we
can recognize $TiO_2$ going into solution in the mullite when cooling.
The higher the flux content of the brick, the greater the proportion
of $TiO_2$ dissolved. On the basis of these investigations we can say

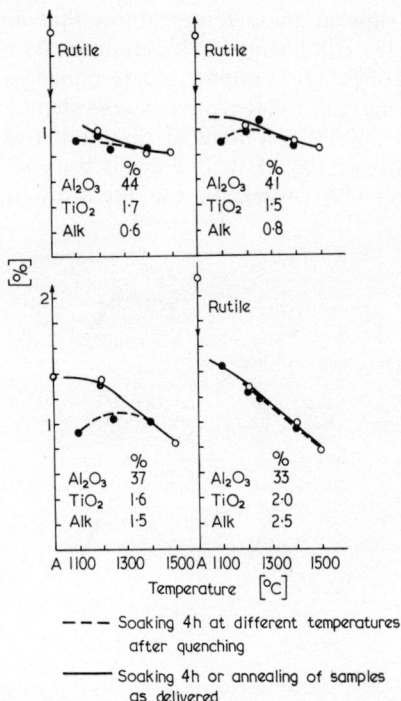

FIGURE 2.—Percentage of $TiO_2$ in mullite as a function of temperature.

that the maximum content of $TiO_2$ in the mullites in fireclay bricks is about 1%. As only a small fraction of $TiO_2$ is incorporated in the structure of mullite during cooling, in bricks with a low alkali content the liquid phase will become rich in $TiO_2$ as a result of segregation of cristobalite.

The total analysis of the isolated mullites shows that, as $Fe_2O_3$ and $TiO_2$ are taken up, the $Al_2O_3$ content changes, but the content of $SiO_2$ remains constant at about 28% (3/2 mullite). This means the $Al_2O_3$ content of mullite in fireclay bricks increases with increasing temperature because $Fe_2O_3$ and $TiO_2$ are released; whereas, on cooling, $Al_2O_3$ is released as other oxides are taken up, which leads to an increase in the mullite content of the brick or to the glass becoming enriched in alumina.

The iron and $TiO_2$ of the mullite affect the intensity of the X-ray interferences. In Figure 3 are compared the $I_{210}$ of mullite isolated by HF/HCl from the quenched and tempered samples and the effect of the foreign oxide in the mullite.

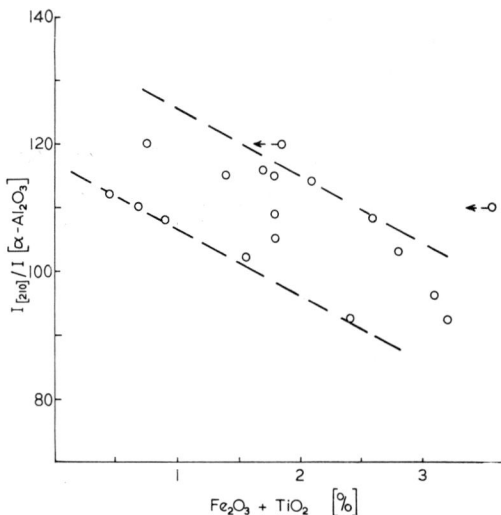

FIGURE 3.—Influence of impurity content on the height of the main peak of mullite (after treatment with HF/HCl).

When mullite is isolated by means of HF, the type of treatment affects the intensity of the mullite peak (Table 1). Boiling the separated residue, especially in the presence of HCl, gives intensities higher than those for samples treated with HF only. These differences result from dissolution of the surface layers of mullite crystals, which increases the intensity. Consequently when a chemically

Table 1

Influence of Chemical Treatment on the Peak Height of the 210 Reflection of Mullite

| Samples | Residue, separated in 10% HF | | | in 40% HF |
|---|---|---|---|---|
| | Boiled in $H_2O$ | Washed with HCl | Boiled in HCl | Boiled in HCl |
| Synthetic mullite | 119 | 114 | 135 | — |
| Fireclay brick:  A | 94 | — | 109 | — |
| B | 92 | — | 116 | 120 |
| C | 86 | — | 96 | — |
| D | 84 | 100 | 114 | 126 |

326    KONOPICKY, ROUTSCHKA AND PATZAK

isolated mullite is used as a standard for quantitative X-ray analysis, the chemical treatment has to be carried out carefully, otherwise the resulting values will be too low.

## REFERENCES

1. BOSE, A. K., MÜLLER-HESSE, H., and SCHWIETE, H. E., *Arch. Eisenhüttenwesen*, **27**, (10), 665, 1956; **28**, (10), 667, 1957.
2. KONOPICKY, K., *Ber. Dtsch. Keram. Ges.*, **36**, (11), 367, 1959.
3. McGEE, T. D., *J. Amer. Ceram. Soc.*, **49**, (2), 83, 1966.

*Part Five*

# DEVELOPMENT OF CERAMICS FOR THE NEWEST APPLICATIONS

# 23.—Defect Structures in Ceramic Materials

By L. E. J. ROBERTS

Chemistry Division, U.K.A.E.A. Research Group,
Atomic Energy Research Establishment, Harwell

## ABSTRACT A511

*Special properties of different classes of ceramics have been exploited in a number of new applications over the past 20 years. The actinide dioxides provide the most important nuclear fuels and doped rare-earth and zirconium oxides are being developed for use in furnaces, M.H.D. electrodes, fuel cells and galvanic cells. The stability of highly defective phases derived from the fluorite structure may depend on the ability of cations having different charges to exist in a favourable local configuration in one host lattice. The phase relationships and electrical properties are determined by the tendency towards short- and long-range order. Some contrasting properties of the semimetallic carbides and nitrides of the transition and actinide elements are also reviewed.*

## Défauts de réseau de matériaux céramiques

*Au cours des vingt dernières années les propriétés particulières de divers groupes de matériaux céramiques ont été mises à profit dans un certain nombre d'applications nouvelles. Des oxydes à conductibilité électrique élevée ont été mis au point pour des fours et des convertisseurs M.H.D. Les oxydes d'actinides fournissent le combustible nucléaire le plus important; les propriétés magnétiques et électroniques de plusieurs systèmes d'oxydes mixtes sont largement exploitées par les industries des télécommunications. Le comportement des ions des métaux de transition et des défauts de réseau est par conséquent extrêmement important. Il existe de plus en plus de données relatives aux facteurs influant sur l'occupation des sites, l'énergie de formation de défauts, l'agglomération de défauts et les mécanismes de migration. La recherche des matériaux utilisables à des températures très élevées, a abouti à de nombreux travaux sur les carbures réfractaires; quelques systèmes à base de carbures, nitrures et sulfures présentent des propriétés électriques et magnétiques inhabituelles qui pourront être mises à profit.*

## Fehlstellen von keramischer Materialien

*Spezielle Eigenschaften verschiedener Klassen keramischer Materialien wurden in einer Anzahl neuer Anwendungsmöglichkeiten im Laufe der vergangenen zwanzig Jahre ausgenutzt. Oxide mit hoher elektrischer Leitfähigkeit wurden für den Gebrauch in Öfen und magnetohydrodynamischen Anordnungen entwickelt; die Oxide der Actiniden erweisen sich als wichtigste Kernbrennmaterialien, die magnetischen und elektronischen Eigenschaften einiger Mischoxidsysteme werden in weitem Maß in der Nachrichtenindustrie verwendet. Das Verhalten von Übergangsmetallionen und von Gitterfehlern hat daher große Bedeutung erlangt. Das beweist die Anhäufung von Faktoren, die die Platzbesetzung, die Fehlstellenbildungsenergie, die Anhäufung von Fehlstellen und den Wanderungsmechanismus beeinflussen; dieser Befund wird behandelt. Die Suche nach Materialien zur Benutzung bei sehr hohen Temperaturen hat zu vielen neuen Arbeiten über Feuerfestkarbide geführt und einige Karbid-, Nitrid- und Sulphidsysteme zeigen ungewöhnliche elektrische und magnetische Eigenschaften, die weiter ausgenutzt werden können.*

# 1. INTRODUCTION

The traditional uses of ceramics exploited their mechanical properties—their high strength at high temperatures—and their chemical inertness. In the last 20 years or so, many new uses of refractory materials have arisen and research into quite different physical properties has already reached a high level of sophistication. Oxide dielectrics, semi-conductors, ferromagnets and ferrimagnets are used throughout the electronic industries; refractory compounds of the actinide elements are highly favoured as nuclear fuels and oxides which are good conductors at high temperatures are being exploited as furnace elements in M.H.D. ducts and in fuel cells.

Many papers to this Conference deal with some aspect of these new uses of ceramic materials. The general field of dielectric, magnetic and semi-conducting oxides is reviewed in the paper by Jonker and Stuijts. The present paper is therefore devoted to some general ideas which underlie the properties and the performance of refractories as nuclear fuels and as electrical conductors. The greater part of the paper is a discussion of the oxide systems, about which most is known at present, though other compounds are rapidly assuming comparable importance, at least for nuclear applications.

Nuclear fuels are required to operate in very large temperature gradients, to withstand temperature cycling without gross distortion and to withstand very heavy radiation damage. A full discussion of this last requirement would be out of place in this paper; irradiation produces very large concentrations of defects in a fuel structure, and results in the introduction of large concentrations of foreign atoms, so a successful fuel material must possess a simple, isotropic (cubic) structure capable of rapid self-healing at the operating temperatures. Ceramic electrodes and electrolytes may also have to operate in temperature gradients and certainly have to withstand the passage of heavy currents and operate in steep gradients of chemical potential, from oxidizing to reducing conditions. Optimum properties are achieved in structures having a very high defect concentration, a condition also common to nuclear fuels after long periods of irradiation. Recent work on the nature and stability of defect structures is accordingly reviewed here.

## 2. NON-STOICHEIOMETRIC OXIDES

The oxides of interest in nuclear and in many electrical applications, $UO_2$, $PuO_2$, rare-earth oxides and doped zirconias, are all based on the fluorite structure. In most cases a cubic phase field of very wide composition range is stable at high temperatures. The fluorite structure can tolerate very large disorder of the anion sub-lattice, either of anion vacancies or of interstitial anions, but the cation sub-lattice remains essentially perfect except, possibly, at the highest temperatures.

### 2.1 Diffusion Measurements

The physical and chemical properties of the fluorite-type oxides at temperatures below 1500°C are dominated by the very high mobility of the oxygen. Some representative values are collected in Figure 1. Recent results on the self-diffusion of oxygen in $UO_2$ crystals close to stoichiometry ($UO_{2.001}$) and containing a large excess of interstitial oxygen ($\sim UO_{2.05}$) are in reasonable agreement with previous values obtained by AUSKERN and BELLE at lower temperatures, using $UO_2$ powders.[1] The results plotted for lime-stabilized zirconia, containing a high proportion of anion vacancies, are those reported by CARTER, using single crystals.[2] The values are close to those reported by other workers and are in good agreement with values calculated from measurements of the electrical conductivity assuming conduction by oxygen ions only with a transport number of unity. The diffusion coefficients for oxygen in pure tetragonal zirconia were calculated from measurements of the electrical conductivity of single crystals reported by ANTHONY[3] on the assumption that, at high temperatures, the conductivity was ionic. The self-diffusion coefficient for Zr in stabilized zirconia was briefly reported by RHODES and CARTER.[4] There is some disagreement concerning the diffusion coefficients of uranium in $UO_2$; the values plotted in Figure 1 were taken from ALCOCK and HAWKINS,[5] who used single crystals of stoicheiometric $UO_2$ and a technique which allowed them to discriminate between lattice and grain-boundary diffusion; the diffusion rate of U down grain boundaries was reported to be higher than the lattice diffusion plotted in Figure 1 by about two orders of magnitude, though the energy of activation was the same.

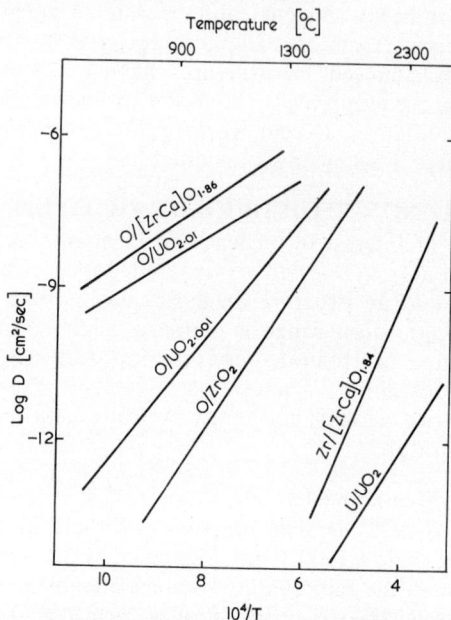

FIGURE 1.—Oxygen and Cation diffusion in fluorite-type oxides.

Diffusion in non-stoicheiometric oxides of this structure, containing either lattice vacancies or interstitial oxygen, is thus very fast and, in either case, the activation energy is 25–30 kcal/mole; this is a fortuitous circumstance, since the mechanisms of vacancy and of interstitial movement are almost certainly different. Close to stoicheiometry, the energy of activation of oxygen diffusion is higher ($\sim$65 kcal/mole) owing to the energy necessary to form intrinsic defects, but oxygen diffusion is still many orders of magnitude faster than metal ion diffusion, and remains so up to the melting point. These figures explain why it is so difficult to quench in the high-temperature structure of the non-stoicheiometric phases: oxygen can diffuse a mean distance of 1 $\mu$m in 1 sec. even at $\sim$900°C and so phase disproportionation reactions depending on variations in oxygen composition are very rapid. Conversely, the cation sub-lattice is virtually "frozen" at temperatures below 1200°C, at which temperature movement of 0·1 $\mu$m would take a given cation about 300 h. Thus it is quite possible for different preparations of the same oxide to contain different concentrations of cation defects, or different degrees of disorder on the cation sub-lattice in the case of a solid solution, dependent upon the method of preparation and thermal history of the sample. Phase separations and ordering processes depending on cation movement will be very sluggish.

## 2.2 Thermodynamic Stability

In common with all transition-metal oxides, the compositions of the actinide oxides and rare-earth oxides such as the Ce and Pr oxides vary according to the oxygen potential of the ambient atmosphere, since more than one valency state is easily accessible. Control of the composition, and hence of many physical properties, during preparation, firing and use demands close control of the composition of the surrounding atmosphere.

The thermodynamic data for the $UO_{2+x}$, $PuO_{2-x}$ and $(U, Pu)O_{2\pm x}$ phases up to temperatures of 1200°–1500°C have been reviewed in detail elsewhere.[6] The phase diagram of the U–O system has been extended to higher temperatures[7] and it has been shown that a sub-stoicheiometric range of $UO_{2-x}$, analogous to $PuO_{2-x}$ and $CeO_{2-x}$, exists at temperatures above 1500°C if the conditions are sufficiently reducing. A selection of the results on the U–O phase is shown in Figure 2; the oxygen potentials in the $UO_{2-x}$ region are estimated from recent results of MARKIN and WHEELER.[8] The full lines in Figure 2 show the variation of the oxygen potential, or partial molal free energy of oxygen ($\Delta \bar{G} = RT \ln p_{O_2}$) with temperature for the compositions shown; the slope of this line at any temperature is the partial molal entropy of solution of oxygen in the solid phase at the defined composition. It is an experimental fact that the value of the partial molal entropy remains constant over several hundreds of degrees, and the linear extrapolation of the experimental results for the partial molal free energy which is used in Figure 2 is then a reasonably good approximation, though not entirely accurate. Also plotted in Figure 2, as broken lines, are the oxygen potentials of different gas mixtures.

The wide separation of the $UO_{2+x}$ and $UO_{2-x}$ regions in Figure 2 illustrates the range of conditions under which $UO_2$ can be prepared as an essentially stoicheiometric phase. Hydrogen containing 1% of water vapour is in equilibrium with $UO_2$ very close to stoicheiometry at all temperatures. Dry hydrogen, containing $10^{-2}$% of water vapour, will reduce $UO_2$ appreciably at 1800°C or above, though U metal will not appear unless the resulting $UO_{2-y}$ phase is quenched to lower temperatures. Very wet hydrogen, or $CO_2/CO$ mixtures rich in $CO_2$, are appreciably oxidizing to $UO_{2.00}$ at high, but not at low temperatures.

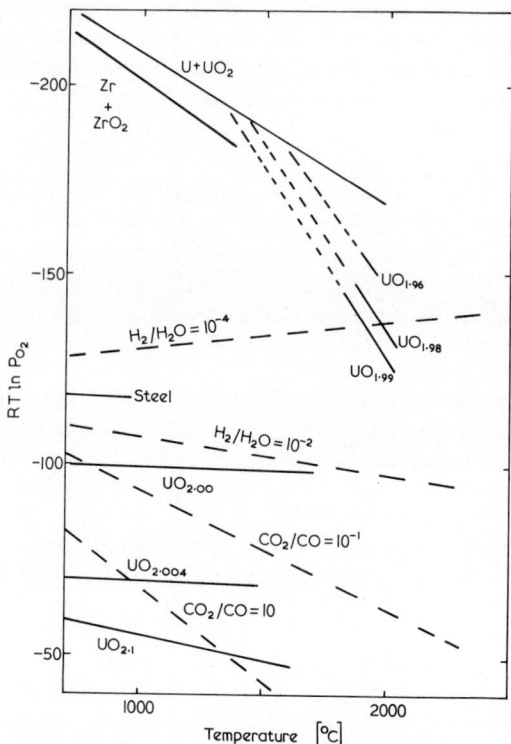

FIGURE 2.—Partial molal free energy of oxygen in equilibrium with oxides of composition $UO_{2\pm x}$ and in equilibrium with various gas mixtures.

As well as defining the conditions of preparation, thermodynamic data of this sort can give some indication of the composition changes that occur in use, in strong temperature gradients. A precise solution to the problem of the redistribution of oxygen in a temperature gradient must be obtained by the methods of non-equilibrium thermodynamics and such a calculation cannot yet be done. However, two effects can be foreseen: (1) oxygen will tend to migrate down a gradient of oxygen potential and (2) in practice, redistribution of oxygen through the gas phase will be important. This is so because a fuel element is sheathed in an impervious can and the inter-connecting residual porosity, together with cracks and crazing due to thermal shock, provides a means of access for the gas in the can to the fuel in both high- and low-temperature regions. MARKIN has shown[9] that the carbon impurity content of the average oxide fuel will be sufficient to generate appreciable pressures of CO and $CO_2$, and so rapid transfer of oxygen through the gas phase is a possibility.

Figure 2 shows that the distribution of oxygen will be very different in the two cases of control by gas–solid equilibration and by the oxygen potential gradient. In the first case, the composition of the gas can be assumed to be constant and the solid composition will therefore tend to be that in equilibrium with a $CO/CO_2$ mixture of constant composition. An oxide that is initially stoicheiometric will remain so in any temperature gradient, particularly in a steel can which will not be appreciably oxidized by a $CO/CO_2$ mixture in equilibrium with $UO_{2\cdot00}$. If the oxide contains excess oxygen, gas equilibration will tend to concentrate the excess oxygen in the hotter regions of the fuel element, i.e. in a direction which is opposite to that of the oxygen potential gradient. If the fuel element is surrounded by an oxygen "sink", such as a can of zirconium metal, then some oxygen will be abstracted from the fuel, to an amount determined by kinetic considerations. Conditions in the fuel then become very reducing, and if the temperature of the hot centre exceeds that at which oxygen can be lost from $UO_2$, it seems likely that $UO_{2-y}$ will be formed, leading to the precipitation of U metal when the fuel cools down.

Some data for the Ce–O system, taken from BEVAN and KORDIS,[10] are shown in Figure 3A. A cerium oxide refractory in a static gas atmosphere will show some redistribution of oxygen in a temperature gradient in the same sense, but not nearly as steeply, as the gradient of oxygen potential would predict. Appreciable reduction of $CeO_2$ occurs at 700° and above, even in a 1/1 mixture of $CO_2$ and CO, and therefore appreciable reduction of ceria electrolytes can be predicted in fuel cell applications, leading to electronic as well as to ionic conductivity. Zirconia-based oxides are, of course, more difficult to reduce. Figure 3B includes the oxygen potentials over the oxide in equilibrium with Zr metal and also, in the hatched section, the oxygen potentials at which some reduction of stabilized lime–zirconia or yttria–zirconia occurs, as shown by the appearance of electronic as well as ionic conductivity. The results in the hatched area were collected by ALCOCK and STEELE,[11] and experience in our laboratories is largely confirmatory, at least of the lower (more negative) oxygen potentials at which reduction can occur. Under conditions commonly found in fuel cells ($CO/CO_2 \sim 10$) no reduction of the zirconia electrolyte should occur below 1000°C, though under more reducing conditions, some reduction may occur at 700°C. At high temperatures, above 1500°C, some reduction should occur even in a good vacuum ($10^{-8}$ atm.).

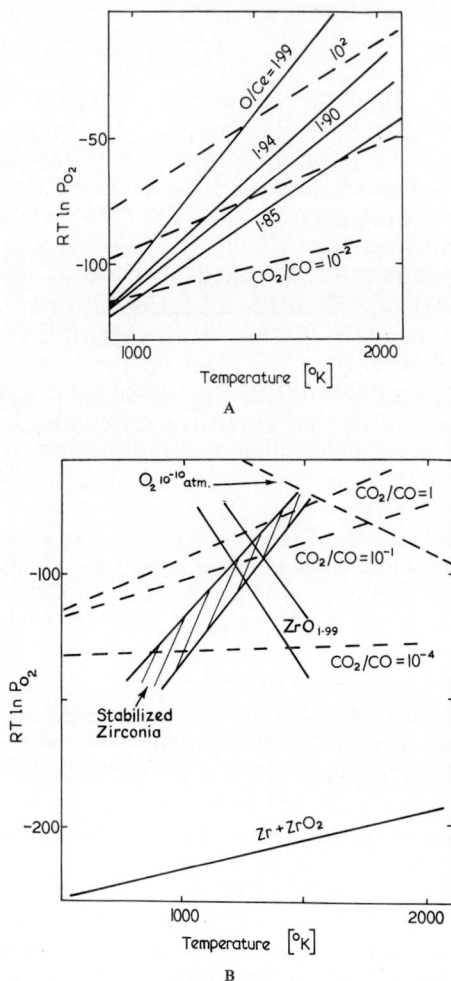

FIGURE 3.—Partial molal free energy data for (A) cerium oxides and (B) the Zr/ZrO$_2$ and cubic stabilized zirconia.

There has been one study of the variation of composition of ZrO$_2$ with oxygen potential[12] and the results reported for ZrO$_{1.98}$ and ZrO$_{1.99}$ are plotted in Figure 3B. These results are not thought to be truly representative of such equilibria in zirconia, and they could not be repeated by KOFSTAD and RUZICKA.[13] As stated, these results for oxygen-deficient zirconias would predict extensive reduction at low temperatures, with a positive value for the partial molal entropy of oxygen, in sharp contrast to the values found for $CeO_{1.99}$ and $PuO_{1.99}$ (see Figure 3A). The reason for these anomalous results on $ZrO_{2-x}$ may be due to the samples used; from some information published in the paper, the experiments were performed using $ZrO_2$ powder of 4·2 m²/g surface area.[12] Previous work in our laboratories has shown that appreciable amounts of oxygen can be removed from the surface of $ThO_2$ at 800°C under mildly reducing conditions where reduction of the bulk oxide is impossible, and Rand and Jackson have characterized the excess oxygen content of samples of "$PuO_{2.09}$" as being due to surface chemisorption on powder samples of 12–15 m²/g surface area. Amounts of oxygen equivalent to changing the O/Zr ratio by 0·01–0·03 may then be characteristic only of surface processes if the surface area is as high as 4 m²/g, and the curious temperature coefficient reported may well be due to some change of the surface characteristics with time; the chemisorption results are not normally as reproducible as are the bulk properties.[13]

The type of conduction shown by $CeO_{2-x}$ and by zirconias at very low oxygen potentials is n-type (electron excess). $UO_{2+x}$, containing an excess of oxygen, is a p-type semi-conductor. P-type conduction has also been found at high oxygen pressures (above $10^{-4}$ atm.) for $ThO_2-Y_2O_3$ solid solutions[11] as well as for a variety of other oxides.[15] The reason for such an observation in the case of $ThO_2$- or $ZrO_2$-based oxides is not known. In the absence of impurity effects, it is possible that some surface-controlled adsorption effect, of the type discussed above, is operative at surfaces, grain boundaries, etc.

### 3. DEFECT STRUCTURES AND ORDERING

There is a growing body of evidence that the model of defect sites introduced at random into a near-perfect matrix is inadequate to account for the properties and behaviour of highly non-stoicheiometric oxides such as the ones considered here. The defects order at low temperatures to give complex superstructures and there is some evidence, reviewed below, of the persistence of local ordering even at high temperatures. The tendency is for a given cation to assume the co-ordination typical of another simple phase in the metal–oxygen system. The terms "vacancy" and "interstitial" are to some extent misleading, though they serve to relate the overall density of the non-stoicheiometric phase to that of the simple structure to which it is related. The best documented examples in oxide chemistry are the so-called "shear" structures of the Mo and W oxides, related to the simple $ReO_3$ structure by the inclusion of planes of 4-co-ordinated metal atoms. The structural relationships in the fluorite systems seem to be rather different.

## 3.1 Uranium Oxides

The structure of the disordered $UO_{2+x}$ phase has been determined by WILLIS by a neutron-diffraction study of crystals held at temperatures where the $UO_{2+x}$ phase is truly stable.[16] The results can be interpreted on the basis of the positions in a fluorite-type cell with a U atom at (0, 0, 0) and lattice oxygens on each $(\frac{1}{4}, \frac{1}{4}, \frac{1}{4})$ position. Interstitial positions of two types were found to be occupied, one (O') displaced from the centre along a $<1, 1, 0>$ direction and the other (O'') displaced from the $(\frac{1}{4}, \frac{1}{4}, \frac{1}{4})$ positions along a body diagonal of the cell. At the same time, vacancies appeared on the normal oxygen sites. Geometrical considerations alone preclude oxygen ions being placed at random on all these positions and the simplest geometrical model that accounts for the diffraction results is that the O' atoms displace the two nearest lattice oxygen atoms into O'' sites. There is some evidence—though not conclusive—that the O' atoms enter in pairs and the local "defect complex" is then that sketched in Figure 4A. At low temperatures, the "defect complexes" order into the complex structure of $U_4O_9$, in which the uranium atoms are also displaced slightly from the fluorite positions. The neutron-diffraction results show clearly that the same type of "interstitial" sites are occupied in $U_4O_9$ as in $UO_{2+x}$, though the complete ordered structure has not been elucidated.

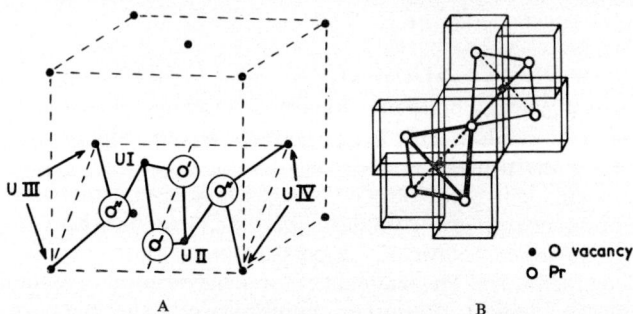

FIGURE 4.—Defect complexes in the fluorite structure: (A) Interstitial oxygen atoms in $UO_{2+x}$ (B) cations surrounding oxygen vacancy pairs in ordered $MO_{2-x}$ systems, such as $PrO_{2-x}$. [Ffg. 4B. is reproduced by permission from ref. 20[

Examination of the structure in Figure 4A reveals that the O' atoms are very close to 2 U atoms which are themselves nearest neighbours; the O'–U distance is 2·18Å compared to 2·37Å in $UO_2$. The asymmetrical structure and the short U–O' distance suggests definite bonding between the O' oxygen and the two nearest U atoms. Additional evidence in support of this comes from thermodynamic data on oxygen dissolved in (U, Th)$O_{2+x}$ solid solutions. The structure of Figure 4A is only possible so long as there are U–U nearest neighbours; hence a large change in energy of bonding might be expected at large dilutions of $UO_2$ in $ThO_2$. This is experimentally true; the partial molal enthalpy of solution of oxygen in (U, Th)$O_{2+x}$ remains similar to that in $UO_{2+x}$ except at concentrations of $UO_2$ in $ThO_2$ less than 10%, when the values decrease sharply.[17]

Some estimate of the energy of disordering of the $U_4O_9$ phase to $UO_{2+x}$ can be made from thermodynamic data values for the heat and entropy of ordering of one mole of $UO_{2\cdot25}$ are $-0\cdot7$ kcal and $-0\cdot5$ e.u.[18] These are not large values and are of the same order as values for many phase transitions.

## 3.2 Rare-earth Oxides

The rare-earth oxides are used in nuclear reactor control rods, in fuel cells and in various laser and electrical applications. A large amount of work has been done on the structural, kinetic and thermodynamic properties of the systems. A series of ordered phases of the general formula $M_nO_{2n-2}$ has been discovered in each of the systems Ce–O, Pr–O and Tb–O, between the end-members $MO_2$ (fluorite) and $M_2O_3$ (hexagonal, A-type or b.c. cubic, C-type).[19] The structure of the b.c.c. $M_2O_3$ oxides may be regarded as derived from the fluorite structure by the ordered omission of oxygen. The oxygen "vacancies" exist in pairs, each pair on the body diagonal of a unit cell, and the structure is made up of strings of these vacancy-pairs arranged in each of the $<111>$ directions.[20] Each cation is then 6-co-ordinated, surrounded by a very distorted octahedron of oxygen. The stable rhombohedral $M_7O_{12}$ structure has also been shown to consist of strings of the body-diagonal vacancy pairs, but all parallel to one of the $<111>$ directions; in this case, $\frac{1}{7}$ of the cations is 6-co-ordinated and $\frac{6}{7}$ are 7-co-ordinated. The maximum electrical conductivity of the Pr–O system occurs at this composition, and it seems likely that the 6-co-ordinated ion is $Pr^{3+}$ with $Pr^{4+}$ and $Pr^{3+}$ equally on the 7-co-ordinated sites.

The structure of all the intermediate phases may be built up of strings of coupled vacancies in this way; removal of 2 oxygen ions from the eight oxygen ions surrounding $1/n$th of the cations means that $1/n$th of the oxygens are missing. The stoicheiometry therefore becomes $MO_{2-2/n}$, or $M_nO_{2n-2}$. The actual structure of the disordered phases stable at high temperatures can only be determined by neutron diffraction carried out at high temperatures, but it does seem plausible that a stable structural unit consists of two oxygen vacancies along a body diagonal, with the local structure of cations that is shown in Figure 4B.

The energy and entropy of ordering cannot be deduced from the available thermodynamic data. The very large energy and entropy changes that occur on reduction of $CeO_2$ or $PuO_2$ may be consistent with the occurrence of comparatively large regions of local order in the disordered $MO_{2-x}$ phase.[17] These may arise in the binary systems, since the $M^{3+}$ and $M^{4+}$ cations can also order readily, merely by the rapid migration of electrons. The entropy changes that occur during the reduction of a solid solution containing 30% of $PuO_2$ in $UO_2$ are far smaller than those found during the reduction of $PuO_2$ itself (Figure 5); it appears that the "defect complexes" in the $MO_{2-x}$ phases involve the local ordering of a comparatively large number of $M^{4+}$ and $M^{3+}$ cations, and the introduction of a $U^{4+}$ cation which cannot be reduced on 70% of the cation sites is enough to prevent their formation.

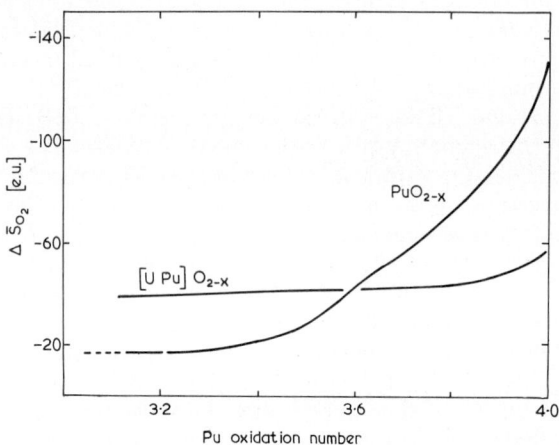

FIGURE 5.—Partial molal entropy of oxygen for the $PuO_{2-x}$ and $(U_{0.7}Pu_{0.3})O_{2-x}$ systems.

The formation of the highly ordered intermediate phases in the binary systems is not always easy, however; complicated and reproducible hysteresis effects have been found in the Pr—O system, and also the occurrence of "pseudo-phases", which are metastable phases of higher symmetry and lower oxygen content than the truly stable low-symmetry phases. These phenomena are thought to be connected with the ordering of the "strings" of coupled vacancies.[20]

In particular, it appears that the transition from a non-stoicheiometric fluorite phase to a non-stoicheiometric b.c.c. (Type C) structure is not continuous except, possibly, at the highest temperatures. A recent study of several $CeO_2-M_2O_3$ systems suggests that a miscibility gap between a fluorite and a b.c.c. phase is the rule and not the exception.[21] Evidence of some ordering within the fluorite phase itself was also obtained. New, diffuse, reflections were observed on the X-ray diffraction patterns of the quenched fluorite phases as the percentage of the $M_2O_3$ phase was increased beyond 10 mole %, and the cell constant showed a non-linear dependence on the $M_2O_3$ content, a property common to many $MO_2-M_2O_3$ systems[18] and not thought to be due to ordering of the cation sub-lattice. Evidence for local ordering of the oxygen vacancies in fluorite phases containing a high proportion of such vacancies is therefore quite strong, although the actual structure has not been determined.

### 3.3 Stabilized Zirconias

It is well known that CaO and a wide variety of $M_2O_3$ oxides can dissolve in $ZrO_2$ to give a cubic solid solution stable over a fairly wide range of composition that can be successfully quenched to room temperature. As the proportion of foreign oxide is increased, many other structures occur—e.g. $CaZrO_3$ (perovskite), $Gd_2Zr_2O_7$ (pyrochlore) and rhombohedral and hexagonal structures;[22] the $ZrO_2-Sc_2O_3$ system shows a particularly large number of intermediate phases.[23] A continuous transition from fluorite to type C structures has been postulated at high temperatures, but is not found at low temperatures;[22] additional lines have been reported in X-ray patterns of the fluorite phases, and further research may indicate a situation not unlike that found recently for the $CeO_2$-based systems. Indeed, the sequence of phases found in the $ZrO_2-Sc_2O_3$ system was compared to those found in the binary cerium oxides.

There is still some doubt on the solubility limits of CaO in $ZrO_2$ and the stability of the cubic solid solution at lower temperatures. DINESS and ROY have redetermined the $CaO-ZrO_2$ phase diagram in the presence of high-pressure water and found that monoclinic $ZrO_2$ could be made to precipitate from all cubic phases containing less than 10 mole % CaO at temperatures of about 1000°C; below 800°C, the phase diagram drawn shows monoclinic $ZrO_2$ and $CaZrO_3$ as the only stable phases.[24] It is pertinent to recall that other investigations in which equilibrium at low temperatures may have been attained or catalysed by a liquid phase have also shown low solubilities in $ZrO_2$ at low temperatures; ROMBERGER et al.[25] report very slight mutual solubility of $UO_2$ and $ZrO_2$ below 1100°C in experiments in which $UO_2$ and $ZrO_2$ were equilibrated with a molten fluoride flux, though previous solid-state studies indicated the existence of stable solid solutions of more than $10^m/_o$ $ZrO_2$ in $UO_2$ and $UO_2$ in $ZrO_2$, persisting to low temperatures.[26] Earlier work on the CaO–$ZrO_2$ system has been summarized by CARTER and ROTH[27] who redetermined the phase limits of the cubic phase on samples which had been sintered at 2050°C and annealed for long periods at lower temperatures. Carter and Roth reported the phase boundaries above 900°C to be 9·8 m/O CaO to 19·6 m/O CaO but found a wide scatter in the values of the lattice parameters of the cubic phase.

The defects present in the cubic zirconias have been proved by many investigators to be anion vacancies. A recent study of the densities and X-ray parameters of equilibrated CaO–$ZrO_2$ solid solutions by DINESS and ROY[28] has confirmed the anion-vacancy model for solid solutions quenched from 1600°C, but higher densities were obtained on some samples quenched from 1800°C, indicating the occurrence of interstitial cations and a profound change of defect type. Further experiments of this type would be of the greatest interest.

Despite some uncertainty regarding the phase diagram, the main features of the electrical properties of the cubic zirconias are clear. Experiments on both single crystals and sintered samples have shown that there is a maximum in the conductivity at any temperature at compositions close to the lower phase boundary of the cubic solid solution; with increasing MO or $M_2O_3$ content, and hence increasing vacancy concentration, the conductivity decreases despite the firm evidence that the transport number of the oxygen ions remains unity. This result has also been found for $ThO_2$–CaO and $ThO_2$–$Y_2O_3$ solid solutions, in which case the conductivity can be measured down to very low vacancy concentrations since the solid solution range extends to 100% $ThO_2$. The conductivity in the solid solutions reached a maximum at between 3 and 4% of anionic vacancies.[11] The major reason for the decline in conductivity at high CaO contents in the $ZrO_2$–CaO solid solutions seems to be the increase in activation energy for migration. There is reasonably good agreement on activation energies found by different authors, though the values for the conductivity vary widely. Figure 6 shows the results obtained by various investigators; the solid line plotted is that reported by CARTER and ROTH,[27] and the individual points are the results of other

authors.[29] This increase in activation energy has been discussed by
TIEN and SUBBARAO in terms of the cation environment of the oxygen
vacancy but, although undoubtedly dependent on the cation distribution in the crystal, it is not markedly dependent on the ordering
process discussed below.

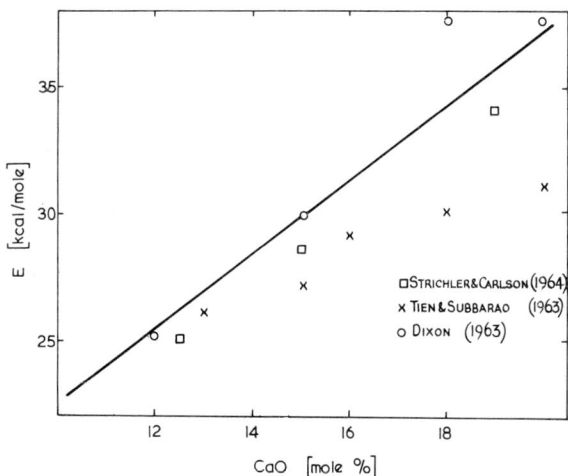

FIGURE 6.—Activation energy for oxide ion conductivity of lime-stabilized
cubic zirconia.

A decrease of conductivity with time of cubic zirconias held at
temperatures around 1000°C has been reported by many authors.
This phenomenon has been studied in great detail by CARTER and
ROTH,[27] who showed that the conductivity reached a stable minimum
after 36 h at 1100°C, but not until 2700 h at 904°C; the conductivity
remained 100% ionic and the activation energy of the conduction
process was not sensibly altered. The original value could be restored
by heating above 1400°C. Together with this change in conductivity
on annealing, distinct changes in X-ray reflections are observed[27,30]
and a diffuse background that always exists between the sharp
Bragg peaks on neutron-diffraction patterns becomes resolved into
new, distinct, reflections.[27] The neutron-diffraction results are not
consistent with the precipitation of $ZrO_2$ or $CaZrO_3$. It is clear that
a new structure is being produced on slow annealing below 1300°C,
and the neutron-diffraction results are consistent with the appearance
of ordered zones in the cubic solid solution phase field. Recent
neutron-diffraction and electron-microscope results reported by

CARTER[30] reinforce this view and show that the oxygen ions in the CaO–ZrO$_2$ solid solution have moved away from the fluorite positions.

So long as there is an abundant supply of oxygen, the current-carrying capacity of the cubic zirconias is high and outputs of 300 mA/cm$^2$ have been reported from operating fuel cells using zirconia electrolytes.[31] However, surface reduction of ZrO$_2$ can be achieved in the absence of oxygen and there is a possibility of cation disproportionation in large voltage gradients, though this has not been found experimentally.[32] Grain-boundary conduction by oxygen ions has been demonstrated, but this is not thought to make a large contribution to the conductivity of a dense electrolyte.[33]

## 4. CUBIC SEMI-METALLIC COMPOUNDS

In addition to non-stoicheiometric oxides, most of the transition, rare-earth and actinide elements form compounds with the non-metallic elements of Groups IV, V and VI with wide ranges of stoicheiometry. A very large family of such compounds have the rocksalt structure and the "ideal" formula is then MX$_{1.00}$. The applications of these compounds is limited to low temperatures in air because of their poor oxidation resistance at high temperatures. Many of the carbides are very hard and are used in the "hard metal" industry; their high melting points (e.g. TaC, $> 4000°C$) and low volatilities are exploited in high-temperature metallurgical operations when air can be excluded, and many of the actinide compounds are being assessed as nuclear fuel materials. A quite new field of application was envisaged in the recent report that some of the cubic nitrides are good Type II superconductors with relatively high transition points.[34] It is impossible to give an adequate survey of such an enormous number of compounds in this paper; a few examples are chosen to contrast with the oxide systems discussed in the preceding paragraphs.

The application of the actinide compounds as nuclear fuels follows from the high densities and high thermal conductivity of the NaCl-type compounds. The Table below lists some properties of the uranium compounds; UO$_2$ is included as a comparison

|  | UO$_2$ | UC | UN | US | UP |
|---|---|---|---|---|---|
| U density (g/cm$^3$) | 9·6 | 12·9 | 13·5 | 9·6 | 9·1 |
| Thermal conductivity (c.g.s.) |  |  |  |  |  |
| 200°C | 0·014 | 0·055 | 0·043 | 0·029 | 0·03 |
| 1000°C | 0·007 | 0·055 | 0·062 | 0·040 | 0·05 |

Because of favourable nuclear properties, the carbides are the chief competitors to oxides as nuclear fuels and a great deal of work has been done on their preparation, thermodynamic properties, and physical and chemical properties.[35] One feature of this research which is common to the studies of all the transition-metal carbides is the attention that must be paid to excluding oxygen and nitrogen, in order to establish the properties of the pure carbide phases.[36] There are three stable uranium carbides: UC, $U_2C_3$ and $UC_2$. $U_2C_3$ is the thermodynamically stable phase in contact with carbon for the temperature range $1000°–1750°C$ but $UC_2$ can dissolve oxygen to form a $UC_xO_y$ composition which is more stable than $U_2C_3$. Further, UC can dissolve oxygen to form a cubic $UC_xO_{1-x}$ phase and UC and UN can form a complete range of solid solutions; a cubic phase $UC_{0.64}N_{0.36}$ has been calculated to be in equilibrium with carbon at $1750°C$. Indeed, the use of oxycarbides or carbonitrides has been advocated in place of the pure carbide and this would certainly be feasible except at very high temperatures where CO or $N_2$ would be lost.

In common with many systems, the uranium carbides are less volatile than the oxides; this is illustrated in Figure 7, where the vapour pressure of $UO_2$[37] is compared with the uranium pressure over the congruently-evaporating composition of the carbide.[38] Even so, the uranium carbides are more volatile than many of the transition-metal carbides: ZrC is appreciably more stable.[39] For use at temperatures around $2000°C$ as cathodes in thermionic diodes employing nuclear heating UC has been dissolved in ZrC in order to lower the evaporation rate.

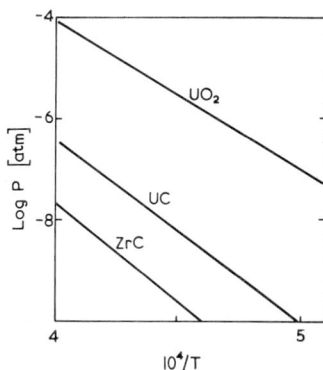

FIGURE 7.—Vapour pressures of $UO_2$, UC and ZrC.

Of the other uranium compounds, the nitride is the most obvious nuclear fuel because of the high density and the convenient rise in thermal conductivity with increasing temperature. Furthermore, both UN and UP can be simply made in a chemically-defined state by heating the higher nitride or phosphide in a vacuum at high temperatures. Neither the nitride nor the phosphide dissolves oxygen to any extent and both are considerably more resistant to oxidation and hydrolysis than is the carbide. A limitation, however, is the comparatively high dissociation pressure of $N_2$ gas at 1700°C and above. It is interesting to note that none of the uranium compounds UC, UN, US or UP show any appreciable range of homogeneity at low temperatures; PuC, ThC and ThN are very variable in composition but PuN and PuP are nearly stoicheiometric.

In general, the NaCl-type carbides and nitrides show very wide ranges of stoicheiometry (e.g. $ZrC_{0.95}$ to $ZrC_{0.55}$; $VN_{1.0}$ to $VN_{0.7}$) with very great differences in physical properties as the composition is altered. For example, the melting points can change by over 1000°, the hardness may be halved and the transition temperature at which superconductivity is observed can alter by many degrees.[34] In most cases where densities have been determined, the "defect" causing the stoicheiometry has been found to be non-metal "vacancies", but some reports of structures defective on both sub-lattices have been made; such a situation is well-established for the $TiO_{1\pm x}$ phase. The variation of lattice constant with composition is often linear but in some cases a maximum value has been found within the homogeneity range; the maximum lattice constant in the $ZrC_{1-x}$ phase occurs at $ZrC_{0.8}$.[39] Even density measurements are not easy in systems like these, with the ever-present trouble of oxygen and nitrogen contamination, and it is impossible to speculate on the defect structure of these phases. Electrical conductivity measurements show that they are metallic in nature, and a logical approach to their structure would be to build up a picture of the band population and structure. Specific heat and magnetic measurements to this end are in progress in many laboratories.

## ACKNOWLEDGMENTS

The author is indebted to many of his colleagues for help during the preparation of this paper and to Dr T. L. Markin for permission to include some unpublished results.

## REFERENCES

1. AUSKERN, G. B., and BELLE, J., *J. Nuclear Materials*, **3**, 267, 1961.
   WHEELER, V. J., and ROBERTS, L. E. J. Unpublished results.
2. SIMPSON, L. A., and CARTER, R. E., G.E.C. Report No. 65-RL-3985M (1965).
3. ANTHONY, A. M., GUILLOT, A., and NICHOLAU, P., *C.R.Acad.Sci. Paris*, **262B**, 896, 1966.
4. RHODES, W. H., and CARTER, R. E., *Bull. Amer. Ceram. Soc.*, **41**, 283, 1962.
5. ALCOCK, C. B., HAWKINS, R. J., HILLS, H. W. D., and McNAMARA, P., "Thermodynamics" Vol. II (I.A.E.A., Vienna, 1966) p. 57.

6. "Thermodynamics and Transport Properties of Uranium Dioxide", Technical Reports Series No. 39 (International Atomic Energy Agency, Vienna, 1965). MARKIN, T. L., and RAND, M. H., "Thermodynamics" Vol. 1 (I.A.E.A. Vienna 1966) p. 145. MCIVER, E. J., MARKIN, T. L. Third International Conference on Plutonium (Inst. Metals, London 1965).
7. EDWARDS, R. K., and MARTIN, A. E. "Thermodynamics", Vol. 1 (I.A.E.A. Vienna, 1965) p. 3.
8. MARKIN, T. L., and WHEELER, V. J. Private communication.
9. MARKIN, T. L. Private communication.
10. BEVAN, D. J. M., and KORDIS, J., J. Inorg. Nucl. Chem., 26, 1509, 1964.
11. STEELE, B. C. H., and ALCOCK, C. B., Trans. Met. Soc., A.I.M.E., 233, 1359, 1965.
12. ARONSON, S., J. Electrochem. Soc., 108, 312, 1961.
13. KOFSTAD, P., and RUZICKA, D. J., J. Electrochem. Soc., 110, 181, 1963.
14. RAND, M. H., and JACKSON, E.E., U.K.A.E.A. Report AERE-R 3636, 1963.
15. STEELE, B. C. H., POWELL, B. E., and MOODY, P. M. R. To be published in Proc. Brit. Ceram. Soc. No. 10, 1967.
16. WILLIS, B. T. M., Nature, 197, 755, 1963.
17. ROBERTS, L. E. J., and MARKIN, T. L. Proc. Brit. Ceram. Soc. No. 8, p. 201.
18. ROBERTS, L. E. J., "Advances in Chemistry" Series No. 39, 1963, p. 66.
19. BRAUER, G., Progress, Science & Technology of the Rare Earths, 2, 312, 1966.
20. HYDE, B. G., and EYRING, LEROY, Proc. 4th Conference Rare Earth Research (1964), p. 623.
21. BEVAN, D. J. M., BAKER, W. W., MARTIN, R. L., and PARKS, T. C., Proc. 4th Conference Rare Earth Research (1964), p. 442.
22. COLLOUGUES, R., LEFEVRE, R., PEREZ Y JORBA, M., and QUEYROUX, F., Bull. Soc. Chim. France, p. 149, 1962.
23. LEFEVRE, J., Ann. Chimie, 8, 117, 1963.
24. ROY, R., MAYAKE, H., and DINESS, A. M., Bull. Amer. Ceram. Soc., 43, 255, 1964; Bull. Soc. Chim. France, p. 1149, 1965.
25. ROMBERGER, A., SEARS, D.R. STONE, H. H., THOMA, R. E., and BAES, C. F., U.S.A.E.C. Report ORNL-3789, 1964, p. 243.
26. COHEN, I., and SCHANER, R. E., J. Nucl. Mat., 9, 18, 1963.
27. CARTER, R. E., and ROTH, W. L., General Electric Research Laboratory Rep., 63-RL-3479M, 1963.
28. DINESS, A. H., and ROY, R., Solid State Comm., 3, 123, 1965.
29. DIXON, J. M., et al., J. Electrochem. Soc., 110, 276, 1963. STUCHLER, D. W., and CARLSON, W., J. Amer. Ceram. Soc., 47, 123, 1964. TIEN, T. Y., and SUBBARAO, E. C., J. Chem. Phys., 39, 1041, 1963.
30. CARTER, R. E. Nuffield Group Spring Symposium, London, April, 1967.
31. WILLIAMS, K. R. (Ed.), "An Introduction to Fuel Cells", p. 212.
32. BRAY, D. T., and MERTEN, U., J. Electrochem. Soc., 111, 447, 1964.
33. TIEN, T., J. Appl. Physics, 35, 132, 1964.
34. GIORGI, A. L. British Ceramics Society Meeting, London, December, 1966.
35. "Carbides in Nuclear Energy". Conference at A.E.R.E., Harwell, (1965).
36. Technical Reports Series No. 14 (I.A.E.A., Vienna, 1962).
37. ACKERMANN, R. J., and THORN, R. J., "Thermodynamics" Vol. I (I.A.E.A. Vienna, 1966), p. 243.
38. STORMS, E. K., "Thermodynamics", Vol. I (I.A.E.A. Vienna, 1966), p. 309.
39. STORMS, E. K., U.S.A.E.C. Report LA-2942, (1964). p. 24,

23

# 24.—High-temperature Electrodes using Stabilized Zirconia

By R. E. W. CASSELTON and M. D. S. WATSON

*International Research & Development Co. Ltd, Fossway, Newcastle upon Tyne, 6.*

***ABSTRACT***                    H439/E181

*Zirconia stabilized with yttria has been studied as a potential M.H.D. electrode material, both in the laboratory under controlled current density, temperature, and atmosphere conditions, and also in a 200-kW propane–oxygen flame seeded with potassium. Experimental results are presented for the d.c. electrical conductivity of sintered yttria-stabilized zirconia with platinum contacts as a function of current density and ambient oxygen partial pressure. At current densities below $10^{-2}$ amp./cm², the electrical conductivity between 1000° and 1500°C is primarily ionic, irrespective of oxygen partial pressure down to $10^{-4}$ atm. Above this current density a breakdown occurs at a critical voltage, resulting in an order of magnitude increase in electrical conductivity. However, this also results in blackening in the region of the cathode, which leads to disintegration of the specimens; ways of overcoming this are considered. This electrolysis effect is a d.c. phenomenon and comparable processes are not observed under a.c. conditions. The behaviour of yttria-stabilized zirconia electrodes in a hot seeded propane–oxygen flame has been investigated, and the effect of the electrode porosity on the thermal shock characteristics considered.*

### Electrodes pour hautes températures en zircone stabilisée

*Les possibilités d'utilisation de la zircone stabilisée à l'oxyde d'yttrium pour la réalisation d'électrodes M.H.D. sont étudiées, d'une part au laboratoire dans des conditions contrôlées d'intensité de courant, de température et d'atmosphère et, d'autre part, dans une flamme de propane-oxygène de 200 kW, ensemencée de potassium. Les résultats expérimentaux sont indiqués pour la conductibilité électrique de la zircone stabilisée à l'oxyde d'yttrium, à contacts de platine en fonction de l'intensité du courant et de la pression partielle ambiante d'oxygène. Lorsque l'intensité du courant est inférieure à $10^{-2}$ amp/cm², la conductibilité électrique entre 1000° et 1500°C est principalement de nature ionique, quelle que soit la pression partielle de l'oxygène jusqu'à $10^{-4}$ atm. Au-dessus de cette intensité* *de courant, il se produit une discontinuité à un voltage critique qui est fonction de la pression partielle de l'oxygène et qui conduit à une augmentation importante de la conductibilité électrique et à un noircissement dans la région de la cathode. La résistance aux chocs thermiques d'échantillons de densité variable, contenant diverses proportions d'oxyde d'yttrium est étudiée et leur comportement dans une flamme chaude de propane-oxygène ensemencée est examinée. Pour des intensités de courant de 1 amp/cm² et certaines conditions, on enregistre des variations de la masse des électrodes après un passage prolongé du courant; elles sont discutées en termes d'érosion, de vaporisation et de changements stoechiométriques.*

### Hochtemperaturelektroden aus stabilisiertem Zirkonoxid

*Mit Yttriumoxid stabilisiertes Zirkonoxid wurd als möglicherweise für MHD-Elektroden verwendbares Material sowohl im Laboratorium unter kontrollierten Bedingungen von Stromdichte, Temperatur und Atmosphäre untersucht, als auch in einer 200 kW-Propan-Sauerstoff-Flamme mit Kaliumzusatz. Es werden die experimentellen Ergebnisse der elektrischen Leitfähigkeit von gesintertem, yttriumstabilisierten Zirkonoxid mit Platinkontakten als Funktion der Stromdichte und des umgebenden Sauerstoffpartialdrucks mitgeteilt. Bei Stromdichten unter $10^{-2}$ A/cm² erfolgt die elektrische Leitung zwischen 1000 und 1500°C vorwiegend ionisch, unabhängig vom Sauerstoff-Partialdruck bis herab zu $10^{-6}$ atm. Oberhalb dieser Stromdichte erfolgt ein Zusammenbruch bei einer kritischen Spannung, die ihrerseits eine Funktion des Sauerstoffpartialdrucks ist, mit dem Ergebnis, daß die elektrische Leitfähigkeit um eine Größenordnung ansteigt. Im Bereich der Kathode erfolgt eine Schwärzung. Die Eigenschaften von Proben verschiedener Dichte und verschiedener Molbruchteile Yttriumoxid hinsichtlich eines thermischen Schocks wurden untersucht und ihr Verhaten in einer heißen, dotierten Propan-Sauerstoff-Flamme betrachtet. Bei Stromdichten von 1 A/cm² wurde unter gewissen Bedingungen eine Änderung der Elektrodenmasse nach längeram Stromdurchgang gemessen. Dies wird im Hinblick auf Erosion, Verdampfung und stöchiometrische Änderungen diskutiert.*

## 1. INTRODUCTION

In M.H.D. power generation the motion of a partially ionized gas across a magnetic field produces power which is extracted through electrodes in contact with the ionized gas. This method of power generation depends on the principle that, when a conducting medium moves transversely through a magnetic field, a potential difference is induced at right angles to both the conducting medium and the magnetic field. In the open-cycle system, the ionized gas derives from a flame seeded with a relatively cheap, easily ionizable material such as potassium carbonate. The closed-cycle system relies on an inert gas heated by an external source, such as a nuclear reactor.

The choice of electrode materials for use in an open-cycle M.H.D. system is severely restricted by the high operating temperatures and by the presence of oxidants in the combustion gases. Metallic electrodes must in general be operated at low temperatures to minimize oxidation, resulting in a low thermionic emission. A high thermionic emission is, however, desirable in order to permit charge transfer across the electrode gap. However, oxidation-resistant refractory materials having a reasonably high conductivity (of the order of 1 (ohm-cm)$^{-1}$ at 2000°C) and a low vapour pressure at these high temperatures may be considered, and it has been shown by HEPWORTH and ARTHUR[1] that these criteria reduce the selection of refractory materials to the oxides of zirconium, hafnium and thorium. Of these three ceramic oxides, thoria may be ruled out because of its poor thermal shock properties and radioactivity. Technically there is little to choose between hafnia and zirconia; for economic reasons zirconia is preferred.

Zirconia undergoes a monoclinic–tetragonal phase transformation at 1000°C which is accompanied by a large volume change, thereby causing cracking in sintered specimens. This can be overcome by stabilizing zirconia in the cubic fluorite phase by the addition of aliovalent oxides, such as calcia, magnesia and yttria. In addition, aliovalent oxide doping increases the electrical conductivity of zirconia by several orders of magnitude because of the formation of oxygen vacancies in the fluorite structure. The cubic phase in magnesia-stabilized zirconia exists only above 1400°C and rapidly destablizes between 1400°C and the monoclinic–tetragonal inversion temperature at 900°C. VIECHNICKI and STUBICAN[2] have shown that the decomposition of 20 mole % $MgO:ZrO_2$ at 1000°C is virtually complete after 15 h. In addition, the volatility of the stabilizing agent in calcia- and magnesia-stabilized zirconia at temperatures above 1800°C in atmospheres of low oxygen partial pressure is high,[3] with the consequent risk of partial destabilization of the zirconia. The cubic phase in 12 mole % $Y_2O_3:ZrO_2$ extends down to room temperature, and the vapour pressure of yttria is known to be of the same order as that of pure zirconia.[4] For these reasons zirconia stabilized with yttria was considered to be a promising electrode material.

The use of stabilized zirconia as M.H.D. electrodes demands oxide–metal junctions to carry the current to the external circuit. Since the conduction in doped oxides is primarily ionic, the oxide–metal junction can be expected to act as a partial barrier to the flow of d.c. current from one material to another. This phenomenon would lead to a polarizing field in the sample, resulting in power losses within the M.H.D. generator, and was anticipated to provide the major limitation in the use of ceramic oxides for M.H.D. electrode applications. Accordingly, a programme of research was initiated to study the basic polarization phenomena in yttria-stabilized zirconia and to observe its behaviour as an M.H.D. electrode under operating conditions.

## 2. EXPERIMENTAL DETAILS

### 2.1 Fabrication of Electrodes and Specimens for Conductivity Measurements

Electrical conductivity measurements by DIXON et al.[5] indicated that the maximum conductivity in $Y_2O_3:ZrO_2$ occurred at 8 mole % $Y_2O_3$, which coincided with the low-yttria boundary of the cubic zirconia field, as determined by DUWEZ et al.[6] Consequently in order to attain the highest possible conductivity consistent with complete stabilization in the cubic phase, all specimens used in this work were of the composition 12 mole % $Y_2O_3:ZrO_2$.

All electrodes and specimens were fabricated from 99·5% pure zirconia (obtained from Magnesium Elektron Ltd), the major impurities being 0·2% $TiO_2$, 0·2% $SiO_2$ and 0·02% $Fe_2O_3$. Yttria of 99·99% purity was supplied by B.D.H. Ltd.

The mixed cold-pressed powders were presintered at 1550°C (to bring about the solid-solution reaction), crushed, ground and sieved to below 90 $\mu$m particle size. Trace-iron pick-up from the percussion mortar was removed by leaching with a dilute mineral acid. X-ray analysis at this stage indicated complete stabilization of the cubic solid solution. Cylindrical bars were isostatically cold-pressed at $3·5 \times 10^3$ kg/cm² and machined in the green state into either cylindrical electrodes, 2 cm × 1·3 cm diam., or rods 5 cm × 1 cm diam. for conductivity measurements. The final sinter at 2000°C for 3 h in an oxy–propane furnace gave densities of 92–95% of the theoretical value. Some electrodes were sintered at 1800°C in order to obtain more porous samples of 20–30% porosity. Metallographic examination showed the final specimens to be single-phase and homogeneous.

To ensure good electrical contact to the platinum electrodes, platinum paste was fired onto all zirconia specimens.

## 2.2 Apparatus and Measurement Techniques for Studying the Conduction Characteristics

The electrical conductivity apparatus, shown in Figure 1, consisted of a specimen (a) sandwiched between two platinum disc electrodes (b) and located inside a gas-tight alumina chamber (c). To prevent oxygen from the outside atmosphere diffusing through to the inner assembly, an outer alumina gas chamber (d) containing argon surrounded the inner chamber (c).

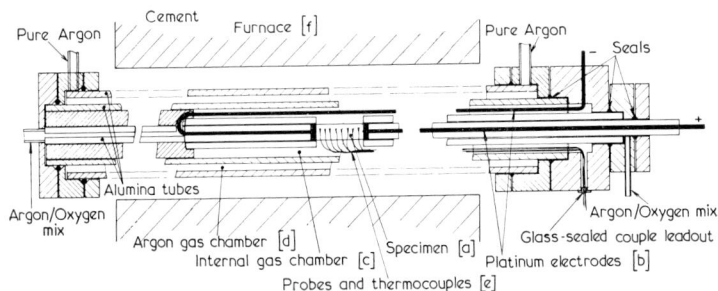

FIGURE 1.—Apparatus for determining potential distributions and electrical conductivity.

To facilitate potential distribution measurements, eight probes (e) made from 0·6-mm diam. wire were inserted into holes drilled radially along the specimen. Four of these probes were of platinum and the remainder were the platinum arms of Pt/Pt-10% Rh thermocouples.

The whole assembly was supported so that the specimen was in the centre of the hot zone of a platinum-wire-wound furnace (f). The temperature gradient over a 5-cm specimen could be minimized to within 4°C by subsidiary heaters located above and below the specimen.

The potential distribution along the specimen was obtained by measuring the potential differences between the end electrodes and each of the eight probes. Accumulators were employed as a d.c. power source, and a Solartron digital voltmeter was used for measuring d.c. potentials.

All potential distribution measurements were made in an oxygen–argon gas stream mixed proportionately to give oxygen partial pressures ranging from $10^{-4}$ to $10^{0}$ atm. No measurements were made at lower oxygen partial pressures, which could only be attained using carbon-dioxide–carbon-monoxide or hydrogen–water-vapour mixtures, and it was found that irregularities occurred in the polarization measurements when using these gases. This was attributed to some modification in the electrode reaction kinetics which probably involves the molecular state of the oxygen gas molecule. The total gas flow was maintained at 1 litre/min., as non-reproducible results were obtained with flow rates less than 400 cm³/min. To ensure accurate and reproducible results, particularly at low oxygen partial pressures, the gases employed were carefully cleaned before entering the conductivity furnace. Argon, initially of 99·995% purity, was cleaned by passing through a molecular sieve to remove water vapour and over zirconium turnings at 800°C in order to remove oxygen and nitrogen. Water vapour was removed from the oxygen gas by passing through a molecular sieve at room temperature.

## 2.3 Experimental Arrangement for Testing Electrodes in the 200-kW Burner

An experimental programme to evaluate the thermal and electro-thermal performance of 12 mole % $Y_2O_3$:$ZrO_2$ was carried out under applied field conditions using a low-velocity 200-kW oxy–propane burner. As shown in Figure 2, current was induced across the gap by an applied electric field and not a magnetic field.

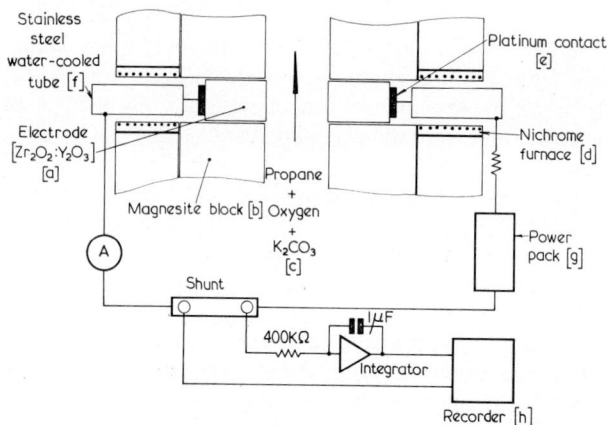

FIGURE 2.—Schematic diagram of burner-testing facility.

The test electrodes (a) were mounted in a segmented magnesite channel (b) and heated to 1850°C by the combustion gases (c) which were seeded with 1% potassium by weight in the form of a potassium carbonate aerosol. In order to minimize stresses arising from thermal shock, the cylindrical electrodes were preheated to

1000°C in nichrome-wound silica furnaces (d) before being gradually introduced into the flame. Platinum discs (e) mounted on water-cooled stainless-steel arms (f) were used for the electrical contacts to the electrode back face, thereby allowing the whole electrode to be operated above 900°C. This is an important criterion, since power losses due to Joule heating become significant below this temperature. The external electrical circuit was completed by a 200-V 10-A d.c. power supply (g) together with an electronic integrator and recorder (h) for monitoring the total charge passed between the electrodes. The platinum–ceramic junctions were maintained in air, and no effort was made to enrich the oxygen supply to the back face junctions.

The electrodes were preheated to 1000°C at 250°C/h in the nichrome furnaces, and the magnesite channel was heated to the same temperature by igniting the oxy–propane burner before slowly moving the electrodes into the flame. Optical pyrometer readings of the electrode front face temperature were corrected for non-blackbody equilibrium using an emissivity value of 0·35 for zirconia.[7] Maximum corrected values were of the order of 1850°C but typical temperatures recorded during a test run, which lasted 1 to 40 min. according to electrode behaviour, were of the order of 1700°–1800°C. At the end of a test run the electrodes were withdrawn from the flame into the preheat furnace and cooled at 250°C/h.

## 3. RESULTS

### 3.1 Conductivity and Polarization Behaviour in Zirconia Stabilized with 12 mole % Yttria

The bulk d.c. electrical conductivity of 12 mole % $Y_2O_3:ZrO_2$ was obtained as a function of absolute temperature from equilibrium potential distribution curves obtained over a 5-cm specimen in a pure oxygen atmosphere using very low currents and is shown in Figure 3 as an Arrhenius plot of log $\sigma$ against $1/T$. Results were reproducible with repeated temperature cycling, and between different specimens of the same composition, after correcting for porosity effects. The activation energies ($Q$) quoted in Figure 3 for 12 mole % $Y_2O_3:ZrO_2$ are derived from a least squares analysis of numerous experimental points, with a probable error of 0·005 eV. Also included for comparison are some earlier measurements made under the same conditions on 17 mole % $MgO:ZrO_2$ and 15 mole % $CaO:ZrO_2$.

Figure 4 (A–C) shows the equilibrium d.c. potential distributions obtained at 1380°C for current densities of $1·29 \times 10^{-3}$, $1·29 \times 10^{-1}$ and $1·29 \times 10^0$ A/cm² and varying ambient oxygen partial pressure

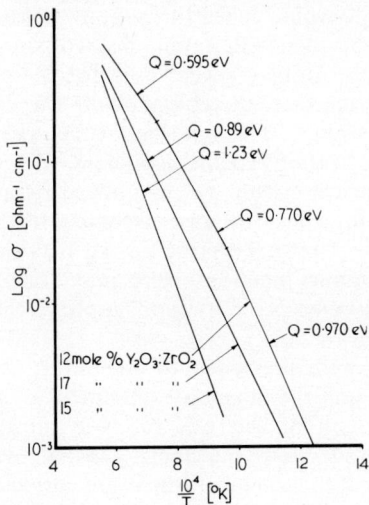

FIGURE 3.—Arrhenius plot of log $\sigma$ against $1/T$ for 12 mole% $Y_2O_3:ZrO_2$, 15 mole % $CaO:ZrO_2$ and 17 mole % $MgO:ZrO_2$.

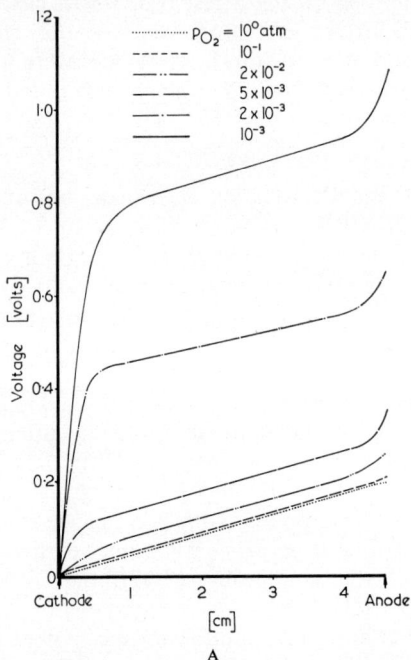

A

FIGURE 4.—Potential distribution at 1380°C in 12 mole % $Y_2O_3:ZrO_2$ at: A. $1·29 \times 10^{-3}$ A/cm² as a function of oxygen partial pressure.

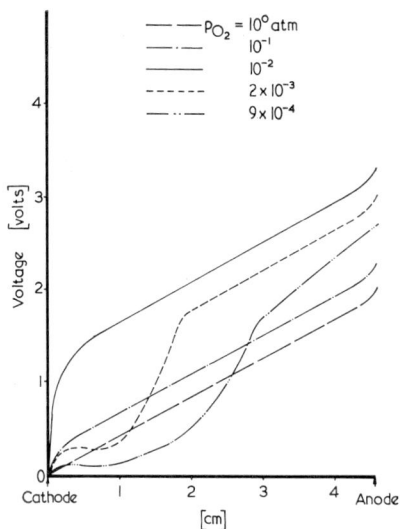

B

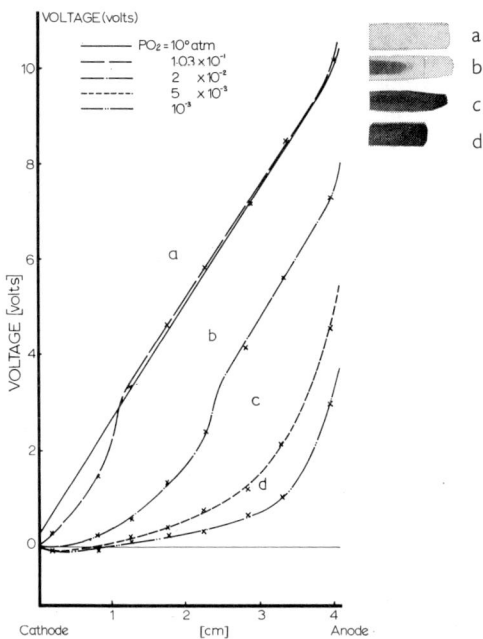

C

FIGURE 4.—*continued*

B. $1 \cdot 29 \times 10^{-1}$ A/cm$^2$ as a function of oxygen partial pressure.

C. $1 \cdot 29$ A/cm$^2$ as a function of oxygen partial pressure, together with associated blackening in quenched specimens.

from $10^0$ to $5 \times 10^{-4}$ atm. The measured potentials are included for reference. Accompanying Figure 4c are photographs of small specimens quenched under the same conditons of temperature, current density and oxygen partial pressure at which the corresponding potential distribution curve was determined, the quenching experiments being performed in a separate apparatus developed for the purpose.

Some of these small specimens are blackened and, on quenching or slow cooling from 1380°C to room temperature in a reduced oxygen atmosphere, the bulk of such specimens tended to break away from the platinum cathode, leaving behind a pip adhering to the cathode. X-ray analysis of the pip material indicated the existence of cubic zirconia, platinum and zirconium metal, together with additional X-ray lines which are as yet unidentified, but may correspond to a platinum–zirconium alloy. X-ray analysis of the bulk of a blackened specimen showed a lattice volume contraction of less than 1 % on blackening, and a manual goniometer count did not reveal the existence of any second phase in either the bulk of a blackened specimen or in the distintegrated remains after re-oxidation.

Weight-change measurements obtained by back oxidation of a totally black sample to stoicheiometry showed a weight loss of about 0·65 % on blackening, indicating the oxygen:metal ratio of the blackened sample to be about 1·85.

The electrolysis phenomena so described is a purely d.c. effect and is not observed under a.c. conditions. For a given temperature and oxygen partial pressure, the a.c. bulk conductivity remained constant with increasing current and frequency.

### 3.2 Electrode Behaviour under Testing Conditions in the 200-kW Burner

Several pairs of 12 mole % $Y_2O_3$:$ZrO_2$ electrodes have been tested in a 200-kW burner at current densities ranging from 0·5 to 3·5 A/cm². Initially it was found that, under testing conditions, the thermal shock resistance of high-density (92–95 % theoretical) electrodes was very poor, despite careful preheating to 1000°C. All these specimens suffered severe cracking and in most cases shattered into fragments, resulting in a test run being brought to a premature close. Further tests using 20–30 % porosity electrodes showed a considerable improvement in thermal shock properties and, although some cracking was experienced, fragmentation did not occur. The size of cracks tended to decrease with increasing porosity.

The electrical behaviour of the cathode during a typical run was completely different from that of the anode in that arc spots were observed at the gas–cathode electrolyte interface but not at the gas–anode electrolyte boundary. These hot ceramic arc spots could be described as scintillations (about 0·5 mm across) randomly distributed over the cathode surface. At the same time large non-scintillating spots appeared above current densities of 0·5 A/cm², and generally only one appeared on the cathode at any one time. They originated at the leading edge of the cathode front face and slowly moved round to the trailing edge. The arc spots sometimes remained in contact with the edge for several minutes before disappearing. Unlike the scintillations, this type of arc spot activity was not always continuous, even though the current remained in excess of 0·5 A/cm.² Owing to the small dimensions of the arc spots, it was not possible to measure their temperature optically.

Examination of the cathode surface after testing showed a glazed and pitted surface with severe glazing in the region of the trailing edge where the large arc spots had resided (Figure 5A). In contrast, examination of the anode showed an unchanged surface apart from cracks (Figure 5B).

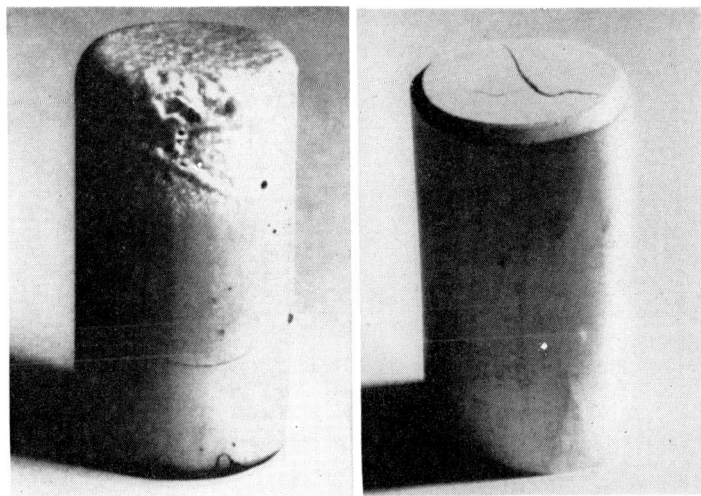

A                                       B

FIGURE 5.—Appearance of cathode and anode electrodes after a test run in the
burner assembly:
A. Cathode (note surface glazing).
B. Anode.

On sectioning the cathode for further examination (Figure 6) a blackened zone in the region of the ceramic–platinum junction was observed. This blackened zone was not uniform in colour, but ranged from black in the zone centre to dark grey at the perimeter. It was not possible to observe the zone colouration at the platinum–ceramic junction, since the surface regions re-oxidized when switching off the current supply and removing the electrode from the combustion flame. However, a general crumbling of the ceramic was observed in the region of the metal–electrode junction. The spiral effect, seen in Figure 6, is thought to be direct evidence of current channelling, which is a phenomenon whereby nearly all the current is conducted through a very narrow channel, and arises from resistance instability due to the negative temperature coefficient of the material. However, it was only observed in a very few cases. As blackening, which is shown in Section 3.1 to depend on the oxygen partial pressure of the gas surrounding the metal–cathode junction, was considered to be detrimental to the testing lifetime of the ceramic electrode, holes were drilled in the platinum disc contact to allow better access of oxygen molecules to the interfacial region. This was fairly successful, although a small amount of blackening was still evident; unfortunately the reduced surface area of contact led to a weaker platinum–ceramic bond.

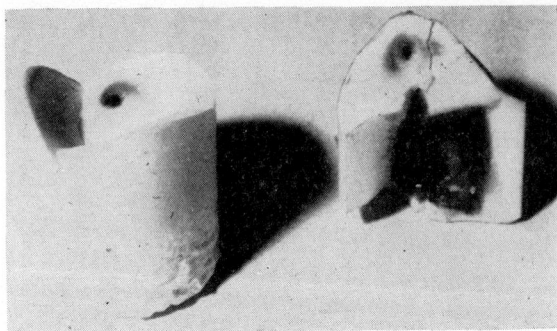

FIGURE 6.—Internal appearance of cathode electrode after a test run in the burner assembly. Note blackening and current channelling.

The electrodes were weighed before and after testing and weight increases of the order of several milligrams were detected. As so many other variables exist (such as average current, time of operation, temperature and fragmentation) it was not possible to deduce unequivocally from our results that the increase in weight was a function of porosity. Chemical analysis of electrodes after testing showed the presence of both potassium carbonate and potassium bicarbonate, originating from the seed, in sufficient quantities to account for much of the observed weight increases.

After removing all traces of the potassium salts by leaching in cold water and subsequent heating to 1400°C, the electrodes generally showed a nett loss in weight of the cathode and a very small nett gain for the anode.

## 4. DISCUSSION

For predominantly ionic conduction the bulk conductivity $\sigma$ is related to the absolute temperature $T$ by the Arrhenius equation:

$$\sigma = A \exp\left(-\frac{Q}{kT}\right)$$

where $A$ is a pre-exponential term which depends on the number of charge carriers contributing to the predominant conduction mode having an activation energy $Q$. Consequently the form of a $\log \sigma : (1/T)$ plot provides information on the conduction modes present, as well as enabling the activation energy to be derived from the slope of the curve.

The results in Figure 3 show that, up to at least 1500°C, the bulk conductivity in 12 mole % $Y_2O_3 : ZrO_2$ is considerably higher than in magnesia- or calcia-stabilized zirconia. Furthermore, whilst zirconia stabilized with divalent ions has a single conduction mode over the temperature range studied, the conduction in trivalent yttria is more complex, since discontinuities were observed in the Arrhenius plot at 820°C and at 1170°C. Dilatometric studies on 12 mole % $Y_2O_3 : ZrO_2$ showed no evidence of any phase transformation between room temperature and 1300°C so that the discontinuities in the Arrhenius plot must correspond to changes in the conduction mode.

It is not possible to make a direct comparison of these results with those of previous workers,[5,8] owing to the different yttria compositions used. However, at first sight it would appear that previous authors determined a constant activation energy for 10 and for 15 mole % $Y_2O_3 : ZrO_2$ over the measured temperature range, but a closer examination of Figure 1 of STRICKLER and CARLSON's paper[8] and of Figure 2 of DIXON et al.'s paper[5] does suggest the existence of a transition point in the region of $1/T = 9 \times 10^{-4}$ °K$^{-1}$, as observed in our work. If due allowance is made for this, the activation energy for the high-temperature part of the 10 mole % $Y_2O_3 : ZrO_2$ conductivity/temperature curve is estimated to be 0·75 eV from Strickler and Carlson's results, and $0 \cdot 8$ eV from Dixon's results, which are in fair agreement with one another and with the activation energy ($0 \cdot 77$ eV) obtained by the present authors for the intermediate conduction mode of 12 mole $\%$ $Y_2O_3:ZrO_2$. More-recent work by one of the present authors[9] suggests that there is little difference in the conductivity of 10 and 12 mole $\%$ $Y_2O_3:ZrO_2$ for the intermediate conduction mode between $820°$ and $1170°C$ but deviations exist above and below these transition temperatures.

The conductivity data in Figure 3 refer to bulk conduction at very low currents, of the order of milliamps, since the potential distributions in Figure 4 (A–C) show quite clearly the increasing non-linearity that results on increasing the current density. The potential distributions at low current densities (Figure 4A) are as expected for predominantly ionic conduction, with a linear distribution in the bulk and increasing polarization voltages at both cathode and anode with decreasing oxygen partial pressure. Analysis of the field in the bulk of the specimen showed that the conductivity decreased slightly with decreasing oxygen partial pressure. Although it would appear that this is a result of increasing oxygen deficiency within the sample with decreasing oxygen partial pressure, this is not so, because of the very low mobility of oxygen ions, which is of the order of $10^{-5}$ cm²/Vsec. A probable explanation of this effect is the existence of a small electron hole contribution arising from the reaction:[10]

$$\tfrac{1}{2} O_2 + \square \rightleftharpoons O^{2-} + 2 h$$

which is superimposed on the predominantly oxygen ion conduction. The law of mass action applied to this reaction predicts a $p_{O_2}$ variation of the hole conductivity with oxygen partial pressure, and this has been observed experimentally in our work.[11]

At intermediate current densities, of the order of $10^{-1}$ A/cm², the potential distribution (Figure 4B) reflects a predominantly ionic conduction process above a critical oxygen partial pressure which is a function of the current. Below this critical partial pressure the potential distribution curve become increasingly non-linear. For example, non-linearity commences at about $10^{-2}$ atm oxygen for $1 \cdot 29 \times 10^{-0}$ A/cm², and $2 \times 10^{-1}$ atm for $1 \cdot 29 \times 10^{0}$ A/cm².

Figure 4C shows that the equilibrium distribution obtained at $1 \cdot 29 \times 10^{0}$ A/cm² in pure argon is very similar to that obtained in $10^{-3}$ atm. of oxygen, and that a small negative space charge, of the order of $0 \cdot 2$ V, is established in this "saturation" state. Comparison of the slopes of the potential curves suggests an order of magnitude increase in the bulk conductivity on reducing the oxygen partial pressure from $10^{0}$ to below $10^{-3}$ atm.

The photographs accompanying Figure 4c show quite clearly that the non-linear potentials are associated with the formation of a blackened zone at the specimen cathode. The origin of the blackening has been ascribed[12] to an electrolysis of zirconia near the cathode arising from an insufficient supply of oxygen from the ambient gas to compensate for the mass transport of oxygen ions away from the cathode interface under the influence of an electric field; i.e. a critical relationship exists between the oxygen partial pressure of the ambient gas and the current passing through the specimen. Analysis of the potential distribution curves shows that in fact a linear relation between critical current for electrolysis and oxygen partial pressure does exist, and furthermore that the onset of electrolysis corresponds to a critical cathode polarization voltage of about 1 V.

As stated in Section 3.2, electrolysis invariably occurred in the cathode electrode of the burner testing facility under normal operating conditions of 2–3 A/cm$^2$ with air (oxygen partial pressure of $2 \times 10^{-1}$ atm) surrounding the cathode metal–ceramic junction. On the basis of the laboratory studies of the electrolysis phenomena, the critical current for the onset of blackening in air is about $4 \times 10^{-1}$ A, which is much smaller than the actual current passing through the electrodes.

It appears highly desirable to inhibit electrolysis in the cathode electrode, owing to the disintegration that occurs on heating or cooling through 900°C. One way of doing this would be to ensure a copious supply of pure oxygen to the cathode metal–ceramic junction but this was not possible with the present experimental arrangement. On the other hand, it was thought that a more porous sample would inhibit blackening by allowing easier access of the oxygen molecules to the substoicheiometric region behind the metal–ceramic junction. This was not found to be so, but it appeared that drilling holes through the platinum contact disc helped to minimize electrolysis. Unfortunately the reduced surface area of contact led to poor zirconia–platinum bonding.

There is a good deal of conflict in the literature[12] over the evidence for current channelling in stabilized zirconia. In this work, blackening has never been observed in the laboratory studies of the conduction mechanisms but, as will be seen from Figure 6, current channelling appears to be a predominant feature of the cathode electrode conduction under applied voltage conditions. A possible reason for this may be the severe axial and radial temperature gradient existing in the electrode, giving a localized channel with the highest temperature and therefore the highest conductivity. As might be ex-

24

pected, this occurs on the upstream side of the electrode i.e. where the flame impinges directly on it.

Arc-spot activity in a ceramic electrode is limited to the cathode, as evidenced by the severe glazing observed on this electrode. The formation of cathode spots is an important process, since it occurs even when the electrode temperature is quite high. Such arc-spot emission suggests that at 1800°C the thermionic emission is insufficient to carry the current. As the temperature within the arc spot will approach the melting point of zirconia (witness the severe electrode glazing), thermionic emission within this localized region is greatly increased and could therefore become a significant current-transfer mechanism.

It will be noticed that electrolysis and arc-spot activity is confined to the cathode electrode, which suggests that the transfer of oxygen from the gas stream into the anode electrode is easily accomplished. The fact that arc spots have not been observed on the anodes tends to confirm that the effect of the boundary layer at the anode is slight.

The adverse thermal shock characteristics of both cathode and anode electrodes under testing conditions is believed to arise from a fluctuating temperature gradient over the electrodes caused by flame instability. The thermal shock characteristics are substantially improved by increasing the electrode porosity, and little visible evidence of cracking was observed on 75% dense samples. On the other hand it would be expected that seed penetration would be greater in the more porous electrodes, but this has not yet been unequivocally ascertained.

## 5. CONCLUSIONS

The two major problems which arise when operating yttria-stabilized zirconia electrodes in an M.H.D. generator are thermal shock and the structural changes which result from electrolysis at the cathode.

Thermal shock may be minimized by adopting a fairly porous sample (about 20–25% porosity) but this will also result in a reduction in the electrical conductivity and a probable increase in seed penetration.

In general, electrolysis limits the effective life of the material considered for an M.H.D. cathode by the destructive structural changes which occur. Electrolysis may be inhibited by a copious supply of oxygen to the cathode metal–ceramic junction and by permitting the oxygen gas to permeate to the junction through holes drilled in the metallic contact disc.

Apart from electrolysis effects, arcing could also be a limitation to the effective life of the cathode. However, it must be stressed that all our work has been carried out under applied-voltage conditions and realistic tests can only be made under generating conditions. Any weight changes observed for the anode electrode on testing may be attributed to absorption of the potassium seed, whereas extraneous factors, such as disintegration or crack formation, prevented an accurate assessment of any weight changes observed in the cathode electrode during testing.

### ACKNOWLEDGMENTS
The authors acknowledge the assistance of Mr S. K. Adams, Mr M. G. Hazelhurst and Mr C. G. Larsen with the experimental work, and Dr G. Arthur and Dr D. Balfour for helpful discussions.

This work was part of a contract sponsored by C. A. Parsons and Co. Ltd, on behalf of the British Collaborative MHD Committee, and the authors thank the sponsors and the Board of Directors of International Research and Development Co. Ltd for permission to publish this paper.

### REFERENCES
1. HEPWORTH, M. A., and ARTHUR, G., "M.H.D. Power Generation" IEE Conference Series No. 4, London, 1963.
2. VIECHNICKI, D., and STUBICAN, V. S., *J. Amer. Ceram. Soc.*, **48**, (6), 292, 1965.
3. SHORT, A. Unpublished work at IRD.
4. ACKERMANN, R. J., and THORN, R. J., Symposium on "High Temperature Technology" (Butterworth: 1964).
5. DIXON, J. M., *et al.*, *J. Electrochem. Soc.*, **110**, (4), 276, 1963.
6. DUWEZ, P., BROWN, F. H., and ODELL, F., *J. Electrochem. Soc.*, **98**, (9), 356, 1951.
7. ELSTON, J., MIHAILOVIC, Z., and ROUX, M., Third International Symposium on M.H.D. Electrical Power Generation, Salzburg. International Atomic Energy Agency, Vienna. 1966.
8. STRICKLER, D. W., and CARLSON, W. G., *J. Amer. Ceram. Soc.*, **47**, (3), 122, 1964.
9. CASSELTON, R. E. W. To be published.
10. KIUKKOLA, K., and WAGNER, C., *J. Electrochem. Soc.*, **104**, 379, 1957.
11. CASSELTON, R. E. W., and SCOTT, J. C., *Physics Letters*, **25A**, (3), 264, 1967.
12. ANTHONY, A. M., *C.R. Acad. Sci. Paris.*, **260**, 1936, 1965.

# 25.—The Use of Ceramics in Molten Carbonate Fuel Cells

By A. C. C. Tseung* and A. D. S. Tantram

*Energy Conversion Ltd, Chertsey Road, Sunbury-on-Thames, Middlesex*

## ABSTRACT BO/E7

*The basic principles of molten-carbonate fuel cells are described. Ceramics are used as insulators, electrodes and electrolyte diaphragms. $Cu_2O$, Ag on alumina or ZnO are used as electrodes. MgO is widely used as the structural material for the electrolyte diaphragm, either in the form of porous discs or as inert filler to form a semi-solid $NaLiCO_3$ electrolyte paste. The influence of electrolyte diaphragm structure on the thermal expansion, ionic conductivity and compressive strength is outlined.*

## L'utilisation des céramiques dans les piles à combustible constituées de carbonate fondu

*Les principes fondamentaux des piles à combustible constituées de carbonate fondu sont décrits. Les céramiques sont utilisées comme isolants, comme électrodes et comme diaphragmes d'électrolyse. $CaO_2$, Ag sur de l'alumine ou ZrO sont utilisés comme électrodes. MgO est largement utilisé comme matériau constitutif du diaphragme d'électrolyse soit sous la forme de disques poreux soit comme charge inerte pour former une pâte d'électrolyte de $NaLiCO_3$ semi-solide. L'influence de la structure du diaphragme d'électrolyse sur la dilatation thermique, la conductivité ionique et la résistance à la compression est indiquée.*

## Die Verwendung von Keramik in Brennstoffzellen mit geschmolzenen Karbonaten

*Beschrieben werden die Grundprinzipien der Brennstoffzellen mit geschmolzenen Karbonaten. Keramische Stoffe werden als Isolatoren, Elektrolyte und elektrolytische Diaphragmen benutzt. $Cu_2O$, Ag auf $Al_2O_3$ oder ZnO werden als Elektroden verwendet. MgO wird viel als Material für elektrolytische Diaphragmen verwendet, entweder in Form von porösen Scheiben oder als inertes Füllmaterial zur Herstellung einer halbfesten $NaLiCO_3$-Paste. Der Einfluß des Aufbaues des*

---

* Now at: Department of Chemistry, City University, St. John Street, London, E.C.1.

*elektrolytischen Diaphragmas auf Wärmedehnung, Ionenleitfähigkeit und Druckfestigkeit wird gezeigt.*

## 1. INTRODUCTION

The molten alkali carbonate fuel cell shows promise as one of the very few cells, capable of using hydrocarbon fuels, that could ultimately be developed into a cheap and reliable system. Molten alkali carbonate is the only fused salt system that would remain invariant in the presence of large amounts of $CO_2$ in the feed gas, since other salts, such as $NO_3^-$, $Cl^-$, $SO_4^{2-}$, etc., would be eventually converted to $CO_3^{2-}$. In practice, hydrocarbon fuels are mixed with steam and passed first into a pre-reformer, which produces a mixture of $H_2$, CO and $CO_2$. The introduction of pure hydrocarbons into a fuel cell at 500°C or above would result in carbon deposition. Figure 1 shows the electrochemical reactions occurring in a molten carbonate cell.

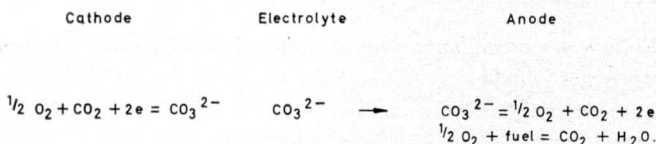

Cathode        Electrolyte        Anode

$$\tfrac{1}{2}\,O_2 + CO_2 + 2e = CO_3^{2-} \qquad CO_3^{2-} \longrightarrow \qquad CO_3^{2-} = \tfrac{1}{2}\,O_2 + CO_2 + 2e$$
$$\tfrac{1}{2}\,O_2 + \text{fuel} = CO_2 + H_2O.$$

FIGURE 1.—Electrochemical reactions in a molten-carbonate fuel cell.

The carbonate ion is the main current carrier and, to provide for its formation at the cathode, carbon dioxide must be supplied with air. This carbon dioxide is released again at the anode. In practical systems the spent anode effluent would be recycled into the cathode air supply.

CHAMBERS and TANTRAM[1] have shown that it is instructive to consider the cell as a combined oxygen and carbon dioxide concentration cell. The oxygen partial pressure at the anode will be kept to a very low value by reaction with the fuel (about $10^{-16}$ atm.). The e.m.f. is given by:

$$\frac{RT}{2F} \ln \frac{p^*_{CO_2}}{p_{CO_2}} + \frac{RT}{4F} \ln \frac{p^*_{O_2}}{p_{O_2}}$$

The $p^*$'s refer to the partial pressures at the cathode and the $p$'s to those at the anode. $p_{O_2}$ can be calculated from the appropriate equilibrium.

$$\text{e.g} \quad \frac{p_{H_2O}}{p_{H_2}\, p_{O_2}^{\frac{1}{2}}} = K_e$$

Looked at in this light, we can see firstly that the cell should work on any fuel that will react rapidly with oxygen. Secondly it is very important that there should be no leak of oxygen on to the anode.

Figure 2 shows a schematic drawing of a laboratory molten-carbonate diaphragm cell. A typical performance is 60–80 watts/ft$^2$ on re-formed methane at 600°C.

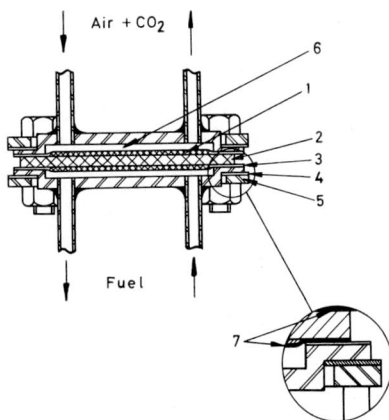

FIGURE 2.—Flame-sprayed disc cell.

We can divide the use of ceramics in molten-carbonate cells into three categories: (1) their use as insulators; (2) their use in electrodes; (3) their use in electrolyte diaphragms. Of these, the latter is the most important and, after some preliminary comment on the first two, the bulk of this paper will be concerned with this subject.

## 2. INSULATORS

The temperature precludes the use of plastics, and ceramics are the only choice. Their selection is governed by the same considerations that apply to electrolyte diaphragms, which are discussed later.

## 3. ELECTRODES

Ceramics may be used in two different ways. A selected electrode material may be ceramic in nature, in which case ceramic techniques are used for electrode fabrication. Examples are the use of NiO and $Cu_2O$ for the cathode. More usually, because of their higher conductivity, metals are used as the active electrode material. In this case, ceramics may still be incorporated, use being made of the ceramic properties to achieve a specific end. One example is with the silver–zinc oxide electrodes used by CHAMBERS and TANTRAM.[1] Here the zinc oxide serves as a sintering inhibitor to the silver catalyst, so that a high surface area may be maintained in operation. Another example, due to the same workers, is the use of alumina as a spacing agent to achieve a controlled-porosity electrode. In this case, they fabricated porous electrodes by flame-spraying silverized alumina under conditions where the silver melted to bond the particles, but the refractory alumina remained solid, so ensuring a porous product.

## 4. ELECTROLYTE DIAPHRAGMS

At the reaction site in a fuel cell electrode there must be easy entry for the reactant gases, easy exit for the products, entry for the current-carrying ion from the electrolyte and exit for the product electrons. These requirements are optimized at the three-phase interface, gas–electrolyte–electrode, and more specifically at the thinly "wetted" part of the electrode. Electrode and cell design must therefore stabilize this interface and preferably maximize this thinly wetted area.

The methods used all depend on surface tension. This stabilization was achieved by using dual-porosity electrodes in a free electrolyte cell,[1] but the approach in which the electrolyte is immobilized or trapped in a diaphragm has found the most favour. In this case, control is effected by making the pore size of the electrodes larger than the effective pore size of the electrolyte diaphragm.

Two distinct approaches may be made to the electrolyte diaphragm. In the first case, a pre-made porous diaphragm of inert non-conducting material is subsequently impregnated with the carbonate electrolyte. In the second case, an intimate mixture of the carbonate electrolyte and inert material is densified to form an impervious electrolyte diaphragm. This is normally referred to as paste or semisolid electrolyte.

The requirements of the inert non-conducting material for both approaches can only be met by ceramics.

The properties required for this material are as follows. Chemical inertness to alkali carbonate melts, in the presence of carbon dioxide and of both highly oxidizing conditions (at the cathode) and highly reducing conditions (at the anode), at temperatures in the range 500°C to 700°C. Good insulating properties; any electronic conductivity will provide an internal short circuit in the cell; for electronic conductivity to be negligible the specific resistance of the material should be greater than $10^3$ ohm-cm. These requirements would seem to restrict the choice of materials to magnesia, alpha alumina, thoria, zirconia, beryllia, some of the rare-earth oxides and possibly some mixed oxides such as magnesium aluminate. Gamma alumina will react with the carbonate melt and there are some doubts about the stability of zirconia. An alternative approach is to allow a reactive starting material to fully react with the melt and to use the reaction product as the inert material. This was done by DAVTYAN [2] with his complicated electrolyte mix and recently by SALVADORI et al.[3]. who used gamma alumina as the starting material. It was fully reacted to lithium aluminate, which then provided the inert material.

By far the most work has been done with magnesia. The properties required for the fabricated diaphragm are as follows. The electrolyte resistance should not be substantially increased, which implies that the inert matrix should have a high porosity and low tortuosity. The effective pore size of the inert matrix must be below that of the electrodes to be used, so that electrolyte is retained within the diaphragm. The diaphragm should have good strength and resistance to thermal shock. Some of these properties will be in conflict and so compromises will be necessary.

A full description of the fabrication of porous pre-sintered magnesia diaphragms will be published elsewhere. In the present paper we will summarize the properties of the products obtained and then describe the paste electrolyte diaphragms in rather more detail.

For the pre-sintered diaphragms, $1\%$ titania was used as a sintering promoter and ammonium carbonate as a porosity promoter. The diaphragms were fired at $1300°C$. The average properties of the diaphragms prepared by the preferred method that was developed are given below.

| | |
|---|---|
| Linear shrinkage on firing | $10\%$ |
| Open porosity | $45\%$ |
| Closed porosity | $2\%$ |
| Maximum pore size | 26 m$\mu$ |
| Pore-size distribution | Figure 3 |
| Tortuosity factor | 2·1 |
| Hardness (Moh's scale, scratch test) | 6 |

For use in the fuel cells these diaphragms were first machined flat and then impregnated with the molten carbonate electrolyte by means of a simple hot-dipping technique.

The main problem found with these cells was loss of electrolyte due to creep. It became clear that a much smaller pore size was desirable, coupled with very accurate control of maximum pore size, which would prove very expensive in practice. An additional economic factor was the necessity for the machining operation on the sintered diaphragms. Gas-sealing problems on the cells were also difficult.

Because of these problems we turned our attention to semi-solid electrolyte diaphragms.

## 5. SEMI-SOLID ELECTROLYTE DIAPHRAGMS

### 5.1 General Properties

In the usual version, an extremely intimate mixture of the carbonate electrolyte and a substantial percentage of very finely divided inert filler (e.g. MgO) is densified to form an impervious electrolyte diaphragm.

At the operating temperature of the fuel cell the carbonate component is molten and in this stage the material is somewhat analogous to a clay–water system over a particular range of composition.

This type of electrolyte derives from the work of DAVTYAN.[2] He described what he called a solid electrolyte made by casting a fused mixture of monazite sand, sodium carbonate, soda glass and tungsten trioxide.

Working independently, both BROERS [4] in Holland and CHAMBERS and TANTRAM [1] in England showed that his electrolyte was in fact of the semi-solid type and they refined the composition down to the essential ingredients of alkali metal carbonates and an inert filler. At this stage BROERS [5] developed this concept further while CHAMBERS and TANTRAM [6] worked on a variant, dependent on using a non-eutectic mixture of the carbonates such that, at the operating temperature, the filler was provided by the excess solid carbonate.

The present data described in this paper refer to the inert-filler type and specifically to the use of MgO in conjunction with the $NaLiCO_3$ melt (m.p. 514°C). Some preliminary comment on the general properties found may prove helpful.

The forces holding these diaphragms together are almost entirely surface tension forces and depend on the presence of an extremely high meniscus length (molten carbonate on magnesia) at the surfaces of the diaphragm. This is demonstrated by the fact that a sample fully immersed in molten carbonate loses virtually all its compressive strength. For this reason the smaller the particle size of the magnesia the better, and we have used "levissima" magnesia (surface area $30 m^2/g$) throughout.

The useful composition range is between about 45 $^w/_o$ MgO and 70 $^w/_o$ MgO. As the percentage of MgO is increased, the strength increases and the power of electrolyte retention increases, but the diaphragm conductivity decreases. We have therefore compromised with a composition of 63·5 $^w/_o$ MgO.

## 5.2 Fabrication Methods

We have described the various fabrication methods in detail elsewhere.[7] The methods tried include cold pressing and liquid-phase sintering, infiltration, hot pressing above the carbonate melting point and hot pressing just below the carbonate melting point.

The last method is preferred. It takes advantage of the high degree of plastic flow exhibited by the carbonates just below the melting point and enables us to mould diaphragms of any desired shape to a high density, without any of the problems of extrusion or sticking to the mould.

## 5.3 Thermal Expansion

Thermal expansion is of importance in cell and battery design, since the diaphragm must be mated with the electrodes, current collectors, separator, plates, etc. A quartz dilatometer, with a transducer to record the expansion on a X–Y recorder, was used to measure the thermal expansion of hot-pressed diaphragms and also of a cast $NaLiCO_3$ sample up to 400°C. Above 400°C this method was not suitable owing to the reaction of the $NaLiCO_3$ with the $SiO_2$ rods and slight compression of the paste electrolyte, caused by the weight of the silica rod above it. Up to 400°C the coefficients do not vary with temperature. Table 1 shows the results for a range of compositions.

**Table 1**
**Linear Coefficients of Thermal Expansion 20°–400°C**

| Material ($^w/_o$) | Coefficient ($cm/cm/°C$) |
|---|---|
| 100% MgO | $11 \cdot 5 \times 10^{-6}$ |
| 80% MgO, 20% NaLiCO$_3$ | $15 \cdot 9 \times 10^{-6}$ |
| 63·5% MgO, 36·5% NaLiCO$_3$ | $19 \cdot 5 \times 10^{-6}$ |
| 50% MgO, 50% NaLiCO$_3$ | $21 \cdot 0 \times 10^{-6}$ |
| 100% NaLiCO$_3$ | $22 \cdot 4 \times 10^{-6}$ |

These results are plotted in Figure 4.

For a binary phase material the coefficient of thermal expansion of the composite will be given by

$$\alpha = \frac{\alpha_1 K_1 \dfrac{W_1}{d_1} + \alpha_2 K_2 \dfrac{W_2}{d_2}}{K_1 \dfrac{W_1}{d_1} + K_2 \dfrac{W_2}{d_2}}$$

where $\alpha$ is the expansion coefficient, $K$ the bulk modulus, $w$ the weight percentage and $d$ the density.[8] The suffix $1$ and $2$ refer to phase 1 and phase 2 respectively.

For our experimental results to show a reasonable agreement with the equation, we would have to assume that $K_1$ for $NaLiCO_3$ was about 3 times $K_2$ for MgO. The curve obtained by making this assumption is the dashed curve in Figure 4. In practice, for single crystals we would expect $K_1$ to be only about $\frac{1}{10}$ of the value of $K_2$.[8] The answer may well lie in the fact that levissima MgO has been shown by density determinations to be microporous.

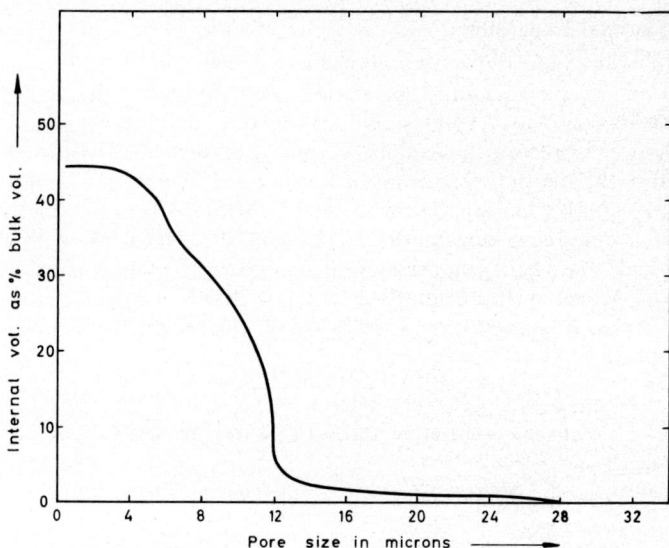

FIGURE 3.—Pore-size distribution of porous magnesia disc as determined by mercury porosimeter.

For example, MURRAY[9] quotes a density of $3.42\,g/ml$ for MgO prepared by thermal decomposition of $Mg(OH)_2$ at $700°C$, compared with $3.58\,g/ml$ for a single crystal. This indicates a microporosity of $7\%$. The bulk modulus for this material may therefore be exceptionally low.

We have no direct measurements for the thermal expansion of the composite material above the melting point of the carbonate, which in practice may not be very meaningful, since the liquid component can move relative to the solid. The linear thermal expansion of the $NaLiCO_3$ melt is approximately $80 \times 10^{-6}$ cm/cm°C.[7] A further figure of practical significance is the volume expansion of $NaLiCO_3$

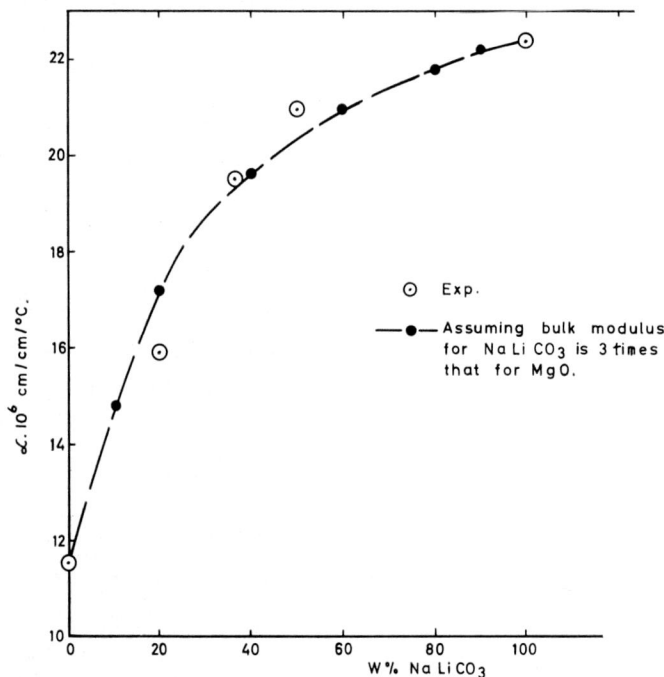

FIGURE 4.—Coefficient of thermal expansion of paste electrolyte discs as a function of composition.

on melting. Using the density/temperature data for the solid and for the melt, we calculate this to be approximately 9%.

### 5.4 Compressive Strength and Creep

The apparatus we used is very similar to that described by NORTON.[10] It is shown diagrammatically in Figure 5.

Figure 6 shows the results of a typical test for 63·5% MgO sample. A given compressive load was applied and the deformation was measured; the load was then removed and the permanent deformation was measured; this procedure was then repeated for a range of loads. In the test illustrated, this procedure was then repeated on the sample at the higher temperature. The base-line point we have taken is zero compression at 6·7 p.s.i., since this is the minimum load unavoidably applied in setting up the apparatus.

We see that the bulk of the deformation is completely elastic. This property is of great practical use, making for easy sealing when constructing cells. We also see that there is a small permanent deformation, increasing slightly with increasing load, which we attribute

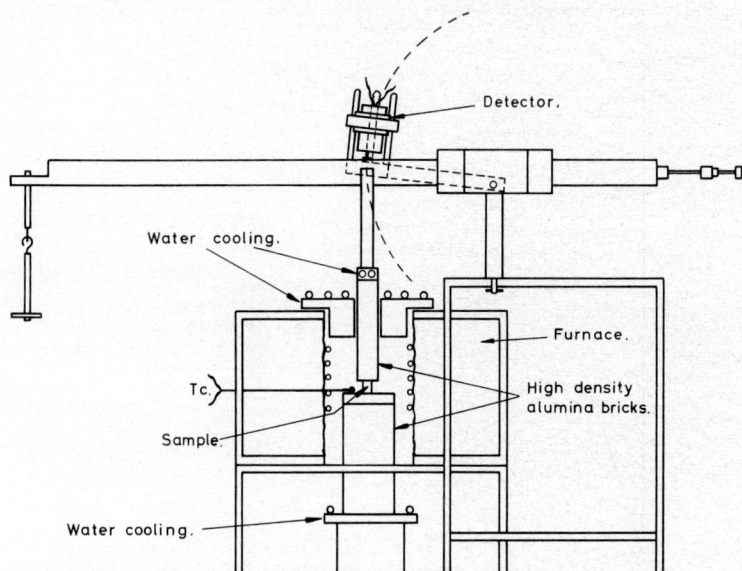

FIGURE 5.—Load-testing apparatus (not to scale)

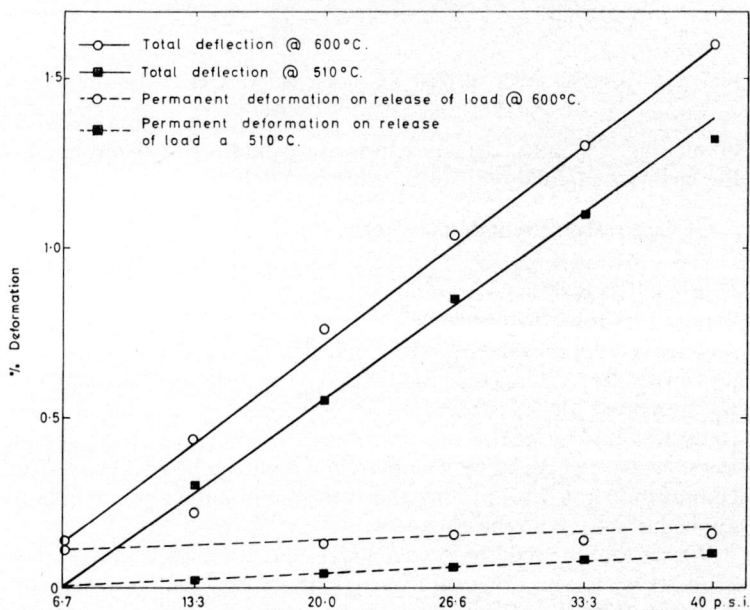

FIGURE 6.—Compression of paste electrolyte discs at elevated temperatures.

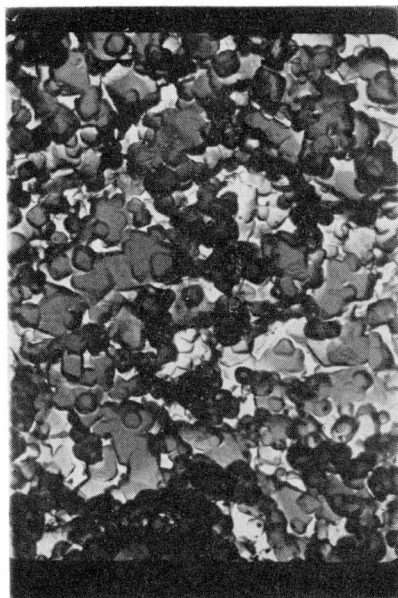

FIGURE 7.—Electron micrograph of a 55% MgO 45% NaLiCO₃ hot-pressed paste electrolyte (carbon replica, ×10,800).

to slight rearrangements in the microstructure. The electron micrograph (Figure 7) shows the open structure of the magnesia and some non-uniformity and hence indicates that such rearrangements would be likely.

A further test in which a sample was held at 40 p.s.i. and 600°C for 48 h showed no further deformation. This lack of creep is confirmed by observations on practical cells using these diaphragms under compression and running for periods up to 60 days.

## 5.5 Ionic Conductivity

This was measured by an A.C. method using a Wayne Kerr Bridge at 1500 Hz. In the experimental set-up the cell constant was inevitably low and hence the accuracy was low.

The relative conductivity $J$ and the tortuosity factor can be calculated using the relationships.

$$J = \frac{\text{Specific conductivity of diaphragm}}{\text{Specific conductivity of pure electrolyte}}$$

$$J = \frac{P}{Q}$$

where $P$ is the fractional volume occupied by the conductive carbonates and $Q$ is the tortuosity factor.

The data available are at present limited, but some results for semi-solid diaphragms are given in Table 2.

Table 2

| Composition (w/o) | Temperature (°C) | J | P | Q |
|---|---|---|---|---|
| 50% MgO | 550 | 0·43 | 0·61 | 1·43 |
| 63·5% MgO | 550 | 0·25 | 0·5 | 2·0 |

The tortuosity factor is a little lower than found with equivalent impregnated pre-sintered magnesia diaphragms.

## 6. FURTHER DATA REQUIRED

The preliminary data obtained have enabled us to choose a suitable composition (63·5% MgO) for these semi-solid diaphragms and to use these diaphragms for an investigation into the development problems of molten-carbonate fuel cells.

For a final optimization, however, we require more data of this type—for example, on compressive strength, yield point, shear strength, conductivity, etc., covering the full range of potentially useful compositions. We would like also to see the effect of alternative inert filters. A further property of great practical significance is the equivalent pore size of the material, which is a measure of its power to retain electrolyte against opposing forces; for example, the surface tension forces arising when the diaphragm is combined with a porous electrode. We would like to make our electrodes of small pore size to increase the active area, but we must not allow them to flood. At present we have no convenient method of measuring this property.

## ACKNOWLEDGMENTS

The authors wish to thank the Board of Energy Conversion Limited for permission to publish.

## REFERENCES

1. CHAMBERS, H. H., and TANTRAM, A. D. S., "Fuel Cells", Vol. 1 (Ed. G. J. Young) (Rheinhold Publishing Corp., 1960) pp. 94–108.
2. DAVTYAN, O. K., *Bull. Acad. Sci. U.R.S.S. Chem. Sci. Tech.*, **107**, 215, 1946
3. SALVADORI *et al.*, "La Pile Gaz pour l'Utilisation des Gaz Industriels." Paper read at Deuxièmes Journèes Internationales d'Etude des Piles Combustibles, Brussels, June 1967.
4. BROERS, G. H. J., and KETELAAR, J. A. A., loc. cit., (1), pp 78–93.
5. BROERS, G. H. J., and SCHENKE, M., "Fuel Cells" Vol. 2 (Ed. G. J. Young), (Rheinhold Publishing Corp, 1963) pp 6–23.

6. CHAMBERS, H. H., and TANTRAM, A. D. S., *Brit. Pat.* 806,592, 1958.
7. TANTRAM, A. D. S., TSEUNG, A. C. C., and HARRIS, B. S., "Hydrocarbon Fuel Cell Technology" (Ed. B. S. Baker) (Academic Press, 1965) pp. 187–211.
8. KINGERY, W. D., "Introduction to Ceramics" (John Wiley & Sons, 1962).
9. MURRAY, P., "Agglomeration" (Ed. W. A. Knapper) (Interscience Publishers, 1961) pp. 93–116.
10. NORTON, F. H., *J. Amer. Ceram. Soc.*, **22**, 334, 1939.

# 26.—Reactions of Oxygen–Potassium Compounds with Some Refractory Oxides

By M. Devalette, C. Fouassier, G. Le Flem, M. Tournoux and
P. Hagenmuller

*Service de Chimie minérale structurale de la faculté des Sciences de Bordeaux,
associé au C.N.R.S.*

## ABSTRACT
E18/A512

*The high melting-point of zirconia makes it a suitable material for magneto-hydrodynamic (M.H.D.) conversion, but its resistance to corrosion by alkali ions in an oxidizing atmosphere is low. The action on zirconia of some oxy-compounds of potassium has therefore been investigated. The less acid the gaseous phase evolved, the richer in $K_2O$ is the compound obtained. Except for the carbonate, the phase resulting depends on the temperature of the reaction, the proportions of the reagents, and even on the shape of apparatus. (1) $K_2CO_3$ reacts only above 950°C: $K_2CO_3 + 3ZrO_2 \rightarrow K_2Zr_3O_7 + CO_2\uparrow$; (2) KOH reacts with zirconia at 600°C. The low-temperature α-form of $K_2Zr_2O_5$ is the phase richest in $K_2O$ obtained: $2KOH + 2ZrO_2 \rightarrow K_2Zr_2O_5 + H_2O\uparrow$; (3) $KNO_3$ begins to react at 800°C. With a sufficient excess, $K_2ZrO_3$ is obtained: $2KNO_3 + ZrO_2 \rightarrow K_2ZrO_3 + oxides$ of nitrogen. For systems poorer in $KNO_3$, α-$K_2Zr_2O_5$ would form between 800° and 850°C; above 850°C these systems lead either to the high-temperature β form or to $K_2Zr_3O_7$: $2KNO_3 + 2ZrO_2 \rightarrow β-K_2Zr_2O_5 + oxides$ of nitrogen; $2KNO_3 + 3ZrO_2 \rightarrow K_2Zr_3O_7 + oxides$ of nitrogen; (4) $K_2O$ mixed in stoicheiometric proportions with $ZrO_2$ forms $K_4ZrO_4$ at 450°C: $ZrO_2 + 2K_2O \rightarrow K_4ZrO_4$. The ranges of thermal stability of the various phases and their crystallographic features are reported. The results obtained are compared with those of other systems studied: Ge–O–K; Sn–O–K, Pb–O–K; Th–O–K. The action of $K_2O$ on $ThO_2$ at 500°C leads only to $K_2ThO_3$, which is isotypical of β-$Na_2ZrO_3$: $ThO_2 + K_2O \rightarrow K_2ThO_3$.*

## Action de composés oxygènes du potassium sur quelques oxydes réfractaires

*La zircone possède un point de fusion dur; elle constitue donc un matériau de choix pour la conversion M.H.D., mais son emploi est conditionné par sa résistance à la corrosion due aux ions alcalins en atmosphères oxydante. Dans cette perspective une étude relative à* *l'action, sur la zircone, de quelques composés oxygénés du potassium, a été menée. Le composé le plus basique obtenu est d'autant plus riche en $K_2O$ que la phase gazeuse formée est moins acide. Sauf dans le cas du carbonate, la nature de la phase obtenue dépend de la température d'attaque, des proportions des constituants de la réaction et même des conditions géométriques. (1) Le carbonate de potassium ne réagit qu'au dessus de $950°C$: $K_2CO_3 + 3\ ZrO_2 \rightarrow K_2Zr_3O_7 + CO_2$. (2) La potasse réagit sur la zircone dès $600°C$. La phase la plus riche en $K_2O$ obtenue est la variété basse température $\alpha$ du dizirconate 2 $KOH + 2\ ZrO_2 \rightarrow K_2Zr_2O_5$-$\alpha + H_2O$. (3) L'action de $KNO_3$ débute à $800°C$. Lorsque $KNO_3$ est en excès suffisant on obtient le métazirconate: 2 $KNO_3 + ZrO_2 \rightarrow K_2ZrO_3$ + vapeurs nitreuses. Entre 800 et 850°C, $K_2Zr_2O_5$-$\alpha$ est mis en évidence pour des mélanges moins riches en $KNO_3$. Au-dessus de $850°C$, ceux-ci mènent soit à la variété haute température $\beta$ du dizirconate, soit au trizirconate: 2 $KNO_3 + 2\ ZrO_2 \rightarrow K_2Zr_2O_5$-$\beta$ + vapeurs nitreuses; 2 $KNO_3 + 3\ ZrO_2 \rightarrow K_2Zr_3O_7$ + vapeurs nitreuses. (4) Pour des proportions stoechiométriques, $K_2O$ sur $ZrO_2$ peut conduire dès $450°C$ à l'orthozirconate: $ZrO_2 + 2\ K_2O \rightarrow K_2ZrO_4$. Les domaines de stabilité thermique de ces diverses phases et leurs caractères cristallographiques ont été mis en évidence. Les résultats obtenus ont été comparés à ceux des systèmes Ge–O–K, Ti–O–K, Sn–O–K, Pb–O–K, Th–O–K. Dans ce dernier cas, le caractère fortement basique de la thorine ne conduit à 500° qu'à un métathorate $K_2ThO_3$ isotype de $Na_2ZrO_3$-$\beta$: $ThO_2 + K_2O \rightarrow K_2ThO_3$.*

### Reaktionen von Sauerstoff-Kalium-Verbindungen mit einigen feuerfesten Oxiden

*Der hohe Schmelzpunkt von Zirkonoxid macht es zu einem geeigneten Material für magneto-hydrodynamische (M.H.D.) Umwandler, aber seine Widerstandsfähigkeit gegen Korrosion durch Alkaliionen in oxidierender Atmosphäre ist gering. Daher wurde die Einwirkung einiger Sauerstoffverbindungen des Kaliums auf Zirkonoxid untersucht. Je weniger sauer die entwickelte Gasphase ist, umso reicher ist die entstehende Verbindung an $K_2O$. Außer für das Karbonat hängt die resultierende Phase von der Reaktionstemperatur, von dem Verhältnis der Partner und sogar von der Form der Apparatur ab. (1) $K_2CO_3$ reagiert nur oberhalb $950°C$: $K_2CO_3 + 3\ ZrO_2 = K_2Zr_3O_7 + CO_2$; (2) KOH reagiert mit Zirkonoxid bei $600°C$. Die Tieftemperaturform ($\alpha$-Form) von $K_2Zr_2O_5$ ist die $K_2O$-reichste Phase: 2 $KOH + 2\ ZrO_2 = K_2Zr_2O_5 + HO_2$; (3) $KNO_3$ beginnt bei $800°C$ zu reagieren. Mit einem ausreichenden Uberschuß wird $K_2ZrO_3$ erhalten: 2 $KNO_3 + ZrO_2 = K_2ZrO_3 + $N-Oxide. Für $KNO_3$-ärmere Systeme würde sich zwischen $800°C$ und $°850C$ $\alpha$-$K_2Zr_2O_5$ bilden; oberhalb $850°C$ führen* *diese Systeme zur Hochtemperaturform ($\beta$-Form) oder zu $K_2Zr_3O_7$:2 $KNO_3+2\ ZrO_2=\beta\text{-}K_2Zr_2O_5+N\text{-}Oxide$: 2 $KNO_3+3ZrO_2=K_2Zr_3O_7$ $+N\text{-}Oxide$; (4) $K_2O$ in stöchiometrischem Verhältnis mit $ZrO_2$ gemischt, bildet $K_4ZrO_4$ bei 450°C: $ZrO_2+2\ K_2O=K_4ZrO_4$. Die Bereiche thermischer Stabilität für die verschiedenen Phasen und ihre Kristallographischen Eigenschaften werden mitgeteilt. Die erhaltenen Ergebnisse werden mit denen anderer untersuchter Systeme verglichen: mit GeO-K, Sn-O-K, Pb-O-K, Th-O-K. Die Einwirkung von $K_2O$ auf $ThO_2$ bei 500°C führt nur zu $K_2ThO_3$, das mit $\beta\text{-}Na_2ZrO_3$ isotyp ist:$ThO_2+K_2O=K_2ThO_3$.*

# 1. INTRODUCTION

Zirconia ($ZrO_2$) has been the object of numerous investigations because of its importance in magnetohydrodynamic energy conversion processes (MHD). The reaction tube is exposed to considerable corrosion by alkali ions in oxidizing medium. Zirconia, as well as $HfO_2$, $ThO_2$ and some of the zirconates of the alkaline earths, has proved to be one of the more resistant materials to such treatment. On this account we decided to study the reactions between the carbonate, hydroxide, nitrate, and monoxide of potassium and the oxides of zirconium and thorium ($ZrO_2$ and $ThO_2$).

In these reactions, the less acid the gas formed, the richer in $K_2O$ is the most basic compound obtained. The composition of the phases obtained depends on the reaction temperature, its duration, the proportion of the initial reaction partners and even on geometric factors.

# 2. ZIRCONIA

## 2.1 Potassium Carbonate and Zirconia

Potassium carbonate reacts with zirconia only above its melting point (891°C). Above 950°C the reaction is perceptible, and at 1050°C it becomes rapid. The compound obtained is $K_2Zr_3O_7$ (potassium trizirconate):

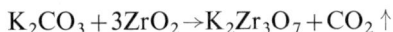

$$K_2CO_3+3ZrO_2\rightarrow K_2Zr_3O_7+CO_2\uparrow$$

Subsequently $K_2Zr_3O_7$ dissociates, forming zirconia.

## 2.2 Potassium Hydroxide and Zirconia

The reaction between potassium hydroxide and zirconia starts at 600°C. Between 600°C and 850°C excess hydroxide gives rise to $\varkappa\text{-}K_2Zr_2O_5$, the low-temperature variety of potassium dizirconate:

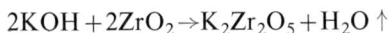

$$2KOH+2ZrO_2\rightarrow K_2Zr_2O_5+H_2O\uparrow$$

For products obtained above 850°C and but otherwise under the same conditions, X-ray diagrams reveal $K_2Zr_3O_7$ coexisting with the high-temperature ($\beta$ form) of $K_2Zr_2O_5$. Pure $\beta$-$K_2Zr_2O_5$ cannot be obtained by this method.

## 2.3 Potassium Nitrate and Zirconia

The thermal decomposition of potassium nitrate under oxygen has been studied by E. S. FREEMAN[1] at temperatures between 650° and 850°C. Between 650°C and 750°C the products are: potassium nitrite ($KNO_2$), nitrogen, oxygen, and traces of nitrogen peroxide. At 800°C the reaction is faster; the nitrite decomposes to form nitrogen, oxygen and potassium oxide ($K_2O$). The minimum temperature at which potassium nitrate could be used would therefore be at least 800°C. This high temperature makes it extraordinarily difficult to prepare single-phase products, because of their relatively low thermal stability: they start to decompose at the temperature at which they are formed, giving off free $K_2O$. The difference between the speeds of formation and of dissociation, however, still allows single-phase products to be prepared if the composition of the starting mixture, the temperature, the reaction time, etc., are carefully chosen. Two cases have to be considered separately, depending on whether the temperature is below or above 850°C.

### 2.31 Reactions below 850°C

Depending on the conditions prevailing, the reaction between potassium nitrate and zirconia can lead to two phases being formed: the metazirconate ($K_2ZrO_3$) and the low-temperature ($\alpha$) modification of the dizirconate $K_2Zr_2O_5$.

To obtain the metazirconate ($K_2ZrO_3$), 10% more of the nitrate is required than the stoicheiometric proportion. Pure $K_2ZrO_3$ is obtained from 5 g $ZrO_2$ after 8 h heating at 820°C.

$$2\,KNO_3 + ZrO_2 \rightarrow K_2ZrO_3 + \text{nitrous gases} \uparrow.$$

When the initial molar ratio $KNO_3/ZrO_2$ is between 2:2 and 1:1, a mixture of $K_2ZrO_3$ and $\alpha$-$K_2Zr_2O_5$ is obtained. At molar ratios below 1:1, in the presence of monoclinic zirconia, dizirconate is formed. Pure dizirconate is obtained by heating 5 g $ZrO_2$ for 8 h at 820°C with 15% excess of nitrate:

$$2\,KNO_3 + 2\,ZrO_2 \rightarrow \alpha\text{-}K_2Zr_2O_5 + \text{nitrous gases} \uparrow.$$

### 2.32 Reactions above 850°C

When the reaction temperature is kept above 850°C, the most basic compound is still the metazirconate $K_2ZrO_3$, but the reaction also produces other compounds poorer in $K_2O$: $K_2Zr_2O_5$ and $K_2Zr_3O_7$. As the temperature is higher, the preparation of these phases calls for more nitrate in order to make up for the larger losses through volatilization.

For 5 g $ZrO_2$ the experimental conditions are:

For $K_2ZrO_3$, 2 h at 900°C, or 30 min. at 1000°C, with 50% excess of nitrate.

For $\beta$-$K_2Zr_2O_5$, 1 h at 1000°C and 30% excess of nitrate.

For $K_2Zr_3O_7$, 1 h at 1000°C and 10% excess of nitrate.

## 2.4 Potassium Monoxide and Zirconia

Potassium oxide starts to react with zirconia at 450°C. The most alkaline compound obtained is the orthozirconate, $K_4ZrO_4$. The constituents are in stoicheiometric proportions:

$$ZrO_2 + K_2O \rightarrow K_4ZrO_4$$

The potassium (mon)oxide was prepared according to a method developed by E. RENGADE:[2] controlled oxidation of potassium metal followed by vacuum distillation in the presence of excess potassium.

## 2.5 Thermal Decomposition of Potassium Zirconates

The orthozirconate of potassium dissociates above 570°C to form $K_2O$ and $\alpha$-$K_2Zr_2O_5$ according to the reaction:

$$2\ K_4ZrO_4 \rightarrow \alpha\text{-}K_2Zr_2O_5 + 3\ K_2O \uparrow$$

The thermal decomposition of the metazirconate $K_2ZrO_3$ also starts at 570°C, but the course of the reaction when the temperature is below 850°C differs from that when it is above.

Between 570° and 850°C the metazirconate forms, $\alpha$-$K_2Zr_2O_5$, which is stable up to 720°C. Between 720° and 850°C, $\alpha$-$K_2Zr_2O_5$ dissociates to produce monoclinic zirconia:

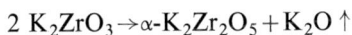

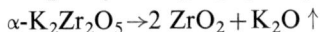

$$2\ K_2ZrO_3 \rightarrow \alpha\text{-}K_2Zr_2O_5 + K_2O \uparrow$$
$$\alpha\text{-}K_2Zr_2O_5 \rightarrow 2\ ZrO_2 + K_2O \uparrow$$

Table 1

| $t < 570°C$ | $K_2ZrO_3$ |
|---|---|
| $570°C < t < 720°C$ | $K_2ZrO_3 \rightarrow \alpha\text{-}K_2Zr_2O_5$ |
| $720°C < t < 850°C$ | $K_2ZrO_3 \rightarrow \alpha\text{-}K_2Zr_2O_5 \rightarrow ZrO_2$ |
| $850°C < t$ | $K_2ZrO_3 \rightarrow \beta\text{-}K_2Zr_2O_5 \rightarrow K_2Zr_3O_7 \rightarrow ZrO_2$ |

Above 850°C the final decomposition product is always monoclinic zirconia, but two intermediate phases occur: the dizirconate, $\beta$-$K_2Zr_2O_5$, and the trizirconate, $K_2Zr_3O_7$:

$$2\ K_2ZrO_3 \rightarrow \beta\text{-}K_2Zr_2O_5 + K_2O \uparrow$$
$$3\ \beta\text{-}K_2Zr_2O_5 \rightarrow 2\ K_2Zr_3O_7 + K_2O \uparrow$$
$$K_2Zr_3O_7 \rightarrow 3\ ZrO_2 + K_2O \uparrow$$

## 3. THORIA

The hydroxide, carbonate and nitrate of potassium have no influence on thoria ($ThO_2$). Potassium monoxide, however, reacts at 550°C, forming metathorate $K_2ThO_3$ as follows:

$$ThO_2 + K_2O \rightarrow K_2ThO_3$$

We failed to find evidence of other compounds being formed, richer either in $K_2O$ or in $ThO_2$.

The stability of the metathorate is very low; above 650°C $K_2ThO_3$ dissociates, liberating $K_2O$ and regenerating $ThO_2$:

$$K_2ThO_3 \rightarrow ThO_2 + K_2O \uparrow$$

## 4. THE $MO_2$–$K_2O$ SYSTEMS
### (M = Ti, Zr, Hf, Th, Ge, Sn, Pb.)

The results presented here have been compared with those obtained for homologous systems in which other elements capable of forming tetravalent oxides are concerned: germanium, titanium, hafnium, tin and lead.

Table 2 enables the crystallographic characteristics to be compared of a number of ternary oxygen-bearing compounds that have been recently prepared, most of them in our laboratory.

The ternary oxygen compounds of comparable composition occurring in the different systems show a wide variety in structure: when the ratio $K_2O/MO_2$ is smaller than unity, only the following compounds are isotypic:

$K_2Sn_3O_7$ and $K_2Pb_3O_7$    Orthorhombic
$K_2Zr_3O_7$ and $K_2Hf_3O_7$    Tetragonal
$\beta$-$K_2Zr_2O_5$ and $K_2Hf_2O_5$    Orthorhombic ($D_{2h}^6$)

When the $K_2O/MO_2$ ratio is 1, the influence of the size of the $Me^{4+}$ ion on the lattice structure is less pronounced: $K_2TiO_3$, $K_2SnO_3$, $K_2ZrO_3$ and $K_2PbO_3$, in which the ionic radius runs from 0·68 to 0·84 Å in Ahrens Classification, all belong to the orthorhombic system. Their lattice parameters are given in Table 3.

Potassium metathorate, $K_2ThO_3$, is isomorphous with $Na_2ZrO_3$ and is monoclinic.

Table 2

| $\dfrac{K_2O}{MO_2}$ | Ge-O-K $r_{Ge^{4+}} = 0.53\text{Å}$ | Ti-O-K $r_{Ti^{4+}} = 0.68\text{Å}$ | Sn-O-K $r_{Sn^{4+}} = 0.71\text{Å}$ | Hf-O-K $r_{Hf^{4+}} = 0.78\text{Å}$ | Zr-O-K $r_{Zr^{4+}} = 0.79\text{Å}$ | Pb-O-K $r_{Pb^{4+}} = 0.84\text{Å}$ | Th-O-K $r_{Th^{4+}} = 1.02\text{Å}$ |
|---|---|---|---|---|---|---|---|
| 2 | $K_4GeO_4$ Triclinic | $K_4TiO_4$ Triclinic | $K_4SnO_4$ Triclinic | | $K_2ZrO_4$ Triclinic | $K_4PbO_4$ Triclinic | |
| 1 | $K_2GeO_3$ | $K_2TiO_3$ Orthorhombic | $K_2SnO_3$ Orthorhombic | | $K_2ZrO_3$ Orthorhombic | $K_2PbO_3$ Orthorhombic | $K_2ThO_3$ Monoclinic |
| $\frac{1}{2}$ | $K_2Ge_2O_5$ | $K_2Ti_2O_5$ Monoclinic | | $K_2Hf_2O_5$ Orthorhombic | $K_2Zr_2O_5$ $\alpha$: Hexagonal $\beta$: Orthorhombic | $K_2Pb_2O_5$ | |
| $\frac{1}{3}$ | | $K_2Ti_3O_7$ | $K_2Sn_3O_7$ Orthorhombic | $K_2Hf_3O_7$ Cubic | $K_2Zr_3O_7$ Cubic | $K_2Pb_3O_7$ Orthorhombic | |
| $\frac{1}{4}$ | $K_6Ge_{11}O_{25}$ | $K_2Ti_4O_9$ | | | | | |
| $\frac{1}{5}$ | | $K_2Ti_5O_{11}$ | | | | | |
| $\frac{1}{6}$ | | $K_2Ti_6O_{13}$ Monoclinic | | | | | |
| $\frac{1}{7}$ | $K_2Ge_7O_{15}$ | | | | | | |

**Table 3**

|  | $a$ (Å) | $b$ (Å) | $c$ (Å) |
|---|---|---|---|
| $K_2TiO_3$ | 10·04 | 6·95 | 5·46 |
| $K_2SnO_3$ | 10·31 | 7·11 | 5·72 |
| $K_2ZrO_3$ | 10·32 | 6·97 | 5·70 |
| $K_2PbO_3$ | 10·45 | 6·99 | 5·95 |

All phases that correspond to a $K_2O/MO_2$ ratio of 2 are isomorphous ($K_4GeO_4$, $K_4TiO_4$, $K_4SnO_4$, $K_4ZrO_4$ and $K_4PbO_4$) and crystallize in the triclinic system.

## REFERENCES

1. FREEMAN, E. J., *J. Amer. Ceram. Soc.*, **79**, 838, 1957.
2. RENGADE, E., *C.R.Acad.Sci.*, **144**, 754, 1907.

# 27.—Fusion-cast Carbide–Boride–Graphite Ceramics

By A. M. Alper, R. C. Doman and R. N. McNally

*Research and Development Laboratories, Corning Glass Works, Corning, New York 14830*

*ABSTRACT*　　　　　　　　　　　　　　　　　　　E22/E17

*Discusses fused carbides and/or borides containing free graphite. Phase assemblages, microstructures, and properties are related to their compatibility and phase diagrams. The following systems are discussed: Ti–C–B, Zr–C–B. Hf–C–B, V–C–B, Cr–C–B, and W–C–B. Only the parts of these systems which contain free graphite are considered. All these bodies have excellent thermal-shock resistance. Other properties, such as electrical, thermal, mechanical, and chemical, can be modified by choosing different phase assemblages. Some of these materials have been cast into large shapes more than 18 in. long, and they can be machined into articles.*

## Céramiques de carbure–borure–graphite coulées à l'état fondu

*Les caractéristiques de carbures et/ou borures, contenant du graphite libre, coulés à l'état fondu sont considérées. Les assemblages de phases, les microstructures et les propriétés sont étudiées au point de vue de leur compatibilité et à l'aide des diagrammes de phase. Les systèmes suivants sont examinés: Ti–C–B, Zr–C–B, Hf–C–B, V–C–B, Cr–C–B, et W–C–B. Seules les portions de ces systèmes contenant du graphite libre sont prises en considération. Tous ces produits présentent une excellente résistance au choc thermique. D'autres propriétés, comme les caractéristiques électriques, thermiques, mécaniques et chimiques peuvent être modifiées en choisissant différents assemblages de phases. Certains de ces matériaux ont été coulés en pièces de grande dimension (plus de 45 cm de long). Ces pièces peuvent ensuite être usinées.*

## Schmelzgegossene Cerbid–Borid–Graphit–Keramik

*Im Artikel werden Carbide und/oder Boride diskutiert, die freien Graphit enthalten. Phasenanordnung, Mikrostruktur und Eigenschaften werden auf ihre gegenseitige Verträglichkeit und auf ihre Phasendiagramme bezogen. Folgende Systeme werden besprochen: Ti–C–B, Zr–C–B, Hf–C–B, V–C–B, Cr–C–B, und W–C–B, wobei nur die*

*Teildiagramme betrachtet werden, in denen freier Graphit auftritt. Keramik dieser Art hat eine ausgezeichnete Widerstandsfähigkeit gegen thermischen Schock. Die anderen Eigenschaften wie z.B. elektrische, thermische, mechanische und chemische können durch die Wahl verschiedener Phasenzusammenstellungen variiert werden. Einige dieser Materialien wurden in große Formen von mehr als 45 cm Länge gegossen. Sie lassen sich auch zu verschiedenen Artikeln verarbeiten.*

## 1. INTRODUCTION

The work on fusion-cast graphite–carbide and/or boride materials that is to be discussed is based primarily on research work that has been going on at Corning Glass Works for many years. This work was primarily exploratory in nature. The main purpose was to investigate unusual fused refractory compositions and determine whether they have unusual microstructures and properties. It was found that fusion-cast carbides and/or borides can be made which contain graphite crystals and that microstructure can be altered by changing the ratio of refractory metal to graphite.

For many years numerous scientists have investigated refractory metals. Most of the work has been directed towards sintered bodies, usually without graphite. More information can be obtained about these materials from SCHWARZKOPF and KIEFFER,[1,2] TINKLEPAUGH,[3] and KIEFFER and BENESOVSKY.[4] P. T. B. SHAFFER[5] has compiled a list of properties of high-temperature materials which includes most of these components. In the past few years, KENDALL, LEEDS, ROSSI, and SLAUGHTER,[6–12] at Aerospace Corporation, and FOSTER and HILDEBRAND,[13 14] at Battelle, have published a number of excellent papers on fusion-cast Group IV B carbides and graphite bodies. Much phase-equilibrium work has been done on these materials. An up-to-date and very complete compilation of most of the important work is found in references 4 and 15 to 31.

The following sections discuss some of the more important findings of the authors.

## 2. SYSTEMS STUDIED

### 2.1 Ti–B–C

The constitution diagram of the Ti–C system has been reported by RUDY, HARMON and BRUKL.[32] It shows that: carbon-deficient titanium carbide can be made; the melting point of TiC is $3067° \pm 15°C$; and, as the carbon content is increased, the liquidus decreases to $2776° \pm 6°C$ at $63° \pm 1·0$ $^a/_o$ carbon at the eutectic. As carbon is increased above the eutectic, the liquidus increases.

Figure 1 shows that the microstructure of the fusion-cast TiC–graphite changes as the graphite content of the bodies is increased. There is a substantial increase in thermal-shock resistance from 5 to 18 cycles (1800°C-water) as the free graphite platelets appear, as shown in the 67% Ti–33% C example. It has also been found that, as the amount of free graphite increases, the thermal-shock resistance increases to 50 cycles. These results indicate the importance of the

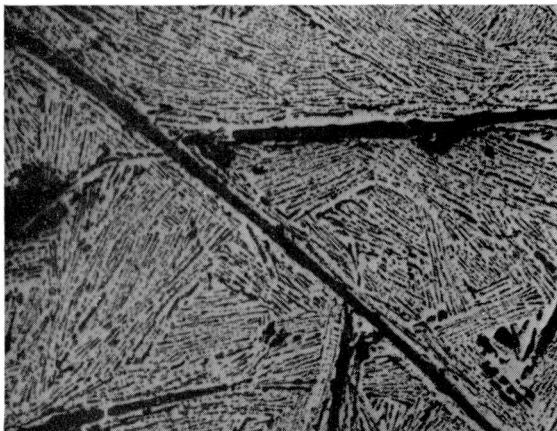

73%Ti–27%C 5 cycles

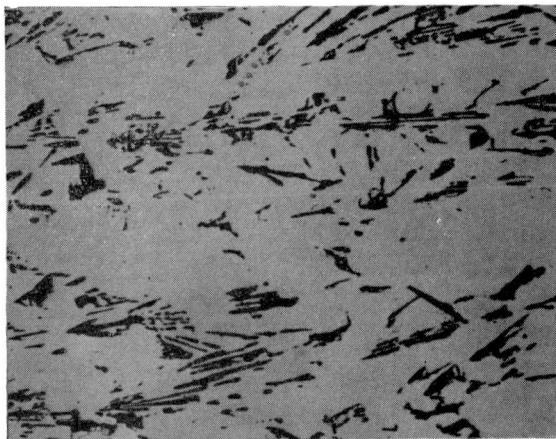

63%Ti–33%C 18 cycles

FIGURE 1.—Photomicrographs of fusion-cast TiC–C alloys.

amount of excess graphite in fusion-cast bodies with regard to the thermal-shock resistance that is desired in the ultimate product.

The strongest bodies have less graphite. Fusion-cast carbide–graphite bodies have been made which have a modulus of rupture between 14,000 and 19,000 lbf/in$^2$. Up to 1340°C no decrease in modulus of rupture occurs with increasing temperature. Measurements were not made at higher temperatures. The modulus of rupture of these materials is inversely proportional to the size of the crystals.

Table 1 summarizes some of the most important properties of a fusion-cast refractory which contains 84% TiC and 16% graphite. The electrical and thermal properties are very metallic in nature. At temperatures above 700°C this material has better oxidation resistance than that of titanium metal or graphite. It also has better resistance than graphite to corrosion by oxide slags (CaO, SiO$^2$ Al$_2$O$_3$, FeO, MgO) or molten metals (Fe, Al, Cu). It has much greater thermal-shock resistance than any refractory oxide or pure carbide or boride.

TiC–C refractories oxidize rapidly at temperatures between 500° and 700°C where a loose, non-adhering anatase oxide forms. At 1000°C, a denser adhering film of rutile forms, which causes oxidation to proceed at a much slower rate. At 1500°C the rate of oxidation again increases, and above 2150°C the material oxidizes rapidly and the film melts and flows off.

Figure 2 is a photograph of a TiC–C brick which was fusion-cast. Experience in the casting of this and similar bricks suggests that large castings (approximately 1000 lb) might be feasible. It is possible to make extremely dense castings with very little porosity. Figure 3 shows that a dense precision-machined piece can be manufactured from these castings, since the graphite phase makes these bodies machinable.

FIGURE 2.—A TiC–C ingot.

**Table 1**

| Properties | 84% TiC / 16% Graphite | 85% ZrC / 15% Graphite | 54% B₄C / 46% Graphite |
|---|---|---|---|
| Number of thermal shock cycles | 20 | 8 | 8 |
| Melting temperature | 2900°C | 3050°C | 2450°C |
| Coefficient of expansion 25°–1000°C | $78 \times 10^{-7}$ | $69 \times 10^{-7}$ | $50 \times 10^{-7}$ |
| Modulus of rupture: | | | |
| Room temperature | 14,500 lbf/in.² | 8000 lbf/in² | 2500 lbf/in² |
| 1340°C | 14,500 lbf/in² | 8000 lbf/in² | 2500 lbf/in² |
| Oxidation resistance: | | | |
| at 1000°C | Poor | Very poor | Good |
| at 1500°C | | | |
| Thermal conductivity: (cal/cm/sec/°C) | | | |
| at 700–1100°C | 0·075 | 0·05 | 0·065* |
| at 25°C | | | |
| at 20°C | | | |
| Specific heat (cal/g°C) | 0·14 at 25°C 0·19 at 200°C 0·24 at 1000°C | | |
| Degree of cracking in manufacture | Very, very slight | | |
| Electrical resistivity: | | | |
| at 25°C | $105 \times 10^{-6}$ ohm/cm | $63 \times 10^{-6}$ ohm/cm | $15{,}500 \times 10^{-6}$ ohm/cm |
| at 1000°C | $124 \times 10^{-6}$ ohm/cm | $155 \times 10^{-6}$ ohm/cm (500°C) | $5·78 \times 10^{-3}$ ohm/cm |
| Specific gravity | 4·1 | 5·3 | 2·3 |
| Young's modulus | $8·1 \times 10^{+6}$ lbf/in² | — | — |
| Shear modulus | $3·9 \times 10^{+6}$ lbf/in² | — | — |
| Poisson's ratio | 0·06 | — | — |
| Resistant to some molten oxides and metals | | | |

*For B₄C after BRADSHAW, W. G., and MATHEWS, C. O., "Properties of Refractory Materials: Collected Data and References" LMSD–2466 (June 24, 1958).

FIGURE 3.—A TiC–C machine nozzle.

The liquidus projection in the Ti–B–C system and the phase assemblages that exist under equilibrium conditions at liquidus temperatures when the composition is varied is reported by RUDY.[15]

The work at Corning indicates that bodies composed of the following phase assemblages can be formed: TiC–C, $B_4C$–C, $TiB_2$–C, TiC–$TiB_2$–C, $TiB_2$–$B_4C$–C.

The addition of boron to the TiC–C system modifies the properties in the following ways: the oxidation resistance is increased, the coefficient of thermal expansion is decreased, the modulus of rupture is approximately the same, the resistance to siliceous slags is decreased, the resistance to molten iron appears to reach a maximum as the $TiB_2$ phase is increased, and the refractoriness is decreased. Bodies containing $B_4C$ have much higher electrical resistance than TiC or $TiB_2$. It is possible to make Ti–B–C fusion-cast refractories higher in graphite than Ti–C refractories.

The coefficient of thermal expansion, over the range 25° to 1000°C for fusion-cast bodies high in $B_4C$, is about $50–55 \times 10^{-7}$ per °C; for those high in TiC, it is about $80–84 \times 10^{-7}$ per °C; and for others high in $TiB_2$, it is about $60–66 \times 10^{-7}$ per °C.

## 2.2 $B_4C$–C

This system is of interest because it is an extremely lightweight material. The density of pure $B_4C$ is about 2·51 g/cm³ and graphite is 2·25 g/cm³. Fusion-cast $B_4C$–C refractories with more than 50 $^w/_o$ graphite can be made that have much greater oxidation resistance than that of graphite. Up to 1000°C, a protective glassy oxide film forms. Above 1500°C, the material oxidizes rapidly. Figure 4, the

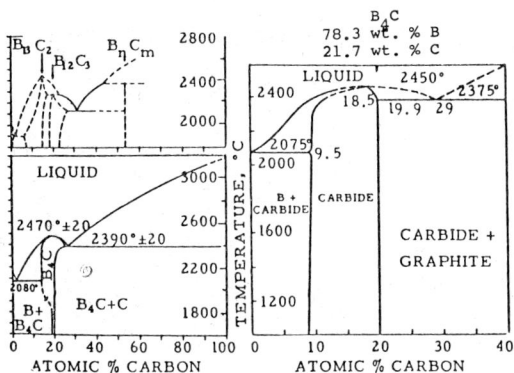

FIGURE 4.—Phase diagrams of the B–C system.[33,34,35]

phase diagram of the B–C system,[33,34,35] shows the liquidus temperatures and the phase compatibility as a function of composition and temperature.

Figure 5 shows how the microstructure changes as the graphite content is increased. Thermal and electrical conductivity increase correspondingly, and resistance to molten siliccous slag and to molten iron, modulus of rupture, Young's modulus, and resistance to oxidation decrease. These refractories can be cast into very large pieces. Since these bodies have some properties which are different from those of pure graphite, they might be used in applications for which graphite is not satisfactory.

Table 1 summarizes the properties of a $B_4C$–graphite (54/46) fused refractory. The most important properties are its high thermal shock resistance, low specific gravity, and refractoriness.

## 2.3 Zr–C–B

Fusion-cast materials containing graphite in the Zr–B–C system are very similar in microstructures, phase assemblages, and properties to those of the Ti–B–C system with free graphite. The following are some of the differences between these two systems: zirconium compounds are more refractory than titanium compounds; the electrical resistivity of the zirconium compounds is less than that of the titanium compounds; the coefficients of thermal expansion of the zirconium compounds are less; zirconium carbide has less oxidation resistance at 1000°C and at 1500°C than titanium carbide; however, above 2000°C they may have greater oxidation resistance than TiC bodies. The specific gravities of zirconium carbide and zirconium boride are greater than similar titanium compounds.

26

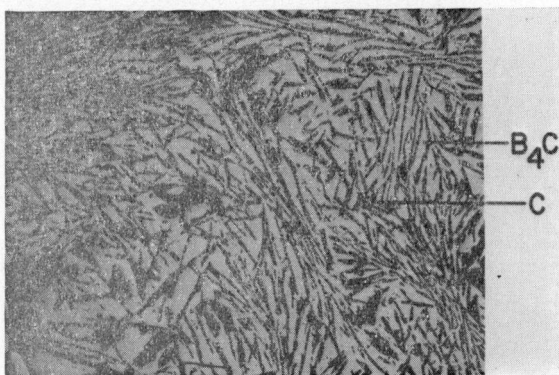

45·6%B 54·4%C

40%B 60%C

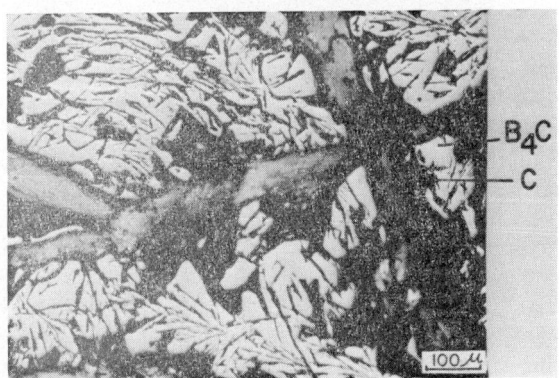

34%B 66%C

FIGURE 5.—Photomicrographs of polished sections of fusion-cast B$_4$C–C alloys.

Figure 6, phase diagram of the zirconium–carbon system,[30,31] indicates that, when carbon is added to zirconium carbide, the liquidus temperature decreases to a minimum at the eutectic, which is 65 $^a/_o$ carbon. When more carbon is added, the liquidus temperature increases rapidly. ADELSBERG, CATOFF, and TOBIN[36] report $2890° \pm 50°C$ for the eutectic. They also discussed the kinetics of the zirconium-carbon reaction at temperatures above $2000°C$,[37] carbon diffusion, and group IVB transition-element monocarbides.[38]

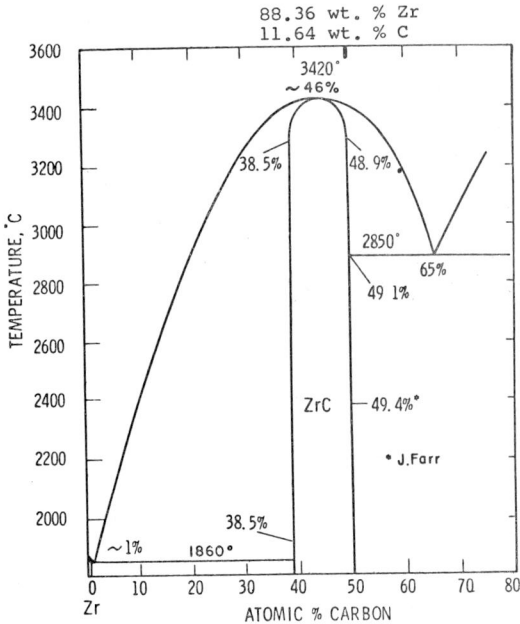

FIGURE 6.—Constitution diagram zirconium–carbon.[30,31]

Figure 7 shows how the microstructure of zirconium carbide–graphite alloys change as the graphite content is increased. Hypo-eutectic bodies generally have large crystals of primary zirconium carbide, which are surrounded by a mixture of fine-grained platelets of graphite–zirconium carbide. Some of these platelets are interconnected. The eutectic material consists almost entirely of fine-grained, partially interconnected, graphite platelets, which are surrounded by zirconium carbide. Hyper-eutectic bodies contain large platelets of graphite, which are surrounded by carbide and intra-granular fine-grained platelets of interlocked graphite. The bodies with more than 5 % graphite have very good thermal-shock resistance.

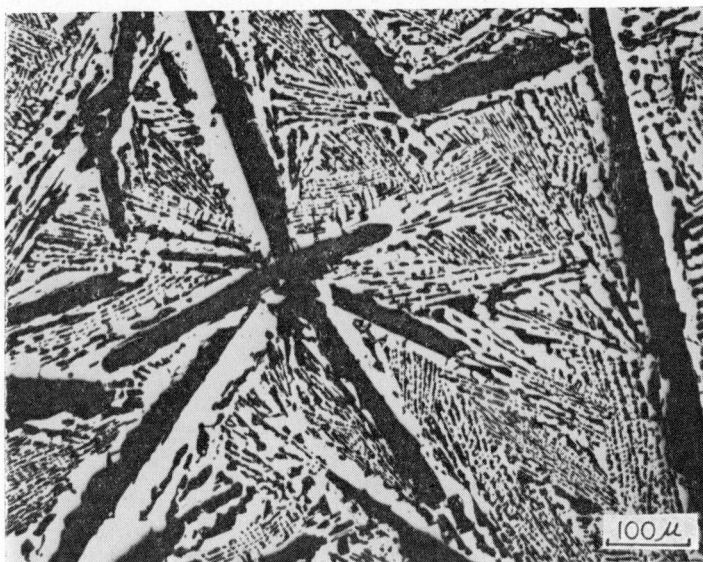

60%Zr   40%C

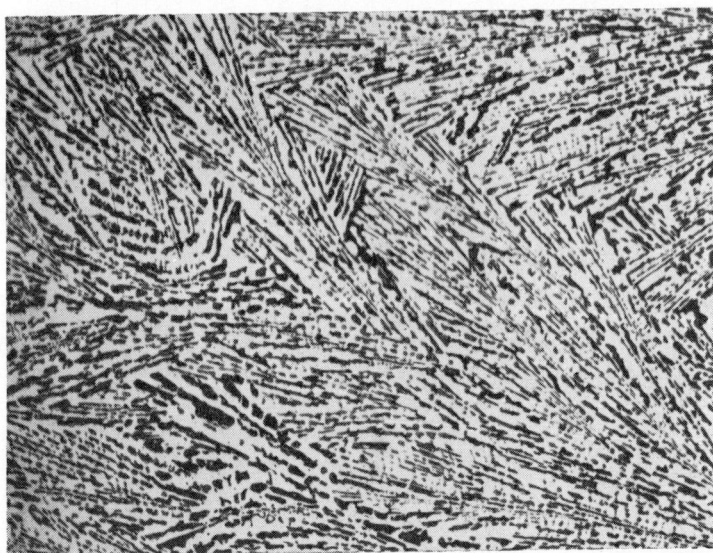

Eutetic

FIGURE 7.—Photomicrographs of polished sections of ZrC–C alloys.

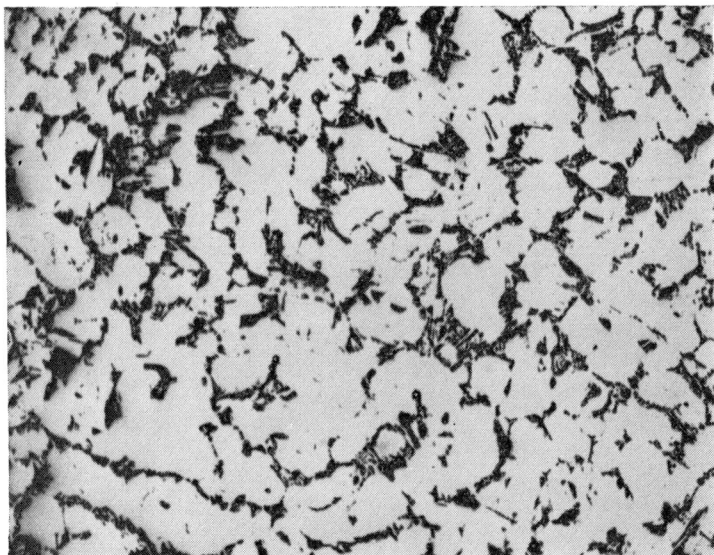

85%Zr  15%C

FIGURE 7.—*continued*

Compared to TiC-containing bodies, zirconium-carbide-containing bodies have poor oxidation resistance up to 1500°C. The oxide coatings which form on ZrC bodies in this temperature range are quite porous and flaky, whereas the oxide coatings on TiC materials are denser and more protective against rapid oxidation. At higher temperatures (above 1800°C) the zirconium oxide coating (melting point $\sim$2700°C) adheres to the carbide whereas the titanium oxide coatings melt (melting point $\sim$1840°C) and lose the protective mechanism.

Zirconium-carbide–graphite has better resistance to siliceous slags and to molten iron than graphite has. When both are exposed to molten iron-oxide-containing slags at 1700°C, zirconium carbide is cut (see Appendix) approximately 15% compared to 30% for graphite. When both are subjected to molten iron at 1700°C, zirconium-carbide-containing bodies are cut 10%, whereas graphite is cut 35%. Their resistance to slag depends on the amount of transition-metal oxide in the slag. Slags that are high (more than 25%) in iron oxide cut carbide–graphite bodies rapidly. Those low in iron oxide (less than 10%) do not attack the carbide-containing refractories very rapidly.

The oxidation state of the iron in the slag is very important. Slags high in $Fe^{3+}$ are much more corrosive than those high in $Fe^{2+}$. When these refractories are put into ferruginous slag, the graphite reacts with the slag and reduces the iron oxide to iron metal. The exposed interface of the refractory becomes enriched in the iron and carbide, and depleted in graphite. The refractory is then corroded by a combined process of oxidation of the carbide phase, solution of the oxide by the slag, and solution of the carbide phase in metallic iron. In certain specialized applications, these materials might replace oxides and graphite in metallurgical furnaces where molten iron or molten steel is being processed. As the graphite content is increased, there is a corresponding increase in electrical resistance; there is also a decrease in coefficient of thermal expansion, resistance to slag and molten iron, modulus of rupture, Young's modulus, and the specific gravity.

Table 1 summarizes the properties of a 85% zirconium carbide–15% graphite refractory. Some of the more important properties of this particular alloy are its high thermal-shock resistance, refractoriness, and low electrical resistivity. The moduli of rupture of refractories in this system were less than those in the TiC–C system, but of the same order of magnitude and, probably with more work, bodies with strengths approaching that of TiC-containing alloys can be made. These materials, like the TiC–C materials, appear to retain their strength at high temperatures.

Though it is more difficult to make large castings of zirconium-carbide–graphite alloys, because of the higher melting point, than of titanium carbide–graphite alloys, large castings (18 in. long) of ZrC–C fusion-cast refractories have been made. However, initial experience indicates it may be possible to make larger castings. Much work has been done on studying the use of these types of materials for aerospace applications.[10,39]

Figure 8 shows the compatibility regions and the liquidus in the Zr–B–C system.[15] Zirconium diboride, $B_4C$ and graphite, and ZrC, $ZrB_2$, and graphite, are compatible. Zirconium carbide has a higher melting point (3440°C) than $ZrB_2$ (3245°C). Consequently the highest-melting materials are the ones highest in zirconium carbide, and the melting point can be decreased by adding boron to the system. Eutectics form between the carbide and graphite and between the boride phases and graphite; and, as the carbon content of hypereutectic compositions is increased, the liquidus is increased. The addition of boron to the ZrC–C system increases the ability to form fused refractories that are high in graphite.

Figure 9 shows photomicrographs of polished sections of fusion-cast Zr–B–C alloys. They represent the microstructures of various

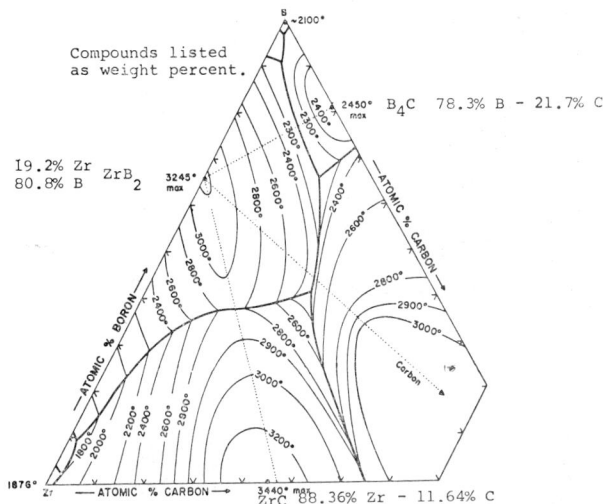

Compounds listed
as weight percent.

B ~2100°

2450° max   B$_4$C   78.3% B - 21.7% C

19.2% Zr   ZrB$_2$   3245° max
80.8% B

1876° Zr   ← ATOMIC % CARBON →   3440° max
ZrC   88.36% Zr - 11.64% C

FIGURE 8.—Liquidus projections in the Zr–B–C system.[15]

alloys in different compatibility fields and in different parts of an individual compatibility region. When boron is added to Zr–C and the whole is melted, the properties are changed in the following way: the oxidation resistance, electrical resistivity, and resistance to attack by molten iron increase, and the coefficient of thermal expansion decreases. Bodies rich in $ZrB_2$ are cut only 4% in molten iron, compared to a cut of 35% of graphite in the same test (2000 g molten iron at 1700°C). These materials have more resistance to ferruginous slag than has graphite. As the graphite phase is increased, the coefficient of expansion, the density, and the strength decrease. The electrical resistivity of these bodies can be modified by changing the ratio of phases and phase assemblages. Zirconium carbide has the lowest electrical resistivity and $B_4C$ has the highest. Zirconium diboride and graphite have an intermediate electrical resistivity. Alloys have been made in this system with greater than 50% graphite. The modulus of rupture of refractories in this system have generally ranged from 6000 to 14,000 lb/in$^2$ Decreasing the grain size increases the strength of these alloys. They might find metallurgical and aerospace applications.

## 2.4 Hf–B–C

The Hf–B–C system is very similar to the Ti–B–C and Zr–B–C systems. The main difference is that the hafnium compounds are

78%Zr, 2%B, 20%C

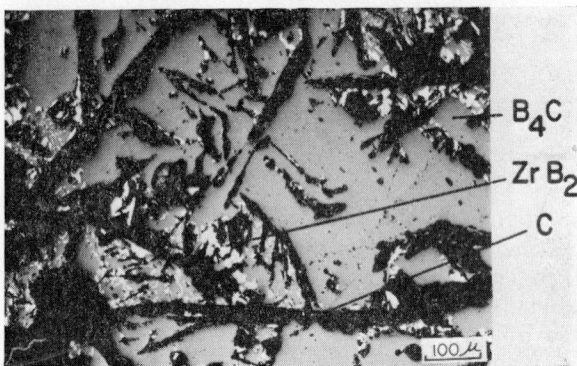

35%Zr, 42%B, 23%C

64·2Zr%, 5·7%B, 29·1%C

FIGURE 9.—Photomicrographs of polished sections of Zr–B–C alloys.

more refractory than zirconium or titanium compounds. The phase assemblages and the microstructures and properties are very similar. Since hafnium is much more expensive than zirconium or titanium, the use of these alloys will be restricted to specialty products.

The constitution diagram of the Hf–C system has been reported by RUDY.[40] A comparison of the phase diagrams of the Ti–C, Zr–C and Hf–C systems shows that the eutectic temperature of HfC–C is $3180° \pm 20°C$ compared to $2850°C$ for ZrC–C and $2776° \pm 6°C$ for TiC–C.

The liquidus of the Hf–B–C system[15] is very similar in appearance to those of the Zr–B–C (Figure 8) and of the Ti–B–C system; however, the system has higher melting temperatures.

Structures of the Hf–C and Hf–B–C alloys are very similar to the other IVB group of Ti–B–C and Zr–B–C (Figure 8). These alloys also have excellent thermal-shock resistance. The specific gravities of these materials are much higher than those of the other materials in the IVB group. HfC alloys have poor oxidation resistance. Oxidation becomes severe above $1000°C$. HfC-containing bodies have a lower electrical resistivity than those containing ZrC and TiC. The electrical resistivity increases slightly with temperature, as it does in alloys high in ZrC or TiC. When measured from $25°$ to $1000°C$, the coefficients of expansion of HfC bodies are slightly less than for ZrC alloys, which are less than for TiC alloys.

More data on group IVB metals–B–C systems can be found in RUDY and ST WINDISCH.[41] A discussion of solubility of boron in the carbide phases and solubility of carbon in the boride phases can be found in KIEFFER and BENESOVSKY[4] and in RUDY and ST WINDISCH.[41]

### 2.5 Group VB Metals–Boron–Carbon

The constitution diagram of the V–C system[42] indicates that VC melts incongruently. Graphite first crystallizes out of the liquid with an atomic ratio of carbon greater than $0·8$; then the graphite reacts with the liquid to form VC. The peritectic is located at $2700° \pm 50°C$. Since VC melts incongruently, the microstructure is different than that of alloys formed by congruent crystallization.

Constitution diagrams of Nb–C[43] and Ta–C[44] show that NbC and TaC melt congruently. NbC has a reported melting point of $3600° \pm 50°C$. As carbon is added to NbC, the melting temperature decreases until a minimum is reached at the eutectic temperature of $3300° \pm 50°C$ at $60·5$ $^a/_o$ carbon. As carbon is added to hyper-eutectic bodies, the liquidus temperature increases.

TaC melts congruently at $3985° \pm 40°C$. With carbon additions, the melting temperature decreases until the eutectic composition

($61 \pm 0.5$ $^a/_o$ C) is reached. The melting temperature of the eutectic is reported at $3445° \pm 26°$C. TaC–C refractories have the highest liquidus of all pure metal-carbide–carbon systems. Consequently they might have promising properties for aerospace applications.

Figure 10 compares the microstructures of NbC–C alloys with those of VC–C alloys. VC–C alloys contain large crystals of graphite surrounded by vanadium carbide, whereas NbC–C alloys contain large crystals of graphite surrounded by the eutectic-type micro-structure which consists of smaller, partially interlocking, intragran-ular platelets of graphite and niobium carbide.

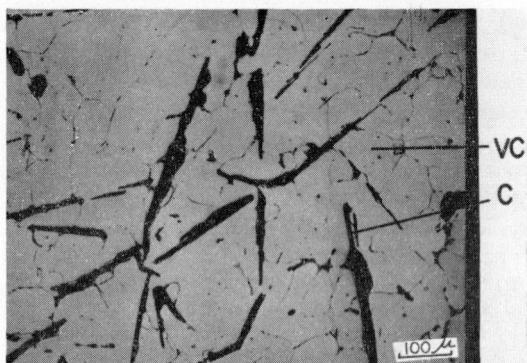

76%V   24%C

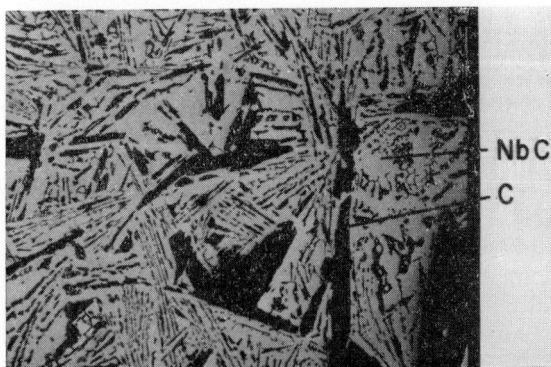

75%Nb,   25%C

FIGURE 10.—Photomicrographs which compare the microstructure of VC–C alloys with NbC–C alloys.

Figure 11 shows the photomicrographs of the polished sections in the Nb–B–C system. These microstructures are very similar to those found in the Ta–B–C system and in the group IVB metal–B–C systems. Similar compatible phase assemblages have been observed in the Ta–B–C and Nb–B–C systems in the following areas of interest: $NbC-NbB_2-C$ and $NbB_2-B_4C-C$ are compatible phases and in the Ta–B–C system, $TaC-TaB_2-C$ and $TaB_2-B_4C-C$ are compatible phase assemblages.

Vanadium borides and/or carbides containing more than $50\%$ graphite have very good thermal-shock resistance. The addition of boron to vanadium-carbide–graphite bodies increases their oxidation resistance and increases the ability to add (more) carbon to these fused refractories. Alloys in the V–B–C system have better oxidation resistance than alloys in the Nb–B–C, Ta–B–C, Ti–B–C, Zr–B–C, and Hf–B–C systems.

The prime use of fused refractories in the group VB metal–boride–carbide–graphite systems will probably be in aerospace applications or in places where materials with high melting points and high resistance to thermal shock are needed.

## 2.6 Group VIB Metals–B–C

Materials in the group VIB metals–B–C systems are not as refractory as those in similar systems containing compounds of the group IVB or VB metals. One other unusual feature of these systems, in contrast to those discussed previously, is that many group VIB metal borides are stable with graphite. Compositions have been made which contain two borides and graphite. Since the borides melt incongruently and peritectic phases are present, equilibrium is not often achieved on solidification, even when the castings solidify slowly. Consequently alloys form with very complex non-equilibrium phase assemblages and with several borides occurring with graphite. In systems containing group IVB and VB metal borides and graphite, however, usually only one boride phase is present, i.e., $MeB_2$, where Me is either Ti, Zr, Hf, Nb, or Ta. Since $Cr_3C_2$ and WC crystallize incongruently, their metal-carbide–graphite alloys often contain other non-equilibrium carbides, such as $W_2C$ and $Cr_7C_3$.

Molten carbide–graphite alloys in incongruent systems, on cooling, first have graphite crystallizing out; then the graphite reacts with the metal-rich liquid to form carbides. Often this reaction is not completed, which results in the non-equilibrium phase assemblages. Consequently alloys containing graphite, carbide, and boride can have more than one carbide and boride present in a non-equilibrium phase assemblage.

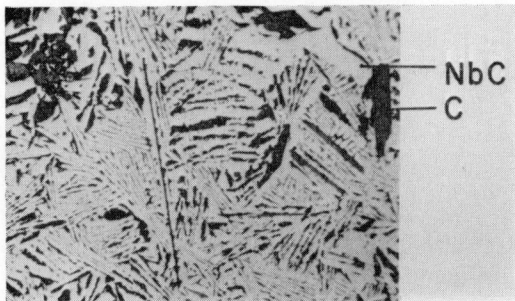

75%Nb,  25%C,  0%B

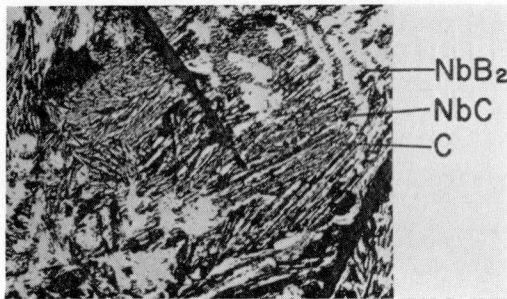

58%Nb,  7%B,  35%C

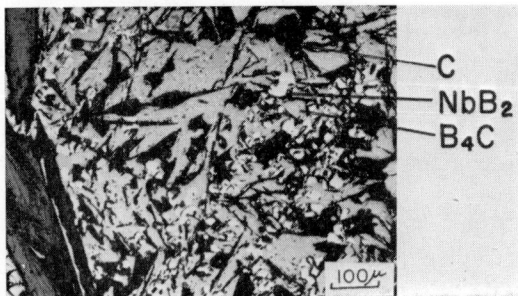

33%Nb,  18%B,  49%C

FIGURE 11.—Photomicrographs of alloys in the Nb–C–B system.

Figure 12 is the constitution diagram of chromium–carbon,[45] which shows that $Cr_3C_2$, $Cr_7C_3$, and $Cr_{23}C_6$ melt incongruently; this helps to explain the non-equilibrium phase assemblages which are often observed in this system. These refractories have excellent thermal-shock resistance. They also have the greatest oxidation resistance, at temperatures from 500° to 1550°C, of all the other simple transition-metal carbides and borides. Consequently, chromium-carbide–graphite and boride–graphite alloys might find applications where there is a need for a moderate refractory material (liquidus above 1900°C) with excellent thermal-shock resistance (samples can be plunged from 1800°C into water repeatedly without breakage) and good oxidation resistance up to 1550°C. The microstructures of some of the Cr–C and Cr–B–C alloys can be seen in Figure 13.

The MoC–C system reported by RUDY et al.[47] contains a low-melting eutectic temperature of 2584° ± 5°C at approximately 45 $^a/_o$ carbon. Hyper-eutectic alloys can be made in that system which contain large primary crystals of graphite surrounded by a eutectic structure consisting of carbide crystals and fine-grained, partially interlocking, intragranular graphite crystals. Several phase changes occur in the solid state as the material is slowly cooled. Alpha $MoC_{1-x}$ converts to $\eta$-$MoC_{1-x}$ at 1960° ± 20°C. At 1655° ± 15°C,

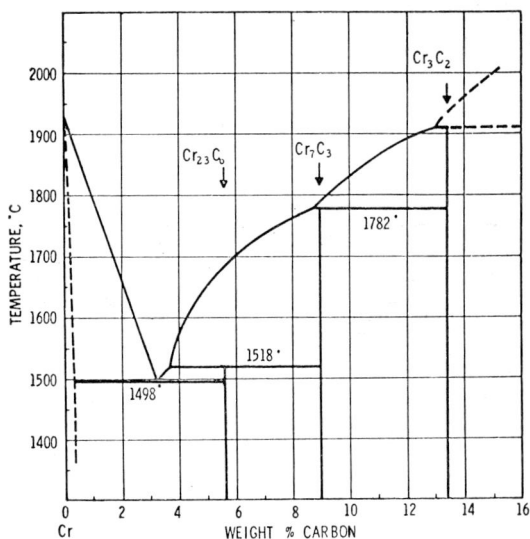

FIGURE 12.—Constitution diagram for chromium-carbon.[45]

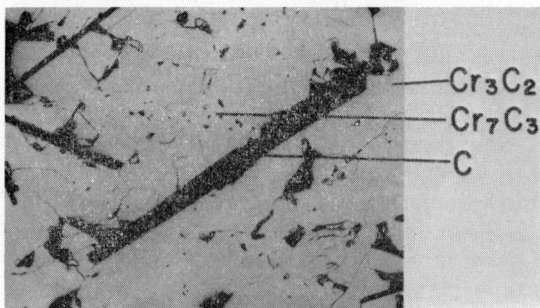

87%Cr,   13%C,   0%B

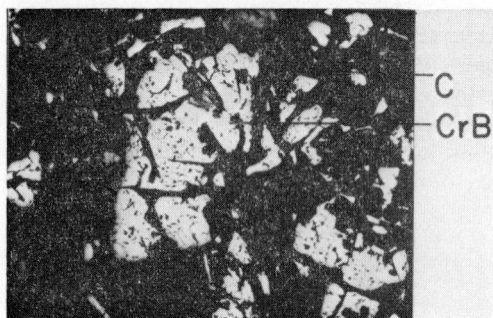

53%Cr,   11%B,   36%C

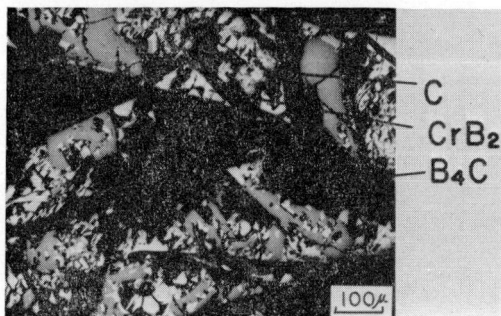

32%Cr,   23%B,   45%C

FIGURE 13.—Photomicrographs of polished sections of fusion-cast Cr–C–B alloys.

$\eta$–$MoC_{1-x}$ breaks down to form $Mo_2C(\beta) + C$; and below $1190° \pm 20°C$, $Mo_2C(\alpha)$ and carbon are the stable phases. Therefore depending on the cooling rate and the type of annealing, different MoC or $Mo_2C$ phases with graphite can be present.

The constitution diagram of the Mo–B system [48] is quite complex and contains numerous incongruently melting compounds and phases that have different structures at different temperatures. Several very interesting alloys have been prepared of a variety of molybdenum borides and graphite.

Figure 14 shows some of the microstructures in the Mo–B–C system. In many cases, fine-grained, partially interlocking intragranular eutectic graphite is formed. These graphite-containing alloys have excellent thermal-shock resistance. The molybdenum–carbon–boron alloys also have less resistance to corrosion by molten iron and molten ferruginous slag than the group IVB carbides and borides. Molybdenum-boride–graphite has greater oxidation resistance than the molybdenum-carbide–graphite, which has only fair resistance.

The constitution diagram of the W–C system shows that WC melts incongruently at $2776° \pm 10°C$.[46] Therefore bodies are made that often contain non-equilibrium phase assemblages.

Figure 15 shows the microstructures of some of the tungsten–carbon and tungsten–boron–carbon alloys. The addition of boron causes the formation of eutectic structure containing tungsten borides and graphite. Samples without boron contain just partially interlocked large crystals of graphite with $W_2C$ and WC. Some samples high in boron contain non-equilibrium phase assemblages of $B_4C$, tungsten boride, graphite, and tungsten carbide.

Tungsten carbides and borides have higher melting points than molybdenum carbides and borides. The tungsten borides have greater oxidation resistance than tungsten carbides. Alloys of tungsten carbides and/or borides and graphite have good thermal-shock resistance. They have poor resistance to ferruginous slag and to molten iron.

The most promising alloys containing group VIB elements are the alloys of graphite with chromium carbides and/or chromium borides which have good thermal-shock resistance, oxidation resistance, and electrical resistivity similar to graphite.

Alloys containing graphite and carbides and/or borides of tungsten or molybdenum have good thermal-shock resistance and moderate oxidation resistance.

Figure 16 compares the microstructure of chromium-carbide–graphite alloys with tungsten-carbide–graphite alloys, molybdenum-carbide–graphite alloys and titanium-carbide–graphite alloys. Since

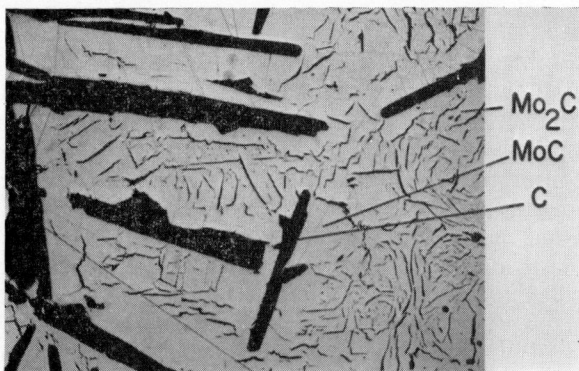

84%Mo,  16%C,  0%B

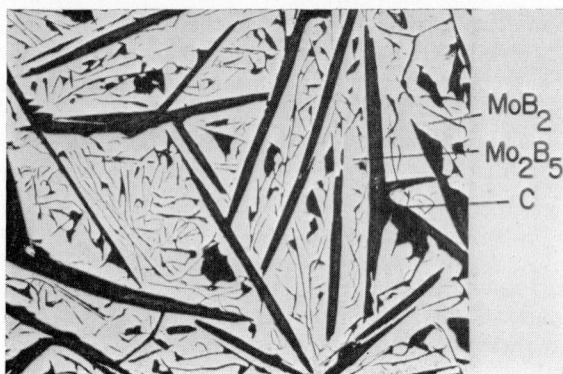

65%Mo,  17%B,  18%C

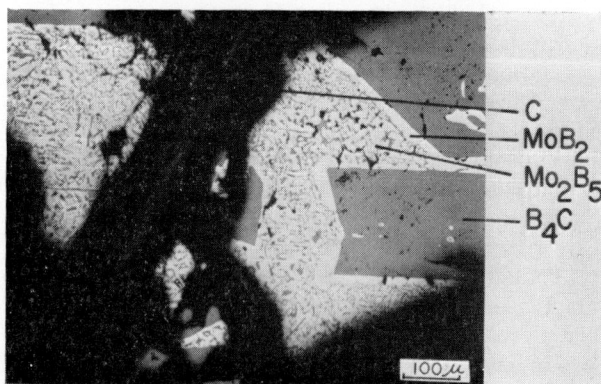

36%Mo,  21%B,  43%C

FIGURE 14.—Photomicrographs of polished sections of fusion-cast Mo–B–C alloys.

87%W, 13%C, 0%B

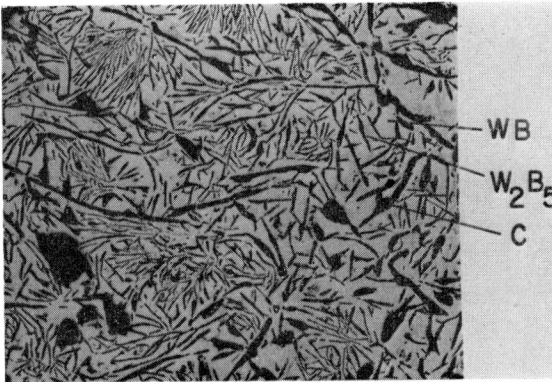

83%W, 6%B, 11%C

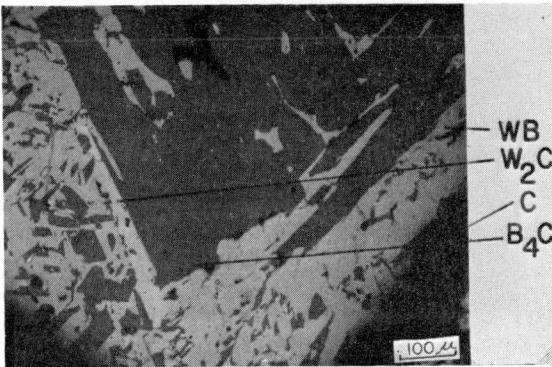

49%W, 13%B, 38%C

FIGURE 15.—Photomicrographs of polished sections of fusion-cast W–B–C alloys.

27

no eutectic exists between the tungsten-carbide–graphite phases in this system, the microstructure consists of large primary graphite crystals surrounded by carbides. Because equilibrium is difficult to achieve, non-equilibrium carbide phases are often found.

Figure 16 also shows that TiC–C alloys have a much simpler microstructure than the group VIB metal carbides–C alloys. Equilibrium is easier to achieve in the TiC–C system.

When chromium is added to TiC–C alloys, the carbide–graphite eutectic structure slowly disappears and bodies high in chromium and graphite do not have the fine-grain graphite phase (see Figure 17).

Peritectic 89%W,   11%C

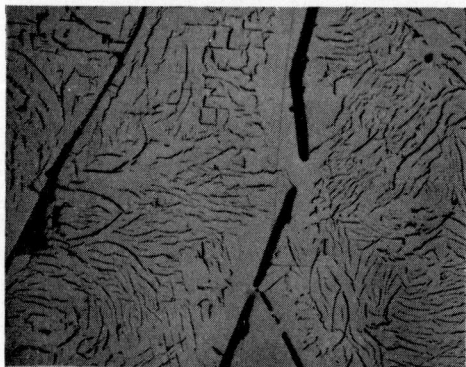

Eutectic 85%Mo,   15%C

FIGURE 16.—Photomicrographs which compare the microstructures of alloys of W–C, Mo–C, Cr–C, and Ti–C.

Peritectic 66%Cr,  34%C

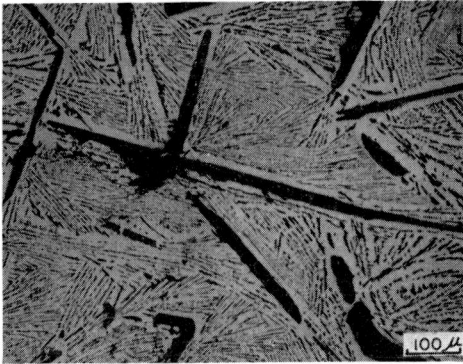

Eutectic 60%Ti,  40%C

FIGURE 16.—continued

Chromium–carbon–boron alloys have very good oxidation resistance. The addition of boron to the chromium–carbon system increases the oxidation resistance even more. The strength of these materials depend on the amount of graphite. Materials high in graphite generally have low modulus of rupture ($< 3000$ lbf/in²). Those with less carbon(approximately 6% graphite) have a modulus of rupture up to 11,000 lbf/in². (These materials exhibit an increase in modulus of rupture when heated to 1340°C.) An increase in modulus of rupture of some samples from 2400 lbf/in² at 25°C to 11,000 lbf/in² at 1340°C has been noted. The $Cr_3C_2$–carbon alloys had higher

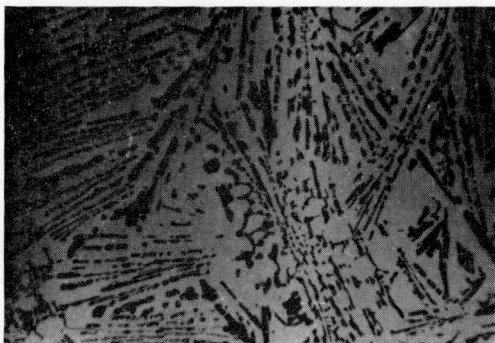

2·8%Cr retained

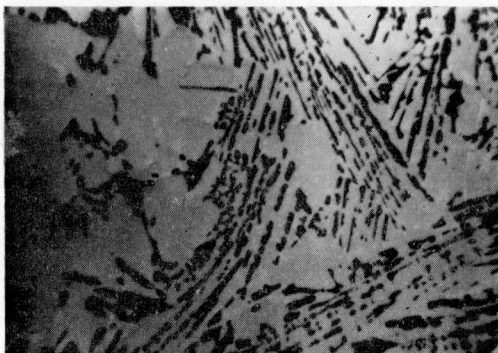

4·6%Cr retained

9·9%Cr retained

FIGURE 17.—Microstructures of Ti–Cr–C alloys.

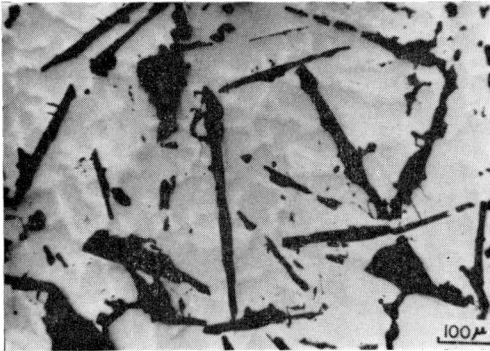

18·0%Cr retained

FIGURE 17.—*continued*

electrical resistivities than the alloys of graphite with the group IVB and the group VB carbides ($659 \times 10^{-6}$ ohm-cm at 25°C and $890 \times 10^{-6}$ ohm-cm at 500°C). Chromium–carbon–boron alloys have very poor resistance to molten iron and to oxide slags. The coefficients of thermal expansion of graphite-containing alloys of chromium carbides and borides were generally higher than alloys with similar phase assemblages of group IVB and group VB carbides and borides.

## 3. CONCLUSIONS

Group IVB, VB, and VIB borides and carbides with substantial graphite content can be fusion-cast. These bodies have excellent thermal-shock resistance. The graphite alloys containing borides generally have better oxidation resistance than those with just carbides and graphite. It is also possible to make interesting fusion-cast refractory materials with unusual phase assemblages and microstructures. Systems containing eutectic compositions have very different microstructures from systems containing peritectic compositions.

Microstructures can be altered by varying the amount or type of metal. The group IVB and VB carbide and boride–graphite alloys look the most promising for applications which require extremely high temperatures. For aerospace applications probably alloys containing carbides and borides of hafnium, tantalum, niobium and zirconium, and graphite are the most promising.

Zirconium diboride and graphite alloys appear to have the best resistance to molten iron. Since titanium–boron–carbon alloys cost much less than other group IVB and VB alloys, they might find application where materials are needed that have greater corrosion resistance to molten iron and slag than graphite, and have equivalent thermal-shock resistance. These alloys can be cast into large shapes and can be readily machined.

Boron-carbide–graphite alloys are lightweight, have good thermal-shock resistance, and show oxidation resistance superior to graphite. Chromium–boron–carbon alloys which contain graphite have excellent thermal-shock resistance and very good oxidation resistance up to 1550°C. These materials might make good furnace parts and kiln furniture.

For additional data on physical properties, references 49–70 should be consulted. Recently three United States Patents have been issued which discuss the composition, properties and manufacturing methods of fusion-cast carbide–boride–graphite materials.[71-73]

## ACKNOWLEDGMENTS

The authors wish to thank Corning Glass Works for support of their research. They also wish to acknowledge the help of the following people: K. E. Zaun, P. R. Smith, N. E. Johnson, and R. W. Denson, Jr., for helping in the preparation of samples; Dr A. A. Erickson, R. G. Ackerman, P. Papa, W. Vine, and W. C. Lewis for their process work; Dr D. E. Campbell and H. L. MacDonell for chemical analyses; H. E. Hagy for physical property measurements; D. L. Zenker for preparation of most of the illustrations, and S. Weisenfeld for photographic work.

## REFERENCES

1. SCHWARZKOPF, P., and KIEFFER, R., "Refractory Hard Metals", (Macmillan Co., N.Y., 1953).
2. SCHWARZKOPF, P., and KIEFFER, R., "Cemented Carbides", (Macmillan Co., N.Y., 1960).
3. TINKLEPAUGH, J. R., and CRANDALL, W. B., "Cermets", (Reinhold Publishing Corp., 1960).
4. KIEFFER, R., and BENESOVSKY, F., "Hartstoffe" (Springer-Verlag, Wien: 1963).
5. SHAFFER, P. T. B., "High-Temperature Materials" (Plenum Press, N.Y.: 1964).
6. KENDALL, E. G., SLAUGHTER, J. I., and RILEY, W. C., AF 04(695)–469 Aerospace Corp., El Segundo, Calif. (1963).
7. KENDALL, E. G., HAYS, C., and RICHARDSON, J. H., TDR–269 (4240–10)–16 Aerospace Corp., El Segundo, Calif. (1964).
8. ROSSI, R. C., and CARNAHAM, R. D., "A Microstructure Approach to Thermal Shock" TDR–669 (6250–10)–4, Aerospace Corp., El Segundo, Calif. (1966).
9. ROSSI, R. C., and CARNAHAM, R. D., TR–669 (6250–10)–10 Aerospace Corp., El Segundo, Calif. (August, 1966).
10. KENDALL, E. G., and McCLELLAND, J. D., TDR–269 (4240–10)–9 Aerospace Corp., El Segundo, Calif. (1964).
11. KENDALL, E. G., "Intermediate Materials" in "Ceramics for Advanced Technologies" (John Wiley & Sons Inc., N.Y.: 1966).

12. KENDALL, E. G., "Rocket Nozzle Technology" in "Ceramics for Advanced Technologies" (John Wiley & Sons Inc., N.Y.: 1966).
13. FOSTER, E. L., Jr, and HILDEBRAND, W., "Arc-Melting and Casting of Refractory Carbides" Battelle Memorial Institute, Columbus, Ohio (1965).
14. FOSTER, E. L., JR, ROUGH, F. A., PRICE, D., HILDEBRAND, W. J., NELSON, S. G., MOAK, D., and ALLEN, J., AF 33(615)–5273 by Battelle Memorial Institute, Columbus, Ohio (1966).
15. RUDY, E., "Ternary Phase Equilibria in Transition Metal–Boron–Silicon Systems" AFML–TR–65–2, Part IV, Vol. II, Wright-Patterson Air Force Base, Ohio (1966).
16. ELLIOTT, R. P., "Constitution of Binary Alloys" (McGraw-Hill Book Co., 1965).
17. STORMS, E. K., "A Critical Review of Refractories" LA 2942, The University of California, Los Alamos, New Mexico (1964).
18. STORMS, E. K., "The Thermo-Dynamics of Refractory Material" Los Alamos, Scientific Laboratory, University of Calif., Los Alamos, New Mexico (1967).
19. STORMS, E. K., "Refractory Carbides" (Academic Press) in the press: London and New York, 1967.
20. KAUFMAN, L., and CLOUGHERTY, E. V., "Investigations of Boride Compounds for High Temperature Applications" in "Metals for the Space Age" (June 1964).
    KAUFMAN, L., and CLOUGHERTY, E. V., "Thermodynamic Factors Controlling the Stability of Solid Phases at High Temperatures and Pressures" Contract AF 33 (657)–9826 and AF 33(657)–8764.
    CLOUGHERTY, E. V., "Research and Development of Refractory Oxidation–Resistant Diborides" Progress Report No. 1, July 1966.
    KAUFMAN, L., "Stability Characterization of Refractory Materials under High Velocity Atmospheric Flight Conditions" Contract No. AF 33(615)–3859 (August 1966).
    KAUFMAN, L., and CLOUGHERTY, E. V., "Investigation of Boride Compounds for Very High Temperature Applications" Report No. RTD–TDR–63–4096, Part III (September 1966).
    KAUFMAN, L., "Stability Characterization of Refractory Materials Under High Velocity Atmospheric Flight Conditions" (December 1966).
21. FARR, J. D., "Refractory Compounds as Structural Materials" Los Alamos Scientific Laboratory of the University of California, Los Alamos, New Mexico (1967).
22. WALLACE, T. C., RUPERT, G. N., and TREIMAN, L. H., "Recent Developments in Equipment and Techniques for the Determination of High Temperature Phase Diagrams", Los Alamos Scientific Laboratory, University of California, Los Alamos, New Mexico (1967).
23. STEINITZ, R., "Physical and Mechanical Properties of Refractory Compounds", presented at short course at UCLA, Los Angeles, Calif. (1967).
24. DePOORTER, G. L., and WALLACE, T. C., "Basic Relationships Involving Phase Boundaries and Homogeneity Ranges of Non-Stoicheiometric Compounds in Refractory Binary and Ternary Systems, Los Alamos Scientific Laboratory, Univ. of Calif., Los Alamos, New Mexico (1967).
25. WALLACE, T. C., "Diffusion and Chemical Kinetics Related to Problems in High-Temperature Chemistry of the Refractory Carbides" Los Alamos Scientific Laboratory, Univ. of Calif., Los Alamos, New Mexico (1967).
26. ANTHONY, F. M., "Design with Brittle Materials—Present and Future" Presented at Univ. of Calif., Los Angeles, Calif. (1967).
27. LAMBERTSON, W. A., "Refractory Compounds as Structural Materials", Univ. of Kentucky (1967).
28. WALKER, P. L., JR., "Carbon—An Old but New Material", Amer. Sci., 50, No. 2, 1962.
29. NIGHTINGALE, R. E., YOSHIKAWA, H. H., and LOSTY, H. H. W., "Physical Properties" in "Nuclear Graphite" (Academic Press, Inc., New York: 1962).

418     ALPER, DOMAN AND MCNALLY:

30. SARA, R. V., LOWELL, C. E., DOLLOFF, R. T., WADD–TR–60–143, Part IV (1963).
31. SARA, R. V., *J. Amer. Ceram. Soc.*, **48**, 243, 1965.
32. RUDY, E., HARMON, D. P., and BRUKL, C. E., AFML–TR–65–2, Part I, Vol. II (May 1965).
33. SAMSONOV, G. V., ZHURAVLEV, N., and AMNVEL, I. G., *Fiz. Metallovi Metallovedenie*, **3**, 309, 1956.
34. DOLLOFF, R. T., WADD Technical Report 60–143, Contract No. AF 33(616)–6286, July (1960).
35. ELLIOTT, R. P., Armour Research Foundation Final Technical Report 2200–12 for U.S. Atomic Energy Comm. Contract No. AT(11–1)–578.
36. ADELSBERG, L. M., CADOFF, L. H., and TOBIN, J. M., "Group IV B and V B Metal Carbide—Carbon Eutectic Temperatures" *J. Amer. Ceram. Soc.*, **49**, (10) 1966.
37. TOBIN, J. M., ADELSBERG, L. M., and CADOFF, L.H. "Kinetics of the Zirconium–Carbon Reaction at Temperatures above 2000°C" *Trans. Metallurgical Soc. AIME*, **236**, July, 1966.
38. TOBIN, J. M., ADELSBERG, L. M., CADOFF, L. H., and BRIZES, W. F., "Nuclear Applications of Non-Fissionable Ceramics", (Amer. Nuclear Soc. Inc., Interstate Printers, Danville, Ill.: 1966).
39. WIEL, S., and BRAMER, S. E., "The Performance of Selected Re-entry Heat Protection Materials in a 200–KW Plasma Jet" Douglas Paper 3942 (1966).
40. RUDY, E., AFML–TR–65–2, Part I, Vol. IV (September 1965).
41. RUDY, E., and ST WINDISCH, Technical Report No. AFML–TR–65–2, Part II, Volume XIII (April 1966).
42. STORMS, E. K., and MCNEAL, R. J., *J. Phys. Chem.*, **66**, 1962.
43. KIMURA, H., and SASAKI, Y., *Trans. Jap. Inst. Met.*, **2**, 98, 1961.
44. RUDY, E., and HARMON, D. P., AFML–TR–65–2, Part I, Vol. V. (December 1963).
45. BLOOM, D. S., and GRANT, N. S., *Trans. Amer. Inst. Met. Engrs*, **188**, 41, 1950.
46. RUDY, E., ST WINDISCH, and HOFFMAN, J. R., AFML–TR–65–2, Part I, Vol. VI (December 1965).
47. RUDY, E., and ST WINDISCH, and CHANG, Y. A., AFML–TR–65–2, Part I, Vol. I (December 1964).
48. RUDY, E., and ST WINDISCH, AFML–TR–65–2, Part I, Vol. III (July 1965).
49. KRIKORIAN, O. H., "Thermal Expansion of High Temperature Materials", UCRL–6132 (September 1960).
50. LOWRIE, R., "Research on Physical and Chemical Principles Affecting High Temperature Materials for Rocket Nozzles," Semi-Annual Report, Contract DA–30–069–ORD–2787 (December 31, 1961).
51. WHITTEMORE, O. J., JR, *J. Canad. Ceram. Soc.*, **28**, 43, 1959.
52. WHITTEMORE, O. J., JR, "Special Refractories for Use Above 1700°C", *Ind. Eng. Chem.*, **47**, 2510, 1955.
53. FINLAY, G. R., "Refractories for 4000°F and Higher", *Chemistry in Canada*, **4**, 41, 1952.
54. LANG, S. M., "Properties of High Temperature Ceramics and Cermets— Elasticity and Density at Room Temperature", N.B.S. Monograph No. 6 (March 1, 1960).
55. NORTON, J. T., BLUMENTHAL, H., and SINDEBAND, S. J., *Trans. AIME*, **185**, 749, 1949.
56. LANGE, N. A., "Handbook of Chemistry", (Handbook Pub. Co., Sandusky, O.: 1946).
57. KIEFFER, R., BENESOVSKY, F., and HONAK, E. R., *Z. anorg. Chem.*, **268**, 191, 1952.
58. STEINITZ, R., *Trans. AIME J. Metals.*, **4**, (9), 983, 1952.
59. WACHTMAN, J. B., JR, SCUDERI, T. G., and CLEEK, G. W., "Linear Thermal Expansion of Aluminium Oxide and Thorium Oxide from 100–1000°K," *J. Amer. Ceram. Soc.*, **45**, (7), 319, 1962.
60. MOERS, K., *Z. anorg. Chem.*, **198**, 262, 1931.

61. GLASER, F. W., "Contribution to the Metal–Carbon–Boron System", *J. Metals*, **4**, 391, 1952.
62. FRIEDERICH, E., and SETTIG, L., *Z. anorg. Chem.*, **144**, 169, 1925.
63. STEINITZ, R., *J. Metals*, **4**, 148, 1952.
64. KINGERY, W. D., "Property Measurements at High Temperature", (Wiley, New York: 1959).
65. BRADSHAW, W. G., and MATHEWS, C. O., "Properties of Refractory Materials: Collected Data and References", LMSD–2466 (June 24, 1958).
66. FINLAY, G. R., "Refractories for 4000°F and Higher", *Chemistry in Canada*, **4**, 41, 1952.
67. MOTT, B. W., "Micro Indentation Hardness Testing" (Butterworth's, London: 1956).
68. CALVERT, E. D., KIRK, M. M., and BEALL, R. A., Bureau of Mines, Report of Investigation No. RI 5951 (1962).
69. SIDGWICK, N. V., "The Chemical Elements and Their Compounds", (Oxford; London: 1951).
70. CAMPBELL, I. E., "High Temperature Technology" (Wiley, New York: 1956).
71. ALPER, A. M., DOMAN, R. C., and MCNALLY, R. N., *U.S. Pat.* 3,340, 076, 1967.
72. ALPER, A. M., DOMAN, R. C., and MCNALLY, R. N., *U.S. Pat.* 3,340, 077, 1967.
73. ALPER, A. M., DOMAN, R. C., and MCNALLY, R. N., *U.S. Pat.* 3,340, 078, 1967.

## Appendix

The materials described were melted in induction furnaces and in electric-arc furnaces.

The compositions referred to in this paper were analysed by wet chemical techniques or by X-ray fluorescence. In addition, some samples were analysed by modal techniques or estimation of the phases from polished sections.

Carbon was determined by a commercially available analyser. Some of the raw materials employed in this study are:

(1) Ti (titanium sponge): 99·3 % Ti, 0·30 % max. Mg, 0·1 % max. Fe, 0·15 % max. Cl.

(2) Zr (zirconium sponge): 99·2 % min. Zr + Hf (Hf 2 %), 0·2 % max. Cr + Fe.

(3) High-purity: 99 + % Ta, Mo, W, Cr, V, Nb.

(4) B: 91 % B, 0·3 % water-soluble B, 4·2 % Mg, 0·3 % $H_2O_2$ insolubles, 0·25 % moisture.

(5) C: Ajax Electrothermic Corp., free-glow Norblack carbon black.

The laboratory tests used in this study are described as:

### Thermal Shock Test

An approximate $1'' \times \frac{3}{4}'' \times \frac{1}{2}''$ sample was cut, heated in an induction furnace immediately to 1800°C, held for 10 min., and then dropped into water at room temperature. This constitutes one cycle. The sample was examined for cracking and spalling. The test was repeated until spalling occurred. Where a sample is referred to as greater than a given number of cycles, this means the test was stopped as a matter of convenience.

## The Corrosion Test

Comprised placing $1\frac{1}{2}'' \times 1'' \times \frac{1}{2}''$ samples in an induction furnace adapted to approximate the temperature and reducing atmosphere of a basic oxygen furnace. At 1700°C, for $2\frac{1}{2}$–3 hs, the samples were passed, with one of their largest surfaces facing upward, through a downwardly directed stream of molten high-lime basic ferruginous slag droplets at a substantially uniform rate of 60 times per hour. The slag was representative of the basic-oxygen-furnace slag developed during the production of a heat of steel and had the following batch composition, by weight: 23·75% $Fe_2O_3$, 25·94% $SiO_2$, 40·86% CaO, 6·25% MgO and 3·20% $Al_2O_3$. At the end of the test, the average thickness of each sample was measured and compared with the original $\frac{1}{2}''$ thickness prior to testing. The results are expressed as a percentage change in thickness (called percent slag cut).

# 28.—Reactions between Nickel and the Carbides and Nitrides of Uranium and their Solid Solutions

By J. C. PASSEFORT, F. ANSELIN and D. CALAIS

*Centre d'Etudes Nucléaires de Fontenay aux Roses (Seine)*

## ABSTRACT                                          E17/K46

*Nickel reacts with uranium monocarbide to give, according to the proportions present, the mixed carbides $UC_2Ni$ or $U_2C_3Ni$ and various compounds in the system U–Ni. For example $2\ UC + 6\ Ni \rightarrow UNi_5 + UC_2Ni$ between 900° and 1300°C. At low temperature this reaction is replaced by $UC + 5Ni \rightarrow UNi_5 + C$, $UC_2Ni$ not being formed for kinetic reasons. The sesquicarbide reacts direct with nickel: $U_2C_3 + Ni \rightarrow U_2C_3Ni$. $UNi_5$, $UC_2Ni$, $U_2C_3Ni$ have been identified by the X-ray pattern and by microanalysis. $UC_2Ni$ is simple cubic ($a = 4·961$ Å, $c = 7·346$Å). $U_2C_3Ni$ has a complex structure. The presence of graphite has been shown by means of carbide labelled with $^{14}C$. Nickel reacts with uranium nitride according to the reaction: $2UN + 5Ni \rightleftharpoons UNi_5 + \frac{1}{2}N_2$ accompanied, depending on the nitrogen pressure exerted, by the formation of sesquinitride: $2UN + \frac{1}{2}N_2 \rightleftharpoons U_2N_3$. An upper limit of the free energy of the compound $UNi_5$ has been evaluated: $G_{UNi_5} - 41,600\ cal. + 5·1\ T°K$. U(CN) solid solutions react with nickel similarly, although some reactions observed with the U(CN)Ni diffusion couples have an incubation period because the free energy of UN or of UC decreases in solid solution. Anisotropic reactions have also been observed in UC–Ni and U(CN)–Ni couples due to the presence of submicroscopic particles of a second phase, $UC_2$, epitaxially on the (100) planes of carbide or carbonitride. Uranium diffuses chemically in nickel according to : $D_{UN} = 137\ cm^2s^{-1}. e^{\frac{-67,000\ cal}{RT}}$.*

## Réactions du nickel avec le carbure d'uranium, le nitrure d'uranium et leurs solutions solides

*Le nickel réagit sur le monocarbure d'uranium pour donner suivant les quantités en présence les carbures mixtes $UC_2Ni$ ou $U_2C_3Ni$ et divers composés du système binaire uranium-nickel. Par exemple, $2UC + 6Ni \rightarrow UNi_5 + UC_2Ni$ entre 900° et 1300°C. A basse température cette dernière réaction est remplacée par la suivante: $UC + 5Ni \rightarrow UNi_5 + C$, la formation de $UC_2Ni$ ne se faisant pas pour des raisons* *de cinétique. Le sesquicarbure réagit directement sur le nickel suivant: $U_2C_3 + Ni \rightarrow U_2C_3Ni$. $UNi_5$, $UC_2Ni$, $U_2C_3Ni$ ont eté identifiés par leur spectre X et microanalyse ponctuelle. $UC_2Ni$ est quadratic simple ($a = 4,961$ Å, $c = 7,346$). $U_2C_3Ni$ a une structure complexe. Le graphite a été mis en évidence par autoradiographie β en utilisant du carbure marqué au carbone 14. Le nickel réagit sur le nitrure d'uranium suivant la réaction: $2UN + 5$ $Ni \rightleftharpoons UNi_5 + \frac{1}{2}N_2$, accompagné suivant la pression d'azote exercée par formation de sesquinitrure: $2UN + \frac{1}{2}N_2 \rightleftharpoons U_2N_3$. On a pu ainsi évaluer une limite supérieure de l'energie libre du composé défini $UNi_5$: $G_{UNi_5} - 41,600$ cal + 5,1 T°K. Les solutions solides $U(CN)$ réagissent sur le nickel de la même façon, cependant certaine réaction observée sur des couples de diffusion $U(CN) - Ni$ présente une période d'incubation due au fait que l'énergie libre de UN ou de UC diminue dans la solution solide. Des réactions anisotropes ont été observées également dans les couples UC–Ni et $U(CN)$–Ni provenant de la présence submicroscopique d'une seconde phase $UC_2$ en épitaxie sur les plans (100) du carbure ou du carboniture. La loi gouvernant la diffusion chimique de l'uranium dans le nickel est la suivante: $D_{UN} = 137$ cm².s⁻¹ $e^{\frac{-67,000 \text{ cal}}{RT}}$.*

## Reaktion von Nickel mit Urankarbid, Urannitrid und ihren festen Lösungen

*Nickel reagiert mit dem Monokarbid des Urans und bildet je nach den vorliegenden Verhältnissen die gemischten Karbide $UC_2Ni$ oder $U_2C_3Ni$ und verschiedene Verbindungen des binären Systems U–Ni, z.B. $2UC + 6Ni \rightarrow UNi_5 + UC_2Ni$ zwischen 900 und 1300°C. Bei niedriger Temperatur wird diese Reaktion ersetzt durch $UC + 5Ni \rightarrow UNi_5 + C$, $UC_2Ni$, welches nicht aus kinetischen Gründen entsteht. Das Sesquikarbid reagiert direkt mit Nickel: $U_2C_3 + Ni \rightarrow U_2C_3Ni$. $UNi_5$, $UC_2Ni$ und $U_2C_3Ni$ wurden durch ihre Röntgendiagramme und durch Mikroanalyse identifiziert. $UC_2Ni$ ist tetragonal ($a = 4,961$ Å, $c = 7,346$ Å). $U_2C_3Ni$ hat eine komplizierte Struktur. Die Gegenwart von Graphit wurde durch β-Autoradiographie an ¹⁴C-markierten Proben nachgewiesen. Nickel reagiert mit Urannitrid gemäß der Reaktion $2UN + 5Ni = UNi_5 + \frac{1}{2}N_2$, und ferner, je nach dem $N_2$-Druckdurch Bildung von Sesquinitrid: $2$ $UN + \frac{1}{2}N_2 = U_2N_3$. Für die Verbindung $UnI_5$ wird eine obere Grenze der freien Energie ermittelt: $G_{UNi_5}C - 41600$ cal + 5,1 T°K. Die festen Lösungen $U(CN)$ reagieren mit Nickel ähnlich, obwohl gewisse Beobachtungen an Diffusionspaaren $U(CN)$–Ni eine Inkubationszeit ergaben, weil die freie Energie von UN oder UC in fester Lösung abnimmt. Anisotrope Reaktionen wurden regelmäßig in den Paaren UC–Ni und $U(CN)$–Ni beobachtet, die auf der Gegenwart von submikroskopischen Teilchen* *einer zweiten Phase, UC$_2$ epitaxial auf (100)-Flächen des Karbids oder Karbonitrids, beruhen, Das Gesetz, nach dem die Urandiffusion in Nickel verläuft ist:* $D_{UN} = 137 \, cm^2 s^{-1} \, exp\frac{-67,000 \, cal}{RT}$.

## 1. INTRODUCTION

In connection with combustible elements used in fast reactors the reactions between nickel and UC, UN and U(C,N) are of dual interest:

(1) The possibility of reactions between nickel alloys or stainless steel used as canning materials for nuclear fuels such as (U,Pu)C, U,Pu)(CN), (U,Pu)N.

(2) To gain an understanding of the kinetics of the sintering of nuclear fuels: very small additions (0·3 $^w/_o$) of nickel to the carbides of plutonium and uranium lower the sintering temperatures from 1600° to 1400°C, although they have no effect on the densification of nitrides.[1,2] More knowledge about the reaction products obtained with nickel might provide an explanation for these phenomena.

## 2. EXPERIMENTAL METHODS

The reactions have been studied either by sintering compacted powders or by using diffusion couples. The nitrides carbides and carbonitrides were prepared by sintering according to the techniques developed by PASCARD and ANSELIN.[1,2] The nitrides are pure single-phase products, the carbides often contain UC$_2$ or U$_2$C$_3$. The nickel used was obtained from Johnson-Matthey (impurities in p.p.m. by weight: Fe,5; Al,2; Ca,1; Cu,1; Ag,1). The sintering experiments were carried out in silica-glass ampoules, sealed initially under vacuum.

The ceramic–nickel joints were made by cold pressing on an ADDA[3] press, and subsequently annealing in vacuum for a short time (10 h) at a temperature below the one chosen for the experimental treatment.

The following methods were used for inspection and examination:

(1) Metallographic techniques.

(2) X-ray powder patterns obtained by means of a X-ray goniometer.

(3) X-ray microanalysis: a CAMECA (CASTAING) apparatus was used. Nickel was determined by means of the K$\alpha_1$ radiation obtained at 25 kV, uranium by L$\alpha_1$ or M$\beta$ radiation at 35 kV.

Curves obtained from a systematic analysis of the definite compounds in the U–N system (UNi$_2$ and UNi$_5$) served to relate the concentrations determined experimentally to real concentrations.

## 3. REACTIONS OF NICKEL WITH URANIUM NITRIDE[4]
### 3.1 Heat Treatment of Diffusion Couples

In the couples that had been heated above 700°, four zones can be observed (Figure 1):

(1) A single-phase zone: UN.
(2) A two-phase zone: grey globules of $U_2N_3$ on a lighter background of $UNi_5$ (Figure 2).
(3) A single-phase zone: $UNi_5$.
(4) A single-phase zone: Ni.

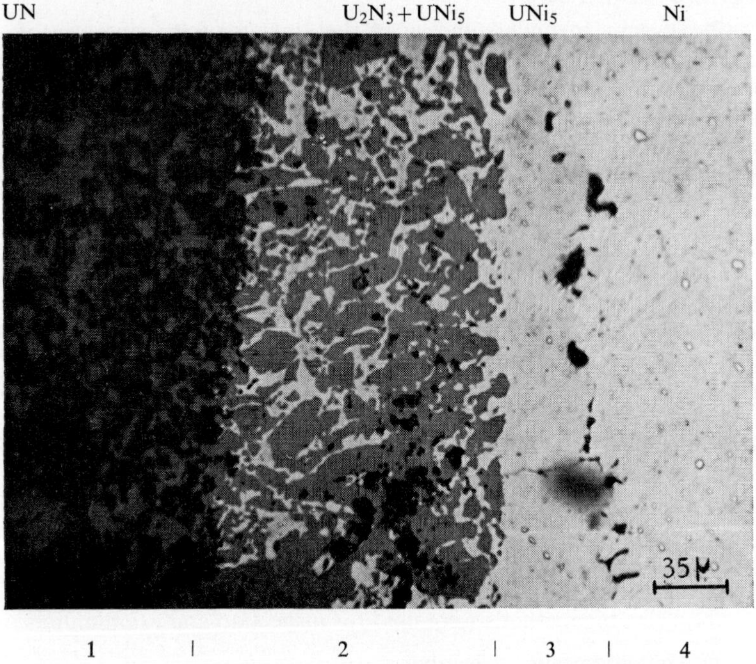

UN                      $U_2N_3 + UNi_5$      $UNi_5$        Ni

1        I        2        I    3    I    4

FIGURE 1.—UN–Ni couple: 840°C. Reaction zone. Chemical etching in a bath of concentrated acetic acid and concentrated nitric acid in equal proportions.

### 3.2 The Reaction $3UN + 5Ni \rightarrow UNi_5 + U_2N_3$

When the analyser crystal of the electron probe micro-analyser apparatus is adjusted for one of the uranium rays, the count is about 1% higher when the beam is directed at UN than when it is directed at the grey globules. As uranium is the only element detected in this phase, the balance must be nitrogen, which suggests a reaction of the type:

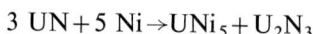

$$3 \text{ UN} + 5 \text{ Ni} \rightarrow \text{UNi}_5 + \text{U}_2\text{N}_3$$

This reaction has been confirmed by direct synthesis by heating compacted powders of UN and Ni at 1000°C (previously mixed in the appropriate proportions) in sealed silica-glass ampoules.

$\text{UNi}_5$ and $\text{U}_2\text{N}_3$ have been identified by their X-ray patterns. When this experiment is repeated, not in a sealed ampoule but in vacuum, the formation of $\text{UNi}_5$ only is observed, because of the complete dissociation of $\text{U}_2\text{N}_3$ at 1000°C and under a pressure of $10^{-5}$ mm Hg:

$$\text{U}_2\text{N}_3 \rightarrow 2 \text{ UN} + \tfrac{1}{2} \text{ N}_2$$

The reactions of nickel with UN can be represented by one of the following two series:

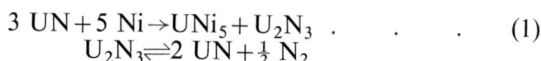

$$3 \text{ UN} + 5 \text{ Ni} \rightarrow \text{UNi}_5 + \text{U}_2\text{N}_3 \quad . \qquad . \qquad . \qquad (1)$$
$$\text{U}_2\text{N}_3 \rightleftharpoons 2 \text{ UN} + \tfrac{1}{2} \text{ N}_2$$

$$\text{UN} + 5 \text{ Ni} \rightleftharpoons \text{UNi}_5 + \tfrac{1}{2} \text{ N}_2 \quad . \qquad . \qquad . \qquad (2)$$

$$\tfrac{1}{2} \text{ N}_2 + 2 \text{ UN} \rightleftharpoons \text{U}_2\text{N}_3 \qquad . \qquad . \qquad . \qquad . \qquad (3)$$

Reaction (1) enables the energy of formation of $\text{UNi}_5$ to be determined.

With $\text{U} + \tfrac{1}{2} \text{ N}_2 \rightarrow \text{UN}$

$$G_{\text{UN}} = -68,500 + 21 \cdot 5T \ ^5$$

and

$$\text{U}_2\text{N}_3 \rightleftharpoons 2 \text{ UN} + \tfrac{1}{2} \text{ N}_2$$
$$\tfrac{1}{2} L P_{N_2} = -L \ k_P$$
$$\log P_{N_2} = 7 \cdot 18 - \frac{11760}{T} \ ^5$$

Pressures are expressed in atmospheres, energies in calories, and temperatures in °K.

It can be deduced that:

$$G_{\text{U}_2\text{N}_3} = -163,871 + 59 \cdot 4T;$$

using reaction (1):

$$G_{\text{UNi}_5} < 3 \ G_{\text{UN}} - G_{\text{U}_2\text{N}_3}$$
$$G_{\text{UNi}_5} < -41,629 + 5 \cdot 1 \ T$$

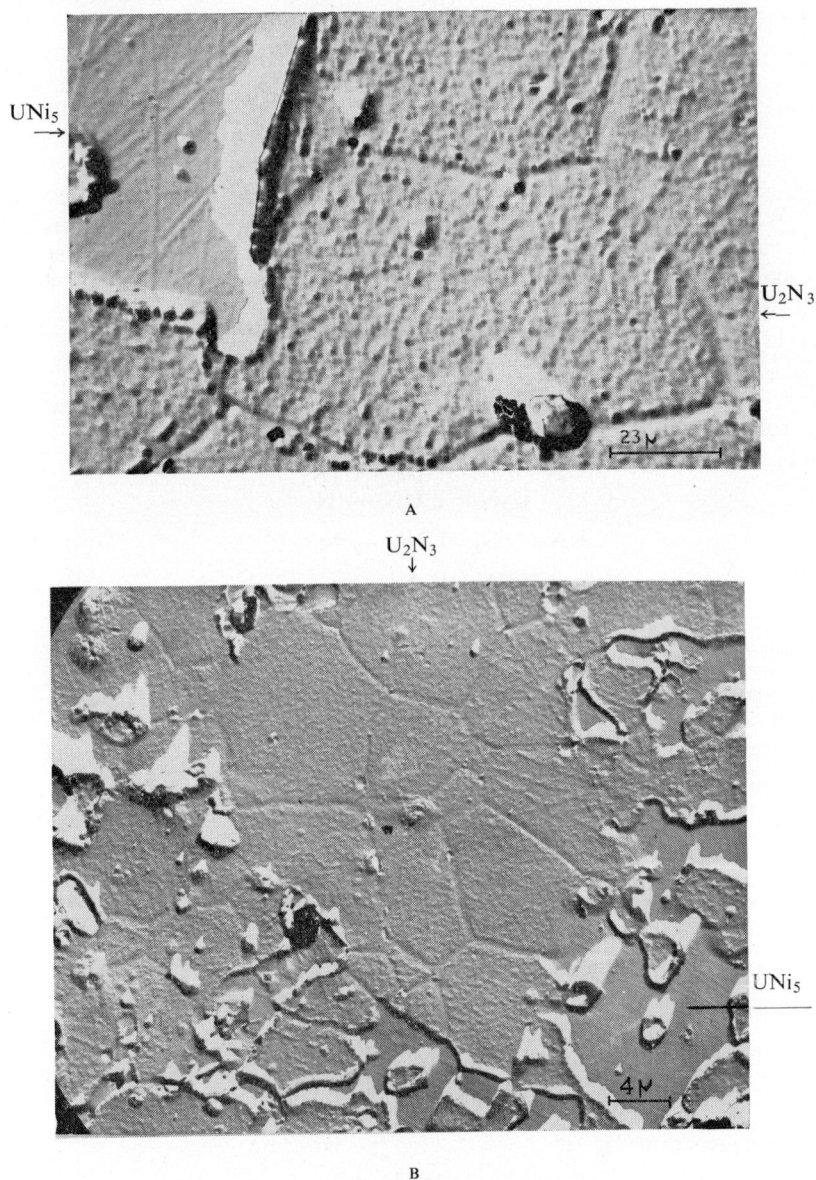

FIGURE 2
Electron micrographs of UN–Ni couple: 910°C. $UNi_5 + U_2N_3$ reaction zone.

$U_2N_3$           $UNi_5$

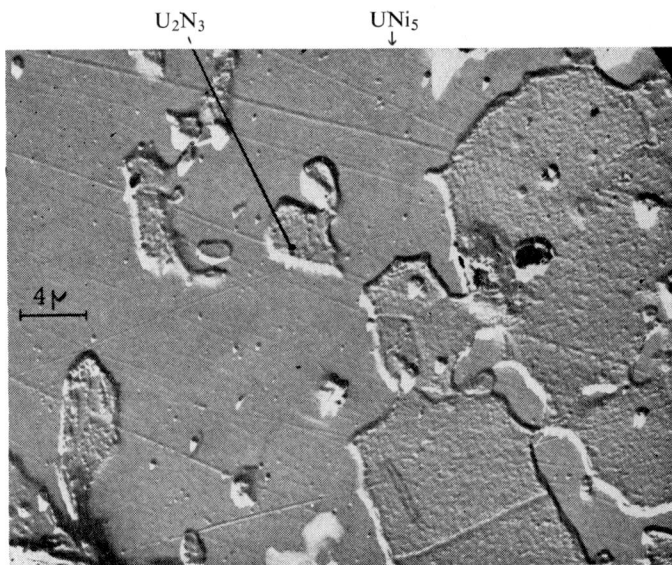

c

FIGURE 2—*continued*

The expression

$$G_{UNi_5} < 3\ G_{UN} - G_{U_2N_3}$$

can be written

$$G_{UNi_5} < G_{UN} - \frac{RT}{2}\ \log p_1$$

where $p_1$ is the dissociation pressure of nitrogen from $U_2N_3$.

Using Equation (2):

$$\Delta G = -\tfrac{1}{2}\ RT \log p_2 = G_{UNi_5} - G_{UN}$$

and

$$G_{UNi_5} = G_{UN} - \frac{RT}{2}\ \log p_2$$

$p_2$ is the nitrogen pressure from reaction Equation (2) and it is clear that

$$p_2 > p_1$$

As $p_2$ is greater than $p_1$, the dissociation pressure of the sesqui-nitride, this compound must be a secondary product of the reaction

28

between nickel and uranium mononitride, i.e. the presence of $U_2N_3$ is not necessary for Ni to react with UN, because reaction (2) is thermodynamically possible, being limited only by the nitrogen pressure in the system. The calculated value for the free energy for the intermetallic compound $UNi_5$

$$G_{UNi_5} < -41,629 + 5 \cdot 1 \ T,$$

is much more negative than the value at 1500°C proposed by BRIGGS,[6]

$$-32,000 < G_{UNi_5} < -25,000$$

who, with his collaborators, considered reaction (2) only.

### 3.3 Width of the Reaction Zones in UN–Ni Couples

The results of our measurements agree with those of KATZ,[7] who experimented at 1000°C. Table 1 and Figure 3 present our values for the width of the zone over the square root of time. At 730°C after 1178 h of heating, small reaction zones were observed dispersed along the interface, the maximum width of which does not exceed 25 $\mu$m. Below 540°C, no effects of reaction between Ni and UN could be traced on photomicrographs.

**Table 1**
**Temperature, Duration of Annealing and Extent of Reaction Zones of UN–Ni Couples**

| Temperature (°C) | Time (h) | Extent of reaction zone | $\dfrac{\theta}{\sqrt{t}}$ cm.s$^{-\frac{1}{2}}$ |
|---|---|---|---|
| 540 | 1560 | | |
| 660 | 1220 | | |
| 730 | 1178 | | |
| 840 | 650 | 170 $\mu$m | $1 \cdot 1 \ . \ 10^{-5}$ |
| 910 | 840 | 225 $\mu$m | $1 \cdot 25 . 10^{-5}$ |
| 988 | 936 | 263 $\mu$m | $1 \cdot 4 \ . \ 10^{-5}$ |

### 3.4 Solubility of Nickel in Uranium Nitride

Measurements by means of electron microprobe analysis at points in the nitride at 1 $\mu$m from the reaction zone, performed on samples that had been heated to 800°C and 1000°C, showed nickel concentrations of $1500 \pm 500$ p.p.m. by weight.

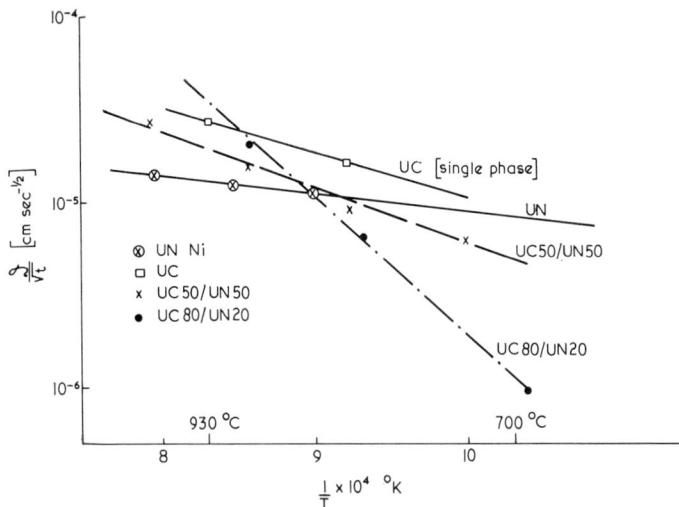

FIGURE 3.—Total size of reaction zones ($\delta$) divided by the square root of the annealing period, $\sqrt{t}$, observed for the couples UC–Ni UN–Ni, U(C,N)Ni as a function of reciprocal absolute temperature.

## 4. REACTIONS OF NICKEL AND URANIUM CARBIDE

### 4.1 Reactions

Between 900° and 1300°, nickel reacts with uranium monocarbide[9] as follows:

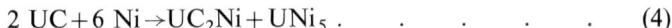

$$2\ UC + 6\ Ni \rightarrow UC_2Ni + UNi_5 . \qquad . \qquad . \qquad . \qquad . \qquad (4)$$

$UC_2Ni$ and $UNi_5$ have been identified by X-ray and electron microprobe analysis. $UC_2Ni$ has a simple tetragonal lattice with the following parameters.

|   | Counting: proportional diffractometer radiation $CuK\alpha$ ($K\alpha$ peak) | Flat plate back-reflection camera $FeK\alpha$ |
|---|---|---|
| $a$ | 4·961 Å | 4·960 Å |
| $c$ | 7·346 Å | 7·346 Å |
| $a/c$ | 1·481 | 1·481 |

The chemical formula has been derived from concentration measurements by X-ray micro-analyses. The results have been corrected by means of calibration curves starting with $UNi_2$ and $UNi_5$. Values so obtained are: $C_{Ni} = 18\% \pm 0·5\%$, $C_U = 74\% \pm 0·5\%$. The values corresponding with $UC_2Ni$ are 18·30 and 74·21. The conclusion has been checked by direct synthesis:

$$UC_2 + Ni \rightarrow UC_2Ni \text{ (Figure 4).}$$

The above reaction yields a single-phase product at 1400°C. Starting with this material we re-checked the micro-analytical results for the U–Ni–C system, allowances being made for the presence of carbon.

FIGURE 4.—The reaction $2UH_3 + 2C + 6Ni$: sintered at 1000°C, annealed at 1200°C. Two phases$_5$ matrix, $UNi_5$; needles, $UC_2Ni$.

$UC_2Ni$ starts to decompose above 1500°C and during cooling several compounds in the U–Ni system are formed (particularly $UNi_5$ and $UNi_2$) and a new phase, $U_2C_3Ni$, develops. The existence of $UC_2Ni$ has been recently admitted by WHITE and DUTTA.[10] The reaction

$$U_2C_3 + Ni \rightarrow U_2C_3Ni \quad \text{(Figure 5)}$$

can take place. $U_2C_3Ni$, as made by direct synthesis, consists of 90% of a phase of this composition; the second phase is uranium carbide, UC, which exists in the initial sesquicarbide. The formula has been deduced from microprobe results, evaluated by comparison with a reference curve obtained with $UC_2Ni$. $U_2C_3Ni$ shows a complicated X-ray pattern, which we were unable to index; in Table 2 we reproduce the interplanar distances and intensities that we found. $UC_2Ni$ and $U_2C_3Ni$ appear as separate phases in the carbides UC and $U_2C_3$ sintered at 1400°C in the presence of 3000 p.p.m. Ni (Figures 6 and 7).

FIGURE 5.—$U_2C_3Ni$ prepared by direct synthesis:
$U_2C_3 + Ni \rightarrow U_2C_3Ni$.
1400°C, 4 h. Matrix $U_2C_3Ni$, light phase, UC.

**Table 2**

**Crystallographic Data for the Compound $U_2C_3$ Ni.**
**(Proportional counter diffractometer.[2,3] CuK$\alpha$ radiation)**

| $d$ ($\mathring{A}$) | Intensity | $d$ ($\mathring{A}$) | Intensity |
|---|---|---|---|
| 3·63 | 5 | 2·593 | 15 |
| 3·424 | 29 | 2·537 | 100 |
| 3·307 | 29 | 2·473 | 26 |
| 3·205 | 14 | 2·445 | 9 |
| 3·158 | 6 | 2·430 | 2 |
| 3·115 | 23 | 2·399 | 2 |
| 3·019 | 40 | 2·377 | 20 |
| 2·941 | 46 | 2·339 | 6 |
| 2·884 | 36 | 2·334 | 4 |
| 2·855 | 36 | 2·302 | 22 |
| 2·807 | 14 | 2·262 | 22 |
| 2·762 | 4 | 2·206 | 6 |
| 2·734 | 2 | 2·200 | 4 |
| 2·630 | 2 | 2·122 | 22 |

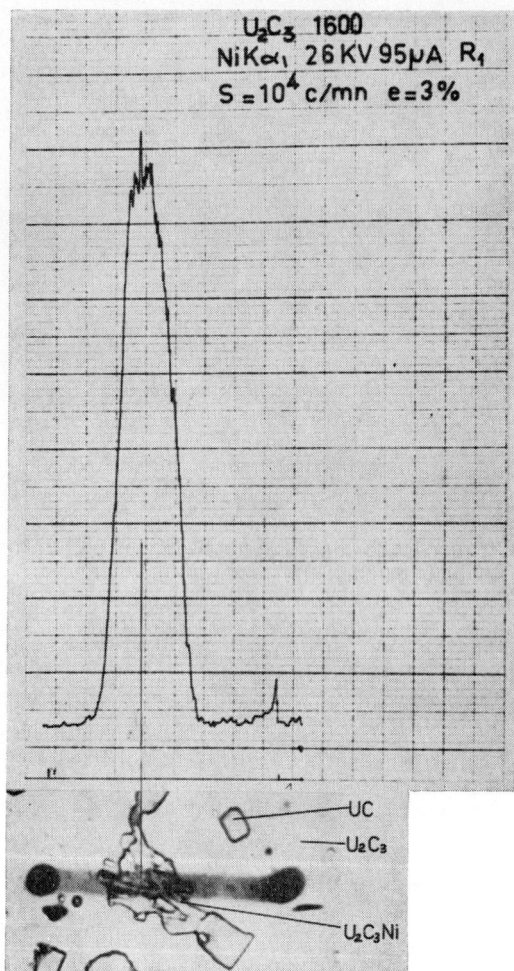

FIGURE 6.—$U_2C_3Ni$ in a uranium sesquicarbide, sintered at 1600°C in the presence of 3000 p.p.m. Ni (X-ray microanalysis).

## 4.2 The Phase Diagram U–C–Ni

This diagram for 1000°C could be constructed from results obtained in sintered compacted U, C, Ni powders. The experimental points are reproduced in Figure 8, which also indicates the examination technique concerned.

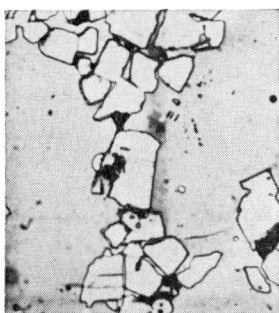

Photomicrograph
—— $U_2C_3$

—— UC
—— $U_2C_3Ni$

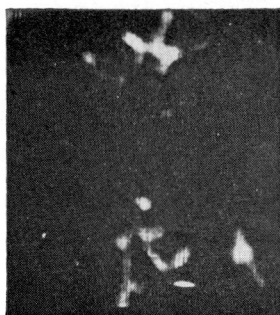

X-ray image (NiK$\alpha_1$) obtained by displacing the sample under the electronic spot in 2 perpendicular directions (scanning); distribution of Ni($U_2C_3Ni$) at the limit of the UC–$U_2C_3$ phases.

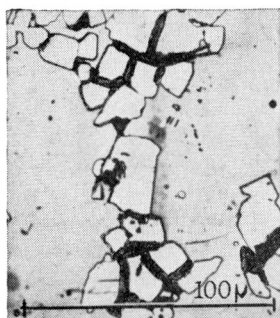

Photomicrograph
The $U_2C_3Ni$ phase is outlined

FIGURE 7.—Localization of Ni in an uranium carbide sintered in the presence of 5000 p.p.m. Ni. Method of analysis: scanning in 2 directions. $U_2C_3Ni$ occurs at the limit of the UC–$U_2C_3$ phase.

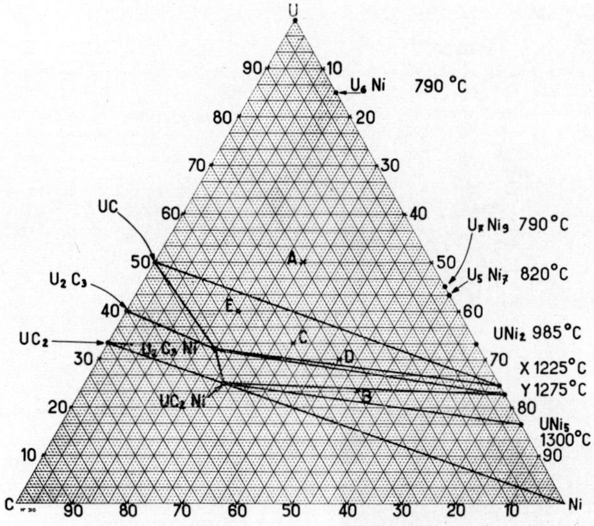

FIGURE 8.—Diagram of the U–C–Ni phase at 1000°C.
A,C,E. X-rays. X-ray microanalysis
B,D. X-rays

### 4.3 U–C–Ni Compatibility

The compatibility has been investigated between 700° and 1000°C by means of diffusion couples.

The thermal treatments and the most important results are indicated in Table 3. From the phase diagram at 1000°C the following reaction zones can be predicted (intersections of the straight line running from the Ni point to the initial carbide composition, with the lines separating the two phases):

(1) UC (single-phase):

$$U_2C_3Ni + X$$
$$U_2C_3Ni + Y$$
$$UC_2Ni + Y \qquad \text{(line } \mathbf{ab} \text{ Figure 9)}$$
$$UC_2Ni + UNi_5$$
$$Ni$$

(2) UC (containing higher carbides)

$$UC + U_2C_3Ni$$
$$U_2C_3Ni + X$$
$$UC_2Ni + Y \qquad \text{(line } \mathbf{cb} \text{ Figure 9)}$$
$$UC_2Ni + UNi_5$$
$$Ni$$

(3)  $U_2C_3$ (containing UC):

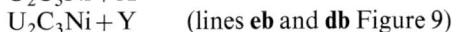

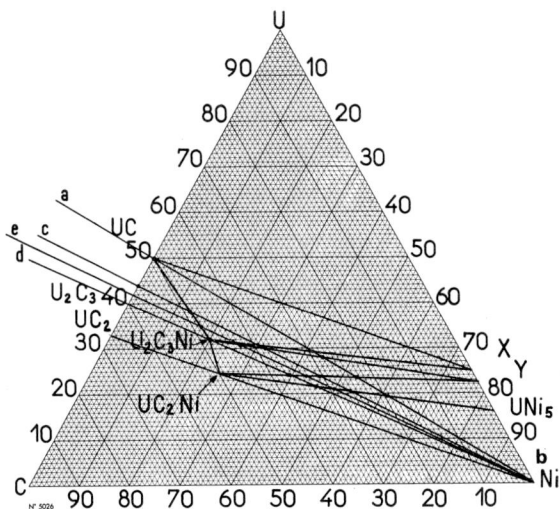

$$U_2C_3Ni + UC$$
$$U_2C_3Ni + X$$
$$U_2C_3Ni + Y \qquad \text{(lines \textbf{eb} and \textbf{db} Figure 9)}$$
$$UC_2Ni + Y$$
$$UC_2Ni + UNi_5$$
$$Ni$$

FIGURE 9.—Prediction of the reaction zones in nickel carbide diffusion couples according to the carbon content of the carbide.

Metallographic examination and micro-probe analysis reveal the following series of diffusion zones (Figure 10).

Zone 1. Single-phase nickel zone.

Zone 2. Two-phase zone consisting from $UNi_5$ (matrix) and graphite (precipitate) (Figures 11 and 12). The precipitate has been identified by $\beta$-autoradiography in a Ni–UC couple labelled with $^{14}C$ (Figure 13); microprobe results show that they contain neither U nor Ni (Figures 14 and 15).

Zone 3. The "needle zone" of multiphase character (Figure 16). This zone is formed by an intergrowth of mixed uranium carbides and nickel and a number of inter-metallic compounds U–Ni. Approximate measurements, made by moving the electron probe spot parallel to the interface, show a steep gradient to exist in the uranium content:

$$C_U \ (\text{UC side}) = 70\%$$
$$C_U \ (\text{Ni side}) = 50\%$$

Table 3

Temperature, Duration of Annealing and Size of Reaction Zones of UC–Ni Couples

| Couples | $\theta$ (°C) | $t$ (h) | $\frac{\theta}{\sqrt{t}}$ $cm.s^{-\frac{1}{2}}$ | $10^4/T°K$ | Solubility ($^w/o$) | Size $UNi_5$ ($\mu m$) | Size of needle zone ($\mu m$) |
|---|---|---|---|---|---|---|---|
| UC (UC$_2$) | 700 | 2100 | $0.4 \times 10^{-6}$ | 10·3 | | 10 | 0 |
| UC (monophase) | 816 | 950 | $1.2 \times 10^{-5}$ | 9·2 | $0.17 \pm 0.05$ | 225 | 0 |
| UC (UC$_2$+U$_2$C$_3$) | 822 | 252 | $3.16 \times 10^{-5}$ | 9·1 | $0.22 \pm 0.07$ | 300 | 0 |
| UC (UC$_2$) | 895 | 640 | $1.6 \times 10^{-5}$ | 8·5 | $0.18 \pm 0.07$ | 235 | 7 |
| UC (UC$_2$) | 930 | 700 | $2.96 \times 10^{-5}$ | 8·3 | $0.17 \pm 0.05$ | 430 | 40 |
| UC (monophase) | 930 | 1850 | $2.71 \times 10^{-5}$ | 8·3 | $0.22 \pm 0.07$ | 600 | 100 |
| UC (UC$_2$+U$_2$C$_3$) | 936 | 1250 | $2.21 \times 10^{-5}$ | 8·3 | | 425 | 45 |
| UC (UC$_2$+U$_2$C$_3$) | 980 | 182 | $3.58 \times 10^{-5}$ | 8 | $0.15 \pm 0.05$ | 230 | 60 |
| U$_2$C$_3$ (UC) | 1010 | 820 | $3.49 \times 10^{-5}$ | 7·7 | $0.0750 \pm 0.05$ | 300 | 300 |

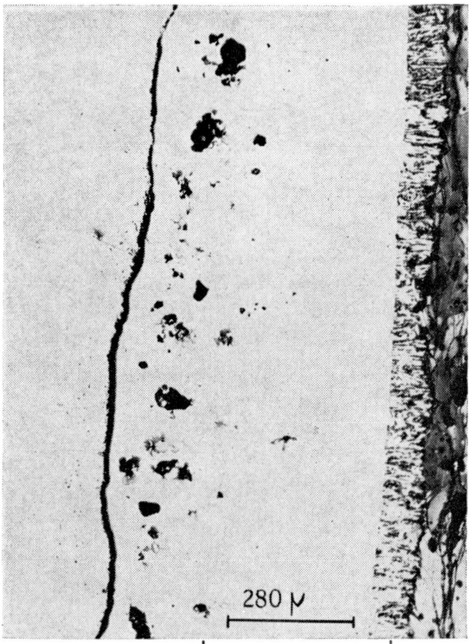
280 μ
Ni
Graphite + $UNi_5$
UC
Needle zone

FIGURE 10.—UC (single-phase)–Ni couple: 930°C, 1850 h. Zone of diffusion.

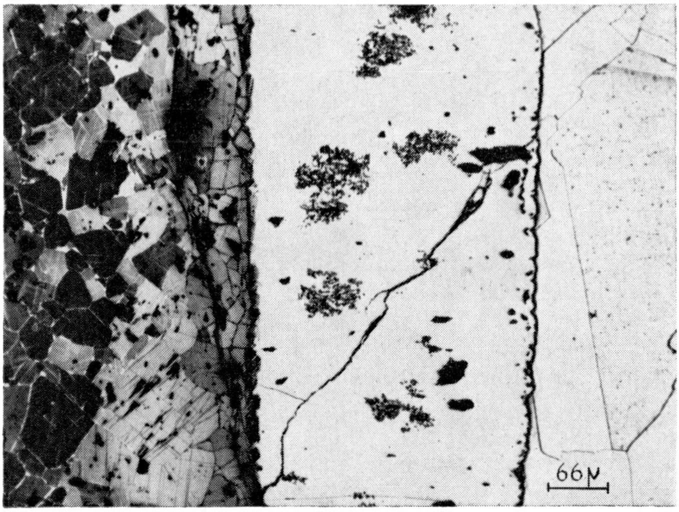
66μ

FIGURE 11.—UC(+$U_2C_3$)–Ni couple: 822°C, 252 h. Zone of diffusion.

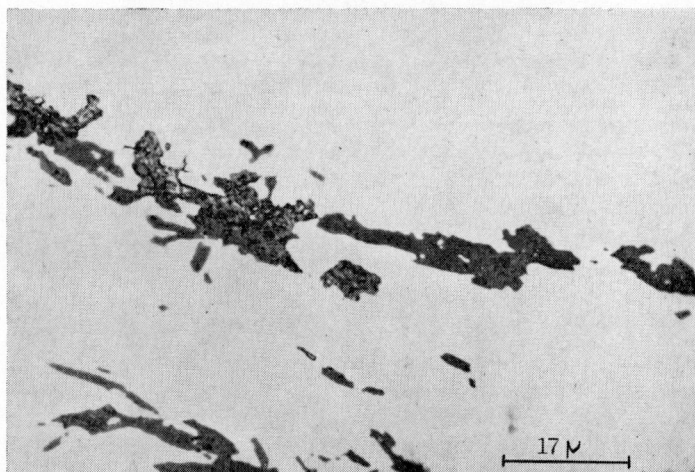

FIGURE 12.—UC(+ UC₂)–Ni couple: 930°C.  Graphite precipitate in the diffusion zone.

although the nickel content is essentially constant, $C_{Ni} = 22 \cdot 5 \%$. These results have been obtained on a single-phase UC couple, sintered at 930°C for 1850 h.

These concentrations do fit very well in the U and Ni contents of the phases:

|  | $C_U$ | $C_{Ni}$ |
|---|---|---|
| UC₂Ni | 74·2% | 18·30% |
| U₂C₃Ni | 83·36 | 10 |
| X (UNi) | 55 | 45 |
| Y (UNi) | 52 | 48 |
| UNi₅ | 44·8 | 55·2 |

These are compounds which, on the basis of the equilibrium diagram, may be expected to exist in the needle zone. At lower temperature the needle zone only develops when higher carbides are present. Between 700° and 900°C the reaction:

$$UC + 5\ Ni \rightarrow UNi_5 + C \qquad . \qquad . \qquad . \qquad . \qquad . \qquad (5)$$

takes the place of the reaction at 1000°C:

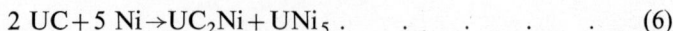

$$2\ UC + 5\ Ni \rightarrow UC_2Ni + UNi_5 . \qquad . \qquad . \qquad . \qquad . \qquad (6)$$

At this lower temperature, carbon is set free; the mixed carbides UC₂Ni and U₂C₃Ni do not form except by a direct reaction between UC₂ or U₂C₃, which are present in the uranium carbide.

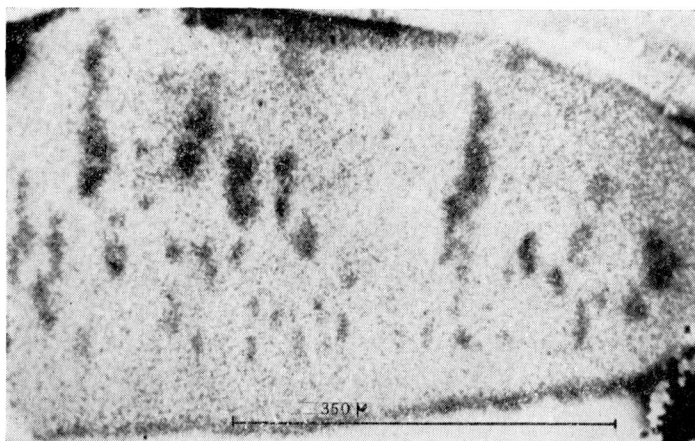

A

B

FIGURE 13.—Identification by $\beta$ radiation of graphite precipitate in the UNi$_5$ zone: UC(labelled with $^{14}$C)–Ni: 936°C, 1250 h.

A. Autoradiography    B. Micrography

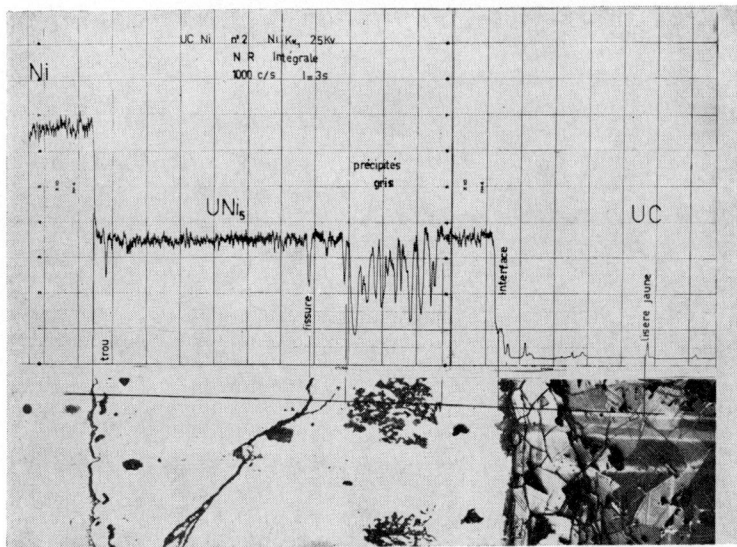

FIGURE 14.—UC–Ni couple: 822°C, 252 h. Record of nickel content in diffusion zone by micro-analysis.

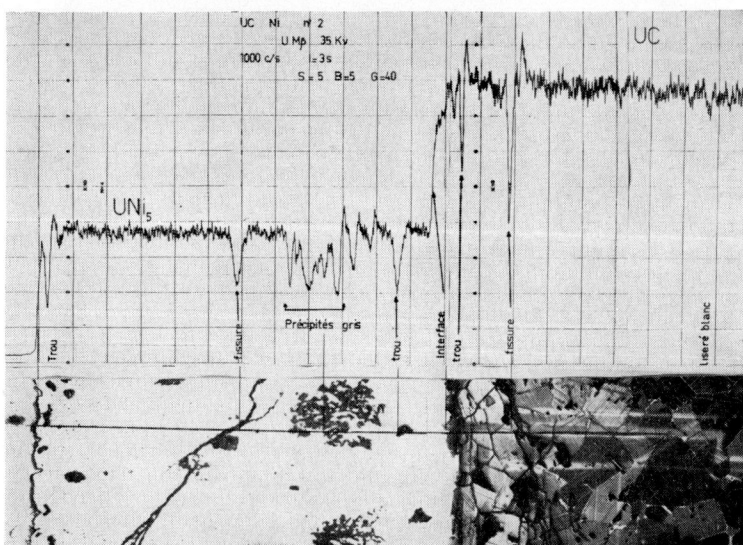

FIGURE 15.—UC–Ni couple: 822°C, 252 h. Record of uranium content in the diffusion zone by microanalysis.

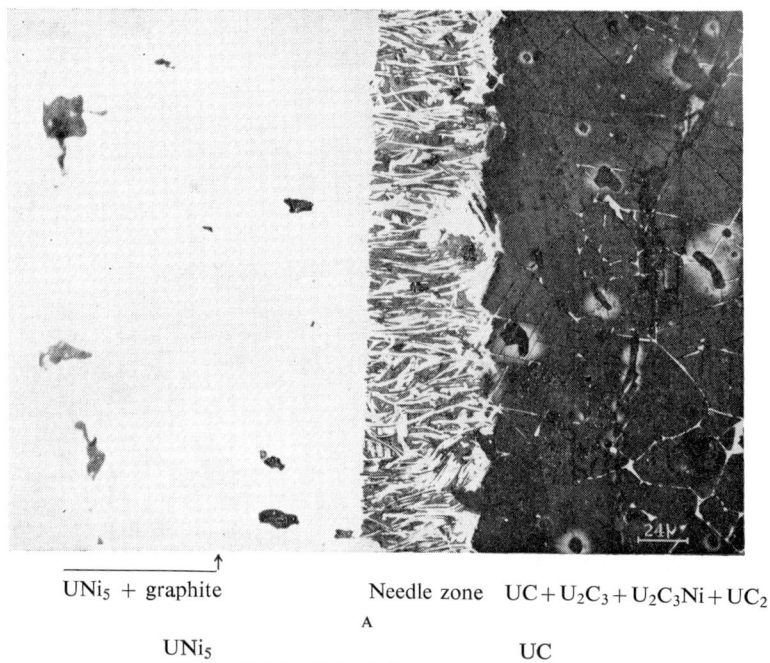

UNi$_5$ + graphite          Needle zone    UC + U$_2$C$_3$ + U$_2$C$_3$Ni + UC$_2$

A

UNi$_5$                              UC

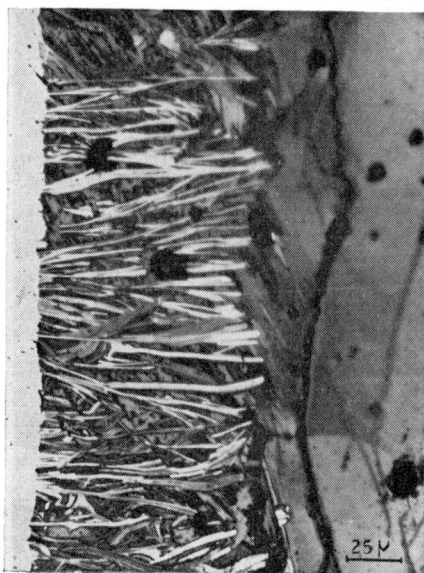

B

FIGURE 16

A. UC($+$UC$_2$$+$U$_2$C$_3$$+$U$_2$C$_3$Ni)–Ni couple: 980°C, 182 h. Needle zone.
B. UC (single phase)–Ni couple. 930°C, 1850 h. Needle zone.

Between $900°$ and $1000°C$ the two reactions take place simultaneously, forming free graphite and direct formation of $UC_2Ni$ by Ni acting on UC. The implication is that, at low temperatures, and for kinetic reasons, the carbon liberated by reaction (5) does not react with UC to form $U_2C_3$. Apart from this, the chemical potential gradient (diffusion couple) favours reaction (6).

This lower-temperature reaction has been reported by BRISI[11] from experiments on compacted powders.

Zone 4. A fringe of $U_2C_3Ni$ existing only in presence of higher carbides (Figure 17).

Zone 5. Carbide.

The mean width of zones $2+3+4$ divided by the square root of the annealing time is virtually independent of temperature between $800°$ and $1000°C$ (1 to $3 \cdot 5 \times 10^{-5}$ cm sec$^{-\frac{1}{2}}$) (Figure 18) but the relative thickness of the needle zone depends both on the annealing temperature and on the contents of higher carbides in the initial carbide mixture. At $700°C$ the reaction zone becomes very narrow: 10 $\mu$m after 2100 h of annealing.

U$_2$C$_3$                          Needle zone

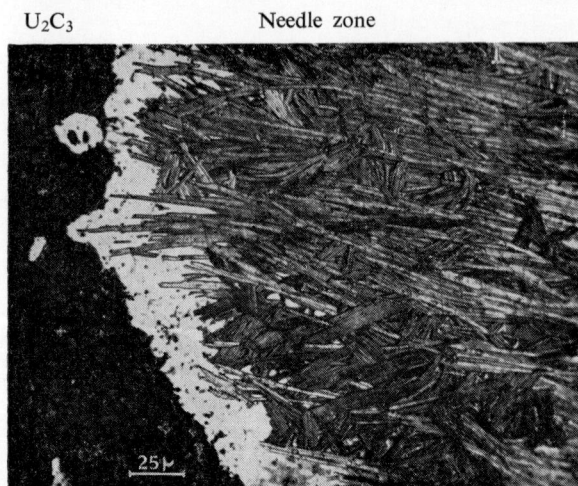

FIGURE 17.—$U_2C_3Ni$ couple: $1010°C$, 820 h. White fringe of $U_2C_3Ni$ between needle zones and carbide.

## 4.4  Solubility of Nickel in the Uranium Carbides

The solubility of nickel in uranium monocarbide and in uranium sesquicarbide has been evaluated by X-ray micro-analysis at $1400°C$.

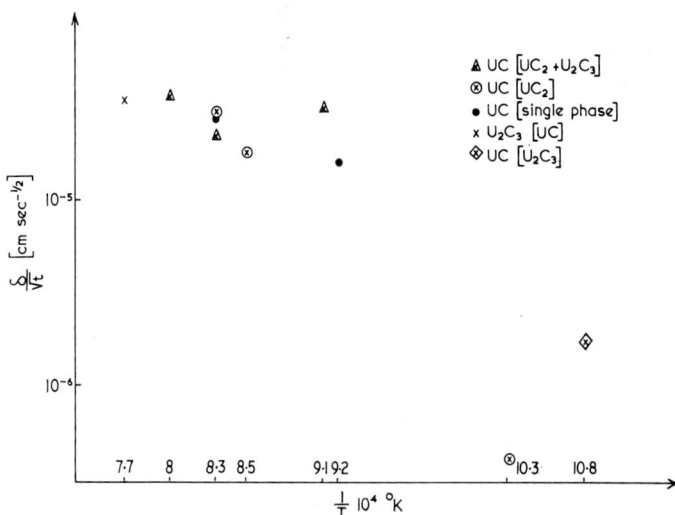

FIGURE 18.—Size of reaction zones divided by the square root of the annealing time for the Ni–carbides couple as a function of the carbon content of the carbides and of the reciprocal absolute temperature.

We found $1000 \pm 300$ p.p.m. in UC and $300 \pm 200$ p.p.m. in $U_2C_3$.

EYRE and BARTLETT[12] found that they could dissolve 1500 p.p.m. of nickel in UC at 1500°C. The difference in solubility of Ni in UC and in $U_2C_3$ is confirmed by STRASSER[13] who at 1500°C found 2000 p.p.m. Ni in UC and 100 p.p.m. in $U_2C_3$.

## 5. REACTIONS BETWEEN NICKEL AND URANIUM CARBONITRIDES

### 5.1 Thermodynamic Considerations

Nickel reacts with uranium nitride according to:

$$UN + 5\ Ni \rightleftharpoons UNi_5 + \tfrac{1}{2}\ N_2 \qquad . \qquad . \qquad . \qquad . \qquad (2)$$

In sealed ampoules the following reaction will also take place:

$$2\ UN + \tfrac{1}{2}N_2 \rightleftharpoons 2\ U_2N_3.$$

From the two preceding reactions it can be deduced that

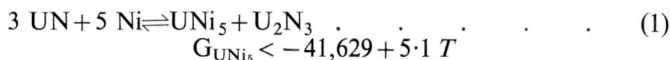

$$3\ UN + 5\ Ni \rightleftharpoons UNi_5 + U_2N_3 \qquad . \qquad . \qquad . \qquad . \qquad (1)$$
$$G_{UNi_5} < -41,629 + 5 \cdot 1\ T$$

When UN is in solution in a carbonitride, the direction of the reaction depends on the nitrogen pressure; if it is supposed that the uranium nitride in the carbonitrides stays in the species state,

29

reaction (1) will take the indicated course if

$$G_{UNi_5} + G_{U_2N_3} < G_{3UN} + 3\ RT \log (UN)$$

or

$$\frac{G_{UNi_5} + G_{U_2N_3}}{3} < G_{UN} + RT \log (UN)$$

(UN) representing the activity of UN.

The same is true for the carbide; it is known that nickel reacts according to:

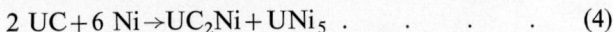

$$2\ UC + 6\ Ni \rightarrow UC_2Ni + UNi_5 \ . \qquad . \qquad . \qquad . \qquad (4)$$

In the case where the carbide is in solution in U(C,N), the direction of the reaction is determined by the following inequality:

$$G_{UC_2Ni} + G_{UNi_5} < \ G_{2UC}(g) + 2\ RT \log (UC)$$

or

$$\frac{G_{UC_2Ni} + G_{UNi_5}}{2} < G_{UC} + RT \log (UC)$$

$G_{UC_2Ni}$ can be evaluated from the thermodynamic balance of:

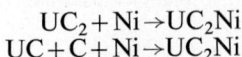

$$UC_2 + Ni \rightarrow UC_2Ni$$
$$UC + C + Ni \rightarrow UC_2Ni$$

$G_{UC_2}$ is very close to $G_{UC}$,

$$\text{so with } G_{UC} + \ -24,400 + 3{\cdot}1\ T$$
$$G_{UC_2Ni} < \ -24,400 + 3{\cdot}1\ T.$$

Figures 19 and 20 represent the variations in free energy of UC and UN as a function of their proportions in the carbonitrides at a given temperature:

$$\frac{G_{UNi_5} + G_{U_2N_3}}{3} \text{ and } \frac{G_{UC_2Ni} + G_{UNi_5}}{2}$$

have been drawn in the same graph.

Two cases may occur:

  (a) (Figure 19)

$$100 < (UC) < A$$

Only UC reacts until the composition of point A is reached, where the reaction stops.

$$A < (UC) < B$$

Neither of the components UC or UNi reacts.

$$B < (UC) < 0$$

Only UN reacts until the composition of point B is reached, where the reaction stops.

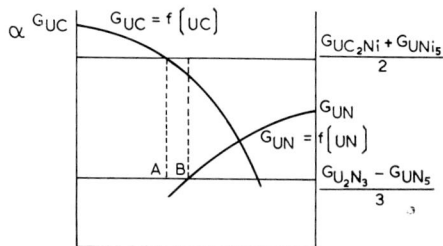

FIGURE 19.—Value of free energy of UC and UN as a function of their concentration in the carbonitrides in relation to the free energies of $UNi_5$, $UC_2Ni$ and $U_2N_3$ (Case a).

(b) (Figure 20)
When

$$100 < (UC) < A$$

UC reacts until the composition of point B is reached; at this point UN starts to react.

$$A < (UC) < B$$

UC and UN react simultaneously.

$$A < (UC) < C$$

Only UN reacts until the concentration of UC reaches the Value A; then the reaction with the nitride begins.

The application of these principles to the diffusion couples U (C,N)–Ni requires a knowledge of the diffusion of nitrogen and carbide in the carbonitride (kinetic considerations).

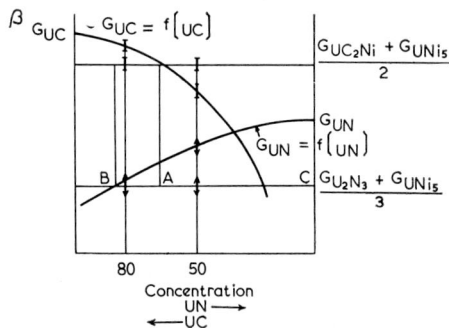

FIGURE 20.—As for Figure 19, but case b.

## 5.2  Observations on Diffusion Couples (50UC/50UN)–Ni

In couples annealed above 640°C, four zones are observed, show-ing aspects comparable with those formed in UN–Ni couples (Figure 21):

(1) Single-phase zone: UC, UN.

(2) Two-phase zone: Grey precipitates of $U_2N_3$ in a lighter matrix of $UNi_5$.

(3) Single-phase zone: $UNi_5$.

(4) Single-phase zone: Ni.

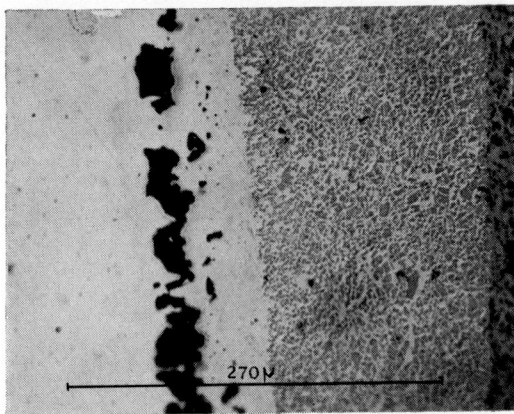

FIGURE 21.—(UC50/UN50)–Ni couple: 895°C, 651 h. Reaction zone. No graphite precipitation.

The absence of both graphite precipitates and a needle zone containing the mixed carbides $UC_2Ni$ and $U_2C_3Ni$ leads to the conclusion that, starting with equimolar composition and heating for the indicated times, only the nitride enters into reaction with Ni; when the annealing time is prolonged, the UC concentration grows.

Two possibilities arise: the carbonitride reaches concentrations of carbon and nitrogen such that no further reaction occurs at all; or the nitride and the carbide react simultaneously; this phenomenon is controlled by the diffusion speed of C and N in U(C,N).

Examination of the three diffusion couples (50UC/50UN)–Ni annealed at 895°C for systematically prolonged periods of time show graphite precipitation in the reaction zones after 4300 h (Figure 22).

This incubation time is a function of the thickness of the samples used: ours were test-pieces 3 mm thick.

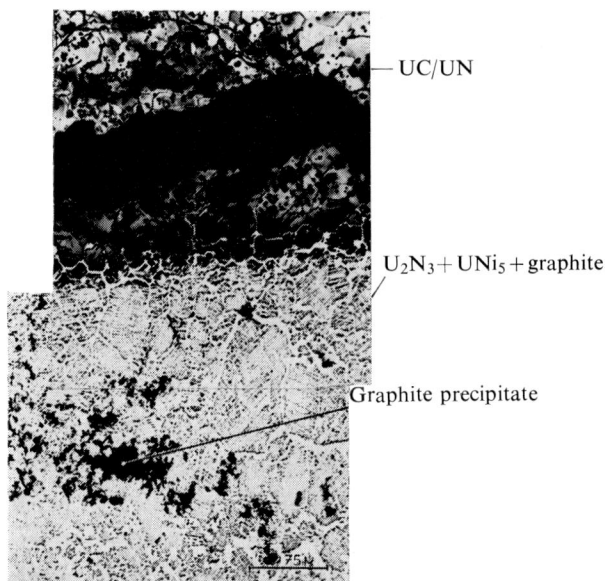

FIGURE 22.—(UC50/UN50)–Ni couple: 895°C, 4200 h. Graphite precipitation in the diffusion zone by the reaction of nickel with UC after a period of incubation.

So the second supposition seems to be the right one; we may conclude that the free energy curves of UC and UN in U(C,N) are in such positions relative to those of $UNi_5$ and $UC_2Ni$ that there will always be a reaction with the carbide and nitride (after a certain period of time).

The growth of the reaction zones $(2)+(3)$ presents no apparent anomaly for the moment at which the carbide reacts; as Table 5 shows $\theta/\sqrt{t}$ is constant.

### 5.3  Observations on the Diffusion Couples (80UC/20UN)–Ni

Examination of these diffusion couples (80–20%) with traces of $UC_2$ confirm our predictions.

In these samples the following reaction zones are found:

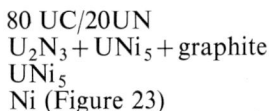

$$80\ UC/20UN$$
$$U_2N_3 + UNi_5 + graphite$$
$$UNi_5$$
$$Ni\ (Figure\ 23)$$

Traces of $UC_2$ are present in the carbonitride. The growth of the reaction zones is slower than in the 50–50 carbonitrides (Table 6).

Table 4

Temperature, Duration and Size of Reaction Zones of Diffusion Couples

| $T(°C)$ | $t\ (h)$ | $\theta\ (\mu m)$ | $\dfrac{\theta}{\sqrt{t}}\ cm.s^{-\frac{1}{2}}$ |
|---|---|---|---|
| 990 | 910 | 470 | $2 \cdot 7 \times 10^{-5}$ |
| 895 | 681 | 235 | $1 \cdot 5 \times 10^{-5}$ |
| 806 | 1850 | 235 | $3 \ \times 10^{-6}$ |
| 736 | 890 | 105 | $6 \cdot 2 \times 10^{-6}$ |
| 636 | 1890 | 40 | |
| 537 | 2170 | 10 | |

Table 5

Increase of Reaction Zones as a Function of time at 895°C in (50UC/50UN)–Ni Diffusion Couples as a Function of Time

| $T(°C)$ | $t\ (h)$ | $\theta\ (\mu m)$ | $\dfrac{\theta}{\sqrt{t}}\ cm.s^{-\frac{1}{2}}$ |
|---|---|---|---|
| 895 | 681 | 235 | $1 \cdot 5 \times 10^{-5}$ |
| 895 | 2080 | 495 | $1 \cdot 8 \times 10^{-5}$ |
| 895 | 4200 | 650 | $1 \cdot 65 \times 10^{-5}$ |

Table 6

Temperature, Duration and Size of Reaction Zones of (80UC/20 UN)–Ni Diffusion Couples

| $T(°C)$ | $t\ (h)$ | $\theta\ (\mu m)$ | $\dfrac{\theta}{\sqrt{t}}\ cm.s^{-\frac{1}{2}}$ |
|---|---|---|---|
| 893 | 1100 | 400 | $2 \ \times 10^{-5}$ |
| 800 | 2100 | 182 | $6 \cdot 6 \times 10^{-5}$ |
| 690 | 1980 | 25 | $9 \cdot 5 \times 10^{-7}$ |

## 5.4 Growth of Reaction Zones

The width of the reaction zones increases with increasing temperature and with time. Results are reported in Tables 4, 5 and 6 and Figures 23 and 24.

UCUN + $\phi$Ni                    UNi$_5$+U$_2$N$_3$+graphite    Ni

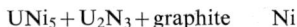

FIGURE 23.—(UC80/UN20)–Ni couple: 1100 h, 893°C. Reaction zones:
U(C,N)–UNi$_5$+U$_2$N$_3$+C–UNi$_5$–Ni.

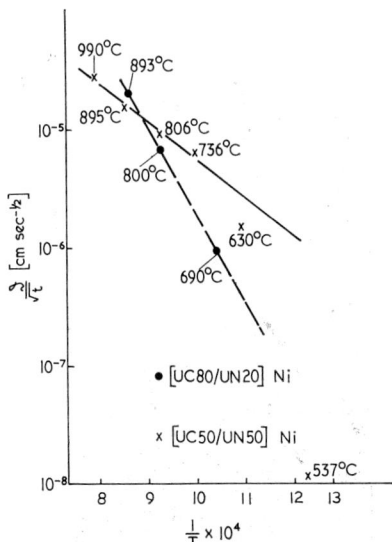

FIGURE 24.—Size of reaction zones divided by the square root of the annealing
time for UCN–Ni couples as a function of reciprocal absolute temperature.

## 6. DECORATION BY NICKEL OF UC$_2$ LAMELLAE
## PRESENT IN URANIUM MONOCARBIDE
## AND CARBONITRIDE

An epitaxial relationship exists between UC$_2$ (tetragonal) and UC
(cubic) for the (100) plane of UC and the (001) plane of UC$_2$.
The Widmanstadten structures of UC$_2$ in a UC matrix can thereby be
explained (lamellae of UC$_2$ parallel to the three (100) planes of the
monocarbide).[15] Examination of the morphology of the U$_2$N$_3$ phase
in the diffusion zone gives rise to the following observations.

### 6.1 The Couple UN-Ni

The $U_2N_3$ phase occurs as precipitates of irregular shape and large dimension in a matrix of $UNi_5$ (Figure 1). The electron microscope permits the observation of grain boundaries; the $UNi_5$ matrixis fully densified (Figure 2).

### 6.2 The Couple (U50/CN50)–Ni

The $U_2N_3$ precipitates are arranged in lamellae parallel to distinct directions in the middle of original U(C,N) grains which are all marked by the presence of $UNi_5$ (Figure 25). The carbonitride is a single phase. Near the interface, these directions corresponding with lines of preferential attack (Figure 26).

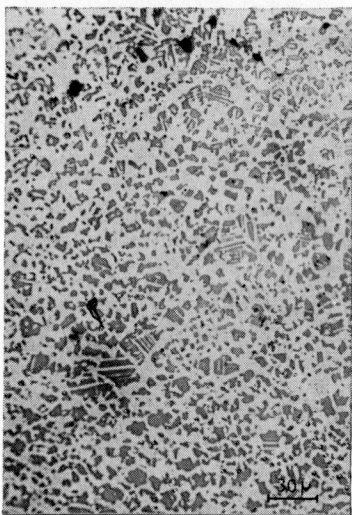

FIGURE 25.—(UC50/UN50)–Ni couple: 895°C, 681 h. Two-phase reaction zone. Geometrical precipitates of $U_2N_3$ in a matrix of $UNi_5$.

### 6.3 The Couple U(C,N)80/Ni20

The effects of anisotropic reactions are very clearly shown in those couples in which the carbonitride contains platelets of $UC_2$ as a second phase; the number of these plates is higher near to the interface and lower in the U(C,N) regions.

The diffusion and the preferential reaction of nickel along the $UC_2$ platelets are evident from the X-ray microanalysis (Figure 27); they explain the lamellar form of $U_2N_3$. The $U_2N_3$ lamellae are parallel to the (100) UCN planes (Figures 28 and 29).

FIGURE 26.—(UC50/UN50)–Ni couple: 895°C, 2,200 h. Interface between UNi$_5$+(U$_2$N$_3$–UC/UN) zones. Note the parallel striae due to an anisotropic reaction in the carbonitride.

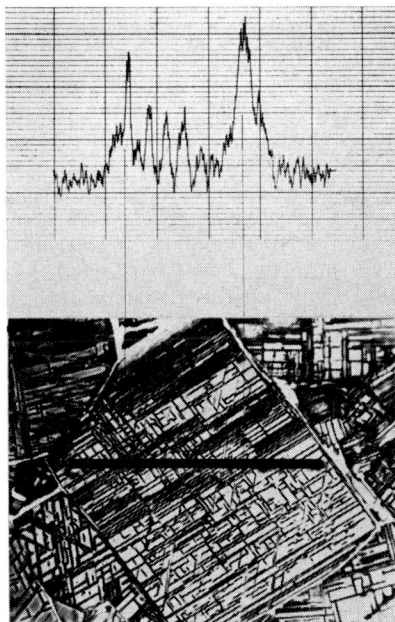

FIGURE 27.—(UC80/UN20)–Ni couple: 893°C, 1100 h. Diffusion and preferential reaction of nickel with the UC$_2$ platelets in UCN in the neighbourhood of the interface.

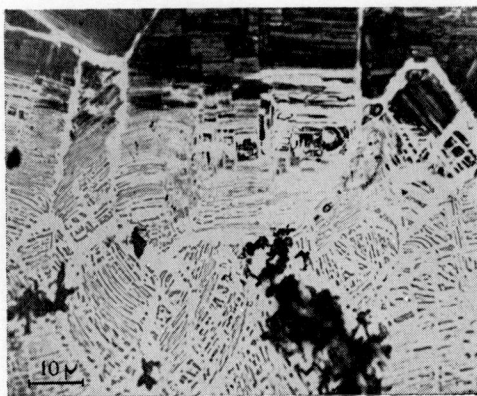

FIGURE 28.—(UC80/UN20)–Ni couples: 893°C, 1100 h. The $U_2N_3$ lamellae are parallel to the (100) planes of the carbonitrides. The orientations of these planes are given by the orientations of $UC_2$ platelets.

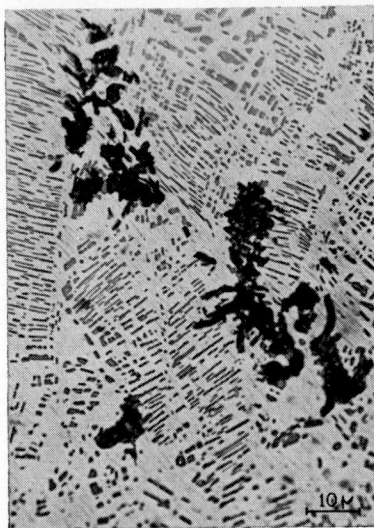

FIGURE 29.—(UC80/UN20)–Ni couple: 895°C, 1100 h. Reaction zones of $UNi_5 + U_2N_3 + C$. Widmandstaten structure of uranium sesquinitride in a matrix of $UNi_5$.

## 6.4 The Couple UC–Ni

The same phenomenon has been observed on monophase uranium carbides[7] (Figures 30 and 31); the conclusion can be drawn that, by preferential reaction with $UC_2$, nickel decorates the (100) planes

UNi$_5$

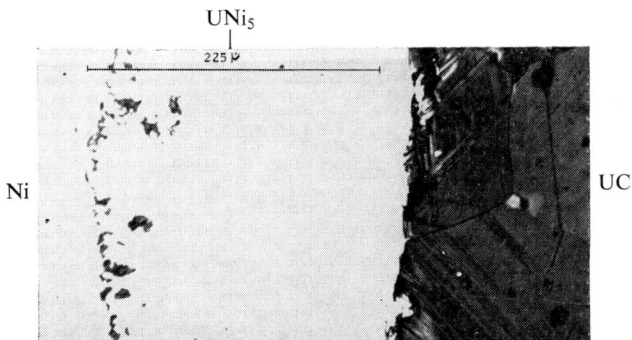

FIGURE 30.—UC (single phase)–Ni couple: 816°C, 950 h. Preferential reaction of nickel along (100) planes of UC (optically single phase). Note the streaks of carbide in the neighbourhood of the interface.

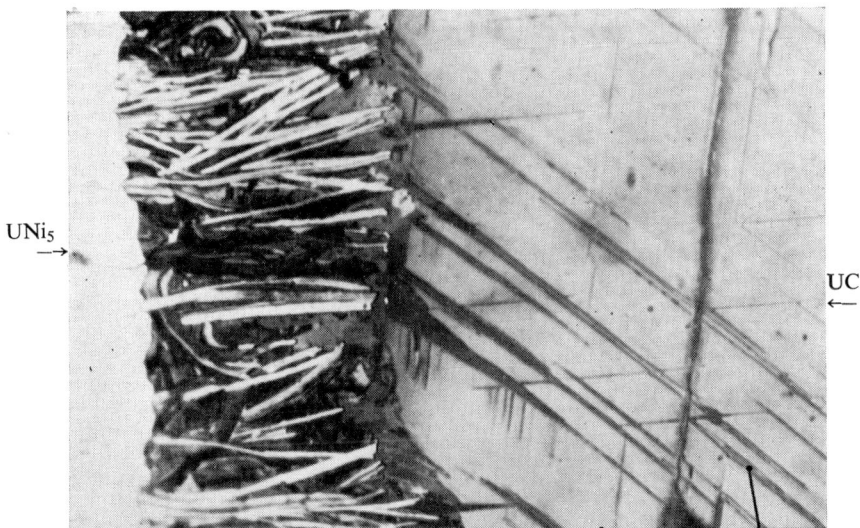

FIGURE 31.—UC(+UC$_2$)–Ni couple: 930°C, 700 h. Preferential reaction of nickel along platelets of UC$_2$ epitaxial with UC along (100) planes.

of the uranium monocarbides and carbonitrides which by micrography are single-phase; we have three reasons that support this hypothesis:

(1) These anisotropic reactions have not been observed on UN–Ni couples; uranium mononitride (isomorphous with UC) when prepared by decomposition in vacuo from U$_2$N$_3$ contains no trace of a second phase.

(2) They have been observed, however, in UC–Fe couples; the carbide was optically single-phase. Fe does not react with UC but only with higher carbides.[9] The presence of submicroscopic lamellae is the only possible explanation for a similar reaction.

(3) No simple crystallographic relations exist between $U_2N_3$ and $UNi_5$ which could lead to epitaxial growth.[10,11]

To sum up, we may say that nickel by a preferential and anisotropic reaction decorates the platelets of $UC_2$ that are found in the uranium monocarbide, although the latter are not discernible by optical microscopy.

## 7. CHEMICAL DIFFUSION OF URANIUM IN NICKEL

This work has been completed by measurements of the coefficient of chemical diffusion $D$ of uranium in solid solution in nickel. WAGNER'S[16] method was used; it consists in measuring the displacement $\varepsilon$ after annealing of an interface between a single-phase and a biphase zone from its original position, which is indicated by tungsten wires. We used nickel as the single phase and nickel saturated

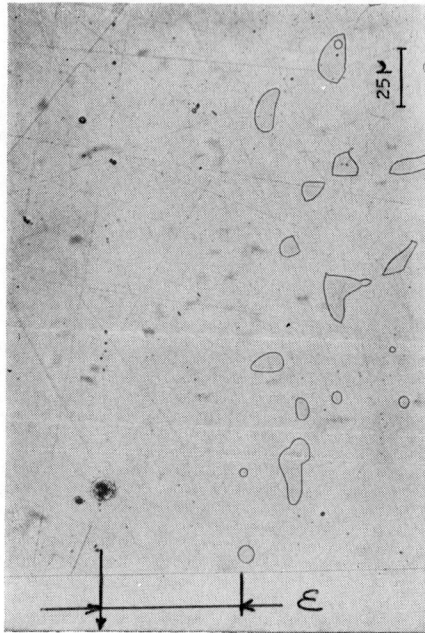

FIGURE 32.—UNi–(U + UNi₅ saturated with Ni)–Ni: 1030°C, 1443 h. Diffusion zone.

with uranium with $UNi_5$ precipitate as the biphase (Figure 32). $D$ is calculated from:

$$\varepsilon = 2\gamma\sqrt{Dt}$$

$$\frac{S_0}{C_0 - S_0} \cdot \frac{1}{\sqrt{\pi}} = \gamma e^{\gamma^2}(1 + erf\gamma)$$

$S_0$ is the solubility of U in Ni measured by X-ray microanalysis; its value varies from 2000 to 7000 p.p.m. by weight at temperatures between 870° to 1030°C. These results are in good agreement with GROGAN's[17] values.

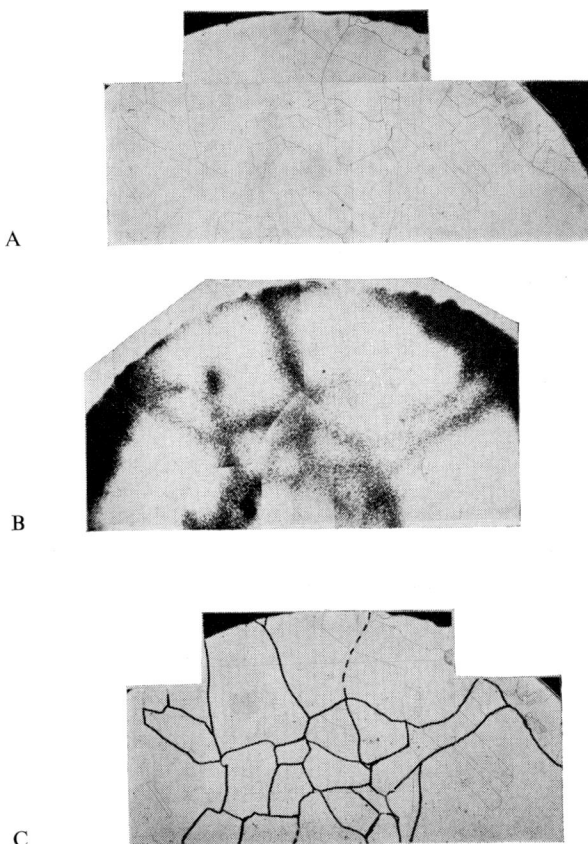

FIGURE 33.—Intergranular diffusion of uranium in nickel shown by fissiography:
A.  Photomicrograph.
B.  Intergranular diffusion.
C.  Boundaries of the grains affected by intergranular diffussion are outlined.

$C_0$ is the atomic concentration in the uranium–nickel biphase used: in our case $C_0 = 2$ at %. At temperatures between 870° and 1030°C. $D_{UNi}$ obeys:

$$D_{UNi} = 137 \text{ cm}^2.\text{s}^{-1} \, e^{\frac{-67,000 \text{ cal}}{RT}}$$

At low temperature the intergranular diffusion of uranium has been demonstrated by "fissiography".[18] This method consists in neutron irradiation of the surface of a diffusion couple after metallographic preparation while it is covered by a thin mica sheet. The mica serves as a detector of fission fragments produced by the superficial irradiation of uranium 235.

By this method, comparable to autoradiography, the distribution of uranium over the sample becomes visible (Figure 33). At higher temperatures surface diffusion takes place (Figure 34).

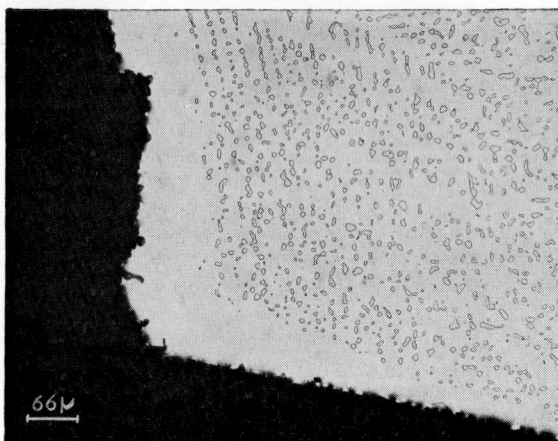

FIGURE 34.—Surface diffusion of nickel at the periphery of the two-phase alloy (Ni-saturated U + UNi$_5$ precipitates), marked by a zone without precipitates. UNi–Ni couple: 1030°C, 1443 h.

## 8. CONCLUSION

Nickel is highly reactive in contact with uranium carbides and nitrides. The following reactions have been observed:

$$2 \text{ UC} + 6 \text{ Ni} \rightarrow \text{UNi}_5 + \text{UC}_2\text{Ni above 900°C.}$$

and

$$\text{UC} + 5 \text{ Ni} \rightarrow \text{UNi}_5 + \text{C}$$
$$\text{U}_2\text{C}_3 + \text{Ni} \rightarrow \text{U}_2\text{C}_3\text{Ni}$$
$$3 \text{ UN} + 5 \text{ Ni} \rightarrow \text{U}_2\text{N}_3 + \text{UNi}_5$$

with

$$2 \, UN + \tfrac{1}{2} \, N_2 \rightleftharpoons U_2N_3$$

from 700°C up wards.

With the uranium carbonitrides the same reactions take place independent of the UC–UN ratio. Some reactions, however, need an incubation period.

## REFERENCES

1. PASCARD, R., Fourth Plansee Seminar. (Reutte. Tirol, 1961), p. 387.
2. ANSELIN, F., *J. Nuc. Mat.*, **10**, (4), 301, 1963.
3. ADDA, Y., and PHILIBERT, Second United Nations International Conference on the Peaceful Uses of Atomic Energy. Geneva, 1958. A/COF. 15 p. 1160.
4. ANSELIN, F., CALAIS, D., LORENZELLI, N., PASSEFORT, J. C., C.E.A. Report 2762. 1965.
5. BUGL, J., and BAUER, A. A., and LAPAT, P. E., and HOLDEN, R. B., International Symposium of Compounds of Interest in Nuclear Reactor Technology IMD. Special Report No. 13. The Metallurgical Society of the American Society of Mining—Metallurgical and Petroleum Engineers. 1964.
6. BRIGGS, G., GUHA, J., BARTA, J., WHITE, J., *Trans. Brit. Ceram. Soc.*, **62**, (3), 221, 1963.
7. KATZ, S., *J. Nuc. Mat.*, **6**, (2), 172, 1962.
8. ANSELIN, F., CALAIS, D., PASSEFORT, J. C., *C.E.A. Report 2845*, 1965.
9. ANSELIN, F., CALAIS, D., DEAN, G., and VAN CRAEYNEST, A., *C.R. Acad. Sci.*, **275**, 3916, 1963.
10. DUTTA, S. K., WHITE, J., Meeting on Nuclear and Engineering Ceramics, A.E.R.E., Harwell, Berks, October, 1965.
11. BRISI, C., and APPENDINO, P., *Anal. di Chim.*, **54**, (2), 661, 1964.
12. EYRE, B. L., and BARTLETT, A. F., *A.E.R.E., Report 5079*, 1965.
13. STRASSER, R., *United Nuclear Corporation, Report 5055*, 1963.
14. GRIEVESON and ALCOCK—quoted by RAND, M. H., and KUBASCHEVSKI, D., *A.E.R.E. Report* 3487, 1960.
15. ANSELIN, F., DEAN, G., LORENZELLI, R., and PASCARD, R., "Carbides in Nuclear Energy", Proceedings of Symposium at Harwell, (MacMillan & Co., London).
16. WAGNER, C.—quoted by JOST, "Diffusion in Solids, Liquids, Gases" (Academic Press, New York, 1960) p. 73.
17. GROGAN, J. D., and PLEASANCE, R. J., *J. Inst. Metals*, **182**, 141, 1953.
18. MORY, J., *C.E.A. Report* 2846, 1965.

# 29.—Properties of Mixtures of Uranium Nitride and Uranium Dioxide

By S. Mönch and N. Claussen

*Max-Planck-Institut für Metallforschung, Institut für Sondermetalle, Stuttgart*

*ABSTRACT*  E21/E173

*Sintering behaviour, electrical conductivity and some mechanical properties of $UO_2$–UN mixtures have been studied. At all concentrations investigated, the system $UO_2$–UN consists of a heterogeneous mixture of both phases which sinter independently. The low electrical conductivity of $UO_2$ can be improved by additions of UN. The Young's modulus and the indirect tensile strength of $UO_2$–UN mixtures does not change much with composition. Sintered samples at concentrations around 30 $^v/_o$ UN and 70 $^v/_o$ $UO_2$ exhibit good electrical properties without loss of mechanical strength, so that the use of this composition may offer some advantage over the pure components.*

### Propriétés de mélanges de nitrure d'uranium et de bioxyde d'uranium

*Le comportement au frittage, la conductivité électrique et certaines propriétés mécaniques de mélanges $UO_2$–UN sont étudiés. A toutes les concentrations, le système $UO_2$–UN est composé d'un mélange hétérogène de deux phases dont le frittage se fait indépendamment. La faible conductivité électrique de $UO_2$ peut être améliorée par des additions de UN. Le module d'Young et la résistance indirecte à la traction de mélanges $UO_2$–UN ne varient pas beaucoup avec la composition. Des échantillons frittés à des concentrations de 30 vol. % UN et 70 vol. % $UO_2$ présentent de bonnes propriétés électriques sans perte de résistance mécanique. L'utilisation de cette compostion peut ainsi présenter certains avantages par rapport aux composants purs.*

### Eigenschaften von Mischungen aus Urannitrid und Urandioxid

*Sinterverhalten, elektrische Leitfähigkeit und einige mechanische Eigenschaften von $UO_2$-UN-Mischungen wurden untersucht. Bei allen Konzentrationen besteht das System $UO_2$-UN aus einem heterogenen Gemisch beider Phasen, die auch unabhängig voneinander sintern. Die niedrige elektrische Leitfähigkeit des $UO_2$ kann durch*

*Beigabe von UN verbessert werden. Der E-Modul und die Zugfestig-
keit von UO₂-UN-Mischungen ändern sich nicht mit der Zusammensetzung. Sinterkörper aus ungefähr 30 Vol-% UN und 70 Vol-% UO₂
zeigen gute elektrische Eigenschaften ohne Verringerung der mechanischen Festigkeit, so daß die Verwendung dieser Zusammensetzung
einige Vorteile gegenüber den reinen Komponenten bietet.*

## 1. INTRODUCTION

In connection with the development of high-temperature nuclear
reactors, the investigation of refractory ceramic fuels is of great
importance. Uranium dioxide ($UO_2$) and uranium mononitride (UN)
play a significant role in this respect. Both compounds consist of
more than 80 $^w/_o$ uranium, and have melting points of 2800° and
2650°C respectively, consequently they are usually prepared by
powder metallurgy techniques. $UO_2$ is already being used successfully as nuclear fuel; UN also seems to be promising, although at the
present time preparation of pure UN is difficult and costly and so the
suggestion has already been made to produce UN from $UO_2$. In this
way, however, UN cannot be obtained oxide-free and so a study of
a $UO_2$–UN mixture is interesting. The $UO_2$–UN diagram is well
known.[2,3] Both components have a very low solubility for each
other and there is a eutectic at 50 mol% and 2200°C. Consequently
all mixtures of the two phases investigated are heterogeneous.

## 2. EXPERIMENTAL PROCEDURE

Nearly stoicheiometric oxide-free UN was prepared by nitriding
uranium powder at 850°C, followed by decomposition of the resulting
$U_2N_3$ in high vacuum. The average particle size was determined on a
Fisher Sub-Sieve-Sizer as 19 $\mu$m. The Fisher value for the nuclear
grade $UO_2$ used in this work was 2·4 $\mu$m. The oxygen content of the
oxide powder corresponded to the composition $UO_{2.057}$.

In order to improve homogenization, mixtures of the two powders
were ball-milled for 30 min. in trichloro-ethylene, which resulted in
the particle size being reduced to one sixth of the original. Metallographic examination of sintered samples showed that both phases
were uniformly distributed.

Samples of different densities were prepared by compacting,
followed by sintering at temperatures between 1500° and 1700°C in
high vacuum (isothermal sintering) and by hot-pressing at 1600°C
under argon. The sintering behaviour was determined dilatometrically
and by density measurements. The weight loss in sintering at 1600°C
was higher on the $UO_2$-rich side of the system, increasing from
0·1 to 0·7%/hour. Comparing the weight losses by evaporation of

multiple annealed samples, one observed an escape of oxygen from the UO$_{2\cdot057}$ initially present, resulting in a UO$_{1\cdot99}$.

Cylindrical samples (4 mm diam., 10 mm long) with gold-plated faces were used for electrical conductivity measurements at room temperature. The indirect tensile strength was determined by diametral compression (dimensions of the samples: 10 mm diam., 8 mm high). A method[4] also based on the diametral compression test served to measure the isothermal Young's modulus and Poisson's ratio. For this procedure, electrical strain gauges with an effective length of 3 mm were fitted in the centre of both faces of the samples (20 mm diam., 4·5 mm high).

Each investigation was carried out with a different set of samples, and the results were adjusted for zero porosity.

## 3. RESULTS

### 3.1 Sintering Properties

Figure 1 shows the sintered densities as a percentage of theoretical density (TD) *versus* concentration for different compacting and sintering conditions.

At all the temperatures the density of the compacts increases with increasing UO$_2$ content, which leads to the assumption that UO$_2$ additions improve the sinterability of UN.[5,6] Comparing the densities of UO$_2$ and UN, and assuming that both components sinter independently in the mixtures, the density should be along the dotted line in Figure 1.

The actual densities achieved, however, were all below this theoretical curve, so the addition of UO$_2$ cannot be regarded as improving the sinterability of UN. Such an improvement by oxidic impurities that may be soluble in UN should, however, not be excluded.

The curves in Figure 1 also show that the sinterability of UO$_2$ is considerably reduced by additions of UN. Therefore if dense compacts are required, UO$_2$–UN mixtures are less suitable than UO$_2$ alone, since, under equal conditions, UO$_2$ can always be sintered to higher densities.

### 3.2 Electrical Properties

The electrical properties of UO$_2$ and UN differ very much. The resistivity of UN (160 $\mu$ cm) is almost comparable to that of metals,[7] whereas UO$_2$ is a relatively bad conductor. The change of the resistivity in the system UO$_2$–UN is seen in Figure 2.

At low UO$_2$ contents, the electrical resistivity of UN hardly changes. Between 20 and 70 $^v/_o$ UO$_2$, a slow increase occurs. Above 70 $^v/_o$ UO$_2$, resistivity increases rapidly, since the UN phase is no longer continuous and the properties of UO$_2$ are dominant.

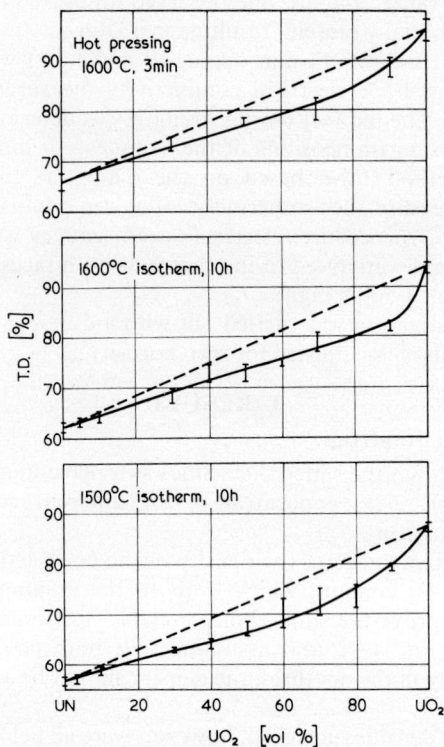

FIGURE 1.—Sintering density of UO₂–UN mixtures.

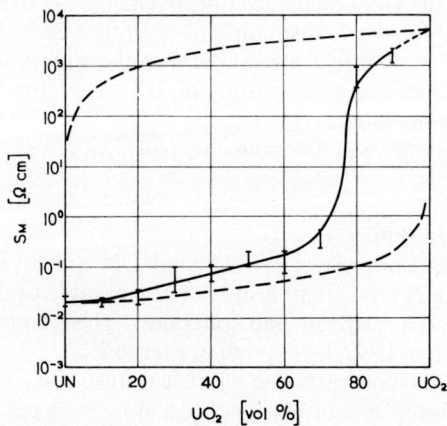

FIGURE 2.—Electrical resistivity of UO₂–UN mixtures.

The two dotted lines represent the theoretical resistivity if the individual resistivities are assumed to act in parallel and in series respectively. In series (upper curve) the bad conductor, $UO_2$, determines the resistivity, whereas in parallel (lower curve) it is UN.

Up to about 30 $^v/_o$ $UO_2$, the measured curve agrees well with the theoretical one for connection in parallel. With further increase of $UO_2$ content, the result is a combination of the two models but parallel connection, however, still predominates; this means that small additions of UN can improve the conductivity of $UO_2$.

### 3.3 Mechanical Properties

Determination of the mechanical properties of brittle sintered compacts involves certain difficulties. On one hand, the results usually show considerable scatter, and on the other hand, for porosity adjustment only empirical relations exist and they differ for different materials.

First, therefore, we examined the dependence of the indirect tensile strength and Young's modulus on $UO_2$ and UN in order to correct the results obtained from the mixtures. For the Young's modulus a linear relation was found. The indirect tensile-strength data (20 samples) were fitted by the method of least squares to an exponential equation proposed by RYSHKEWITCH.[8] The following porosity equations could be determined for the Young's modulus:

$$E_{UO_2} = 23,400 \, (1 - 3 \cdot 1 \, P) \, kp/mm^2$$
$$E_{UN} = 24,000 \, (1 - 2 \cdot 2 \, P) \, kp/mm^2$$

and for the indirect tensile strength

$$S_{UO_2} = 5 \cdot 2 \, . \, e^{-6 \cdot 1 P} \, kp/mm^2$$
$$S_{UN} = 7 \cdot 0 \, . \, e^{-5P} + \, kp/mm^2.$$

The correction constants for the mixtures were obtained from the data of the components by linear interpolation.

Figure 3 shows the corrected Young's modulus ($E$) and Poisson's ratio ($\mu$) as a function of composition. Young's modulus, which is almost the same for both components, exhibits a slight minimum at 20 $^v/_o$ $UO_2$.

Poisson's ratio first increases from 0·25 (UN) with additions of $UO_2$, then remains unchanged at 0·32 ($UO_2$).

The indirect tensile strength ($S$) of various $UO_2$–UN mixtures is given in Figure 4. The bars in the graph represent the standard deviation obtained from 6 samples. The tensile strength of the mixtures decreases from the strength of UN (7 $kp/mm^2$) to the one of $UO_2$ (5·2 $kp/mm^2$) except for mixtures with 70 and 80 $^v/_o$ $UO_2$. So far we have found no explanation for this increase, but small amounts of a

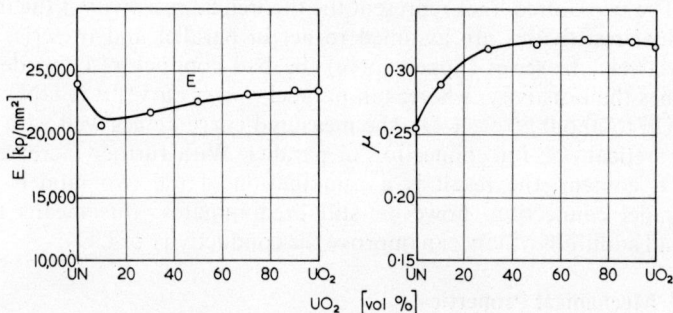

FIGURE 3.—Young's modulus ($E$) and Poisson's ratio ($\mu$) of $UO_2$–UN mixtures.

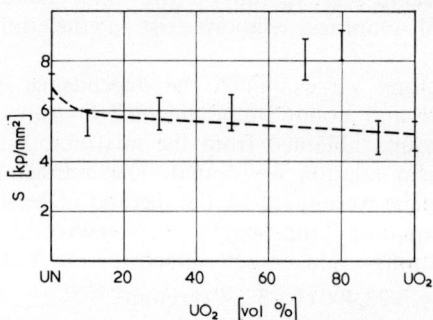

FIGURE 4.—Indirect tensile strength of $UO_2$–UN mixtures.

third, unidentified, phase in this concentration range, found by metallographic examination, could be responsible.

A considerable decrease in the indirect tensile strength, which had been expected originally, does not occur; therefore, in the application of $UO_2$–UN mixtures, their mechanical strength is similar to the strengths of the individual components.

## REFERENCES

1. IMOTO, S., and STÖKKER, H. J., *Ber. Dtsch. Keram. Ges.*, **43**, 130, 1966.
2. BLUM, P., and GUINET, P., *C.R. Acad. Sci.*, **253**, 1053, 1961.
3. KELLY, H. J., Report USBM-RC-1150. (Pratt and Whitney Aircraft, Middletown).
4. CLAUSSEN, N., *Z. Materialprüf.*, **9**, No. 4, 1967.
5. U.S. Patent 3,213,161 (1965).
6. THÜMMLER, F., ONDRACEK, G., and DALAL, K., *Z. Metallkde.*, **56**, 535, 1965.
7. HAYES, B. A., and DE CRESCENTE, M. A., PWAC-481 (1965). Report (Albany Metallurgy Research Center).
8. RYSHKEWITCH, E., *J. Amer. Ceram. Soc.*, **36**, 65, 1953.

# 30.—Preparation and Physical Properties of Perovskite-type Compounds, $[(1-x)\text{La},x\text{Ca}]\text{MnO}_3$

C. M. ISERENTANT* and G. G. ROBBRECHT

*Natuurkundig Laboratorium Verschaffelt der Rijksuniversiteit te Gent*

## ABSTRACT                                                    E730

The preparation of a number of $[(1-x)La,xCa]MnO_3$ compounds shows the great influence of the firing conditions on the composition. Chemical analysis yields the percentage $Mn^{4+}$ ions present in the specimens of which the cell dimensions are determined. From thermal expansion measurements, the behaviour of the compounds during phase transformations is investigated. In the case of macroscopic distortion arising as a consequence of Jahn–Teller type distortions in the immediate environment of the octahedral sites, the configurational energy of the perovskite structure is derived and the structure of the stable low-temperature phase is deduced, in analogy to Wojtowicz's theory for spinels.

## Préparation et propriétés physiques des composés du type perowskite $[(1-x)La,xCa]MnO_3$

La préparation d'un certain nombre de composés du type $[(1-x)$ La, xCa] MnO₃ met en évidence la grande influence exercée sur la composition par les conditions de cuisson. L'analyse chimique fournit le pourcentage d'ions $Mn^{4+}$ existant dans les échantillons pour lesquels on détermine les dimensions de la maille. En se fondant sur des mesures de dilatation thermique, on étudie le comportement des composés pendant les transformations des phases. Dans le cas de distorsions macroscopiques résultant de distorsions du type Jahn–Teller au voisinage immédiat des sites octaédriques, on détermine l'énergie configurationnelle de la structure perowskitique et on en déduit la structure de la phase stable à basse température, par analogie avec la théorie de Wojtowicz relative aux spinelles.

## Anfertigung und physikalische Eigenschaften von Verbindungen des Perowskit-Typs: $[(1-x)LaxCa]MnO_3$

Die Anfertigung einer Anzahl Verbindungen $[(1-x)LaxCa]MnO_3$ zeigt den großen Einfluß der Feuerungsbedingungen auf die

* Research fellow of N.F.W.O.

*Zusammensetzung. Die chemische Analyse ergibt den prozentualen Mn⁴⁺-Gehalt in den Proben, denen Zelldimensionen bestimmt werden. Auf Grund von thermischen Ausdehnungsmessungen wird das Verhalten der Verbindungen während Phasentransformationen untersucht. In den Fällen einer makroskopischen Verdrehung, die als Folge einer Jahn-Teller-Verschiebung in unmittelbarer Nachbarschaft der Oktaederplätze eintritt, wird die Konfigurationsenergie der Perowskitstruktur abgeleitet und die Struktur der stabilen Tieftemperaturphase in Analogie zur Wojtowicz'schen Theorie für Spinelle bestimmt.*

## 1. INTRODUCTION

The interest in compounds of the type $ABO_3$, which have structures closely related to the well-known perovskite structure, stems from the unusual physical properties of these substances. Only their ferroelectric properties have been extensively explored for possible practical applications.

The current trend is to look for new experimental information about the nature of the chemical bond in the transition-metal alloys which crystallize in the perovskite-type structure, and on its relation to their magnetic and crystallographic properties. However, no conclusive work has been done on the nature of the crystal transformations so often found in these substances. Recently it has been recognized that the ideal perovskite-type structure is seldom realized; tetragonal and orthorhombic distortions are frequently observed. In many cases the distortion is highly temperature dependent and transformations to more symmetrical structures can appear at specific temperatures.

It is the purpose of this paper to describe and discuss some physical properties of the mixed series $[(1-x)La,xCa]MnO_3$.

## 2. EXPERIMENTAL

### 2.1 Preparation of the Samples

The compounds included in the present investigation are $LaMnO_3$, $CaMnO_3$ and some solid solutions of the type $[(1-x)La,xCa]MnO_3$ with $x$ equal to 0·2, 0·4, 0·6 and 0·8 respectively. The preparation, according to conventional ceramic techniques, is straightforward in its principle. The starting materials, selected chiefly on the basis of purity and particle size, are the oxide $La_2O_3$ (Johnson & Matthey "Specpure") and the carbonates $CaCO_3$ (Merck, pro analysi) and $MnCO_3$ (British Drug Houses). Stoicheiometric amounts are thoroughly mixed with ethanol as a mixing medium, and milled in an agate mortar for 4 h. The homogeneous mixture is quickly dried under an infra-red lamp to minimize preferential settling of any dense or large particles. The dry material is then pre-sintered in air at a temperature of about 950°C for about 20 h. Experimentally it was established that pre-sintering of the whole mixture was necessary to avoid extensive shrinkage of the final shape. The powder is then again throughly mixed and milled to promote homogeneity. The final sintering takes place by firing at temperatures in excess of 1075°C in an atmosphere ranging from pure oxygen to 99·9 %$N_2$, 0·1 %$O_2$, for periods of 35 h on the average (Table 1). This is a very important process. Apart from the atmosphere in the furnace, which influences the chemical composition to a great extent, the course of the reaction will be determined in a complicated way by many other factors: peak temperature, heating and cooling rates, porosity of the shape, etc. Finally, in order to obtain well-crystallized products the samples were annealed at 800°C for about 90 h.

Relatively large test rods were required with a view to measuring the thermal expansion. Perovskites are very hard and brittle substances, which makes them difficult to machine. One is thus obliged to shape the samples before final firing by pressing the presintered powder, mixed with ethanol to make a paste, in a steel mould.

The procedure for obtaining the relatively large test rods, required with a view to measuring the thermal expansion, has been described in detail in previous papers.[1,2]

### 2.2 Chemical Composition

Because of the difficulties in controlling the various factors governing the final sintering, the end-products often vary somewhat in composition and properties. As a check on the purity of the compounds, chemical analysis for tetrapositive manganese ions was carried out using the method as described by WOLLAN and KOEHLER.[3] The results are given in Table 1.

### 2.3 X-ray Diffraction Analysis

X-ray powder diffraction patterns were taken at room temperature of all the compounds mentioned above, using Cu K$\alpha$ radiation. The diffraction patterns are closely related to those that could be expected from simple perovskite structures. Main differences between the ideal and some of the observed patterns are the split reflections and superlattice reflections shown in the latter. With the help of some previous results[3-5] and by comparison with the diffraction data from gadolinium orthoferrite. GdFeO$_3$,[6] the

31

Table 1

Results of Chemical and Crystallographic Analysis

| Specimen | Firing conditions | $Mn^{4+}$ (%) | Symmetry* | Pseudo-cell dimensions | Pseudo-cell volume (Å³) |
|---|---|---|---|---|---|
| $LaMnO_3$ I | 1200°C, 30 h, in 99·9% $N_2$–0·1% $O_2$ | 8 | O | $a_1 = a_3 = 7·876$ $a_2 = 7·658$ $\beta = 92°\ 12'$ | 474·7 |
| $LaMnO_3$ II | 1200°C, 15 h; 800°C, 72 h; in 99·9% $N_2$–0·1% $O_2$ | 11 | O | $a_1 = a_3 = 7·872$ $a_2 = 7·742$ $\beta = 90°\ 44'$ | 479·7 |
| $LaMnO_3$ III | 1200°C, 30 h; 800°C, 72 h; in 99·9% $N_2$–0·1% $O_2$ | 15 | O | $a_1 = a_3 = 7·829$ $a_2 = 7·761$ $\beta = 90°\ 30'$ | 475·7 |
| $La_{0·8}Ca_{0·2}MnO_3$ | 1400°C, 40 h; 800°C, 90 h; in 99·9% $N_2$–0·1% $O_2$ | 35 | T | $a_1 = a_2 = 7·701$ $a_3 = 7·661$ | 464·3 |
| $La_{0·6}Ca_{0·4}MnO_3$ | 1400°C, 40 h; in 86% $N_2$–14% $O_2$ 800°C, 90 h, in 99·9% $N_2$–0·1% $O_2$ | 67 | C | $a = 7·612$ | 441·1 |
| $La_{0·4}Ca_{0·6}MnO_3$ | 1400°C, 40 h, in 99% $O_2$; 800°C, 90 h, in air | 77 | C | $a = 7·538$ | 428·3 |
| $La_{0·2}Ca_{0·8}MnO_3$ | 1200°C, 30 h; 800°C, 72 h; in 99% $O_2$ | 94 | C | $a = 7·481$ | 418·7 |
| $CaMnO_3$ | 1075°C, 24 h, in 99% $O_2$ | 98 | C | $a = 7·438$ | 411·5 |

* O = orthorhombic    T = tetragonal    C = cubic

diffraction patterns were indexed. The results of the crystallographic analysis are summarized in Table 1. The LaMnO$_3$ samples
are orthorhombic. All the other compositions, except
La$_{0.8}$Ca$_{0.2}$MnO$_3$, which seems to be tetragonal, are cubic. For the
orthorhombic distortion, the pseudocubic lattice is monoclinic.
The true lattice symmetry has of course to be determined from
single-crystal data. The orthorhombic cell contains 16 or, neglecting
a few very weak reflections, four distorted perovskite units in the
true crystallographic cell.

In Figure 1 the lattice parameters $a_1$, $a_2$ and $a_3$ for the pseudocell
of the orthorhombic structures and $a$ for the cubic structures (all
parameters taken at room temperature) are plotted *versus* the Mn$^{4+}$
concentration. The distortion of the LaMnO$_3$ sample with the
fewest Mn$^{4+}$ ions is larger than that of the other samples. The
introduction of Mn$^{4+}$ ions reduces the magnitude of this monoclinic distortion. From the figure, by extrapolating the data listed
in Table 1, we can conclude that the transformation from the orthorhombic structure to the cubic structure takes place when about
20% Mn$^{4+}$ ions are present in the sample, whereas Wollan and
Koehler mention 25%, Wold and Arnott 21% and Jonker 16%:
the ordering effect responsible for the distortion is completely
destroyed. Finally, the [(1 − x)La, xCa]MnO$_3$ series shows a
continuous linear decrease of lattice volume with increasing Mn$^{4+}$
content.

### 2.4 Thermal Expansion

The thermal expansion of all the compounds [(1 − x)La, xCa]
MnO$_3$ was measured with a dilatometer similar to the one already

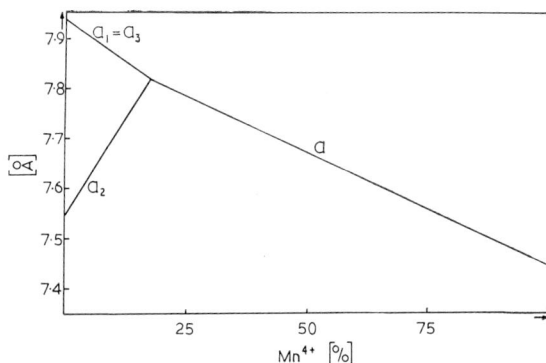

FIGURE 1.—Lattice parameters of [(1−x)La,xCa]MnO$_3$ as a function of Mn$^{4+}$
concentration.

described.[2] However, this instrument was automatized by making use of a transducer and a digital print-out system, whereby it was possible to measure the expansion with an accuracy of 0·2 $\mu$m of a rod of about 100 mm from liquid nitrogen temperature to about 800°C.

The thermal expansion as a function of the temperature shows, although not very pronounced for some compounds, an anomaly at about −150°C. This is a magnetic transition which was not studied in the present investigation. A much greater anomaly, caused by a phase transformation of the crystal structure, was found in all compounds except $La_{0.2}Ca_{0.8}MnO_3$ and $CaMnO_3$. During this transition, which extends over a range of temperature, the expansion of the rods is very small; in the case of $LaMnO_3$ the rods even shrink a little. This unexpected behaviour induced us to submit this lanthanum manganite rod again to heat treatment and measurement, and this procedure was repeated twice. After each treatment the ionic composition of the rod had changed significantly, in particular the $Mn^{4+}$ content had increased appreciably. The results of this additional X-ray and chemical analysis are found in Table 1, indicated as $LaMnO_3$ II and III. The thermal expansion behaviour

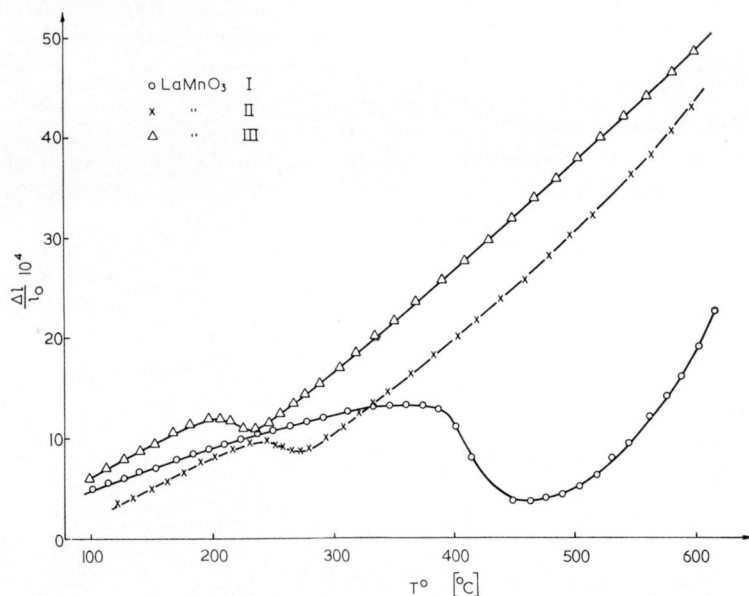

FIGURE 2.—Expansivity of $LaMnO_3$ I, II and III *versus* temperature.

of rods II and III is similar to that of rod I, but the transition is less pronounced and it shifts to lower temperatures with increasing $Mn^{4+}$ content: this is clearly seen in Figure 2. This phenomenon was confirmed by the measurements on the other samples, which contain a much higher percentage of $Mn^{4+}$ ions. As the $Mn^{3+}$ ions are responsible for the distortion, the introduction of more $Mn^{4+}$ ions dilutes the available number of ions participating in the co-operative distortion mechanism, which is reflected in a lowering of the transition temperature.

## 3. THEORETICAL

### 3.1 Theoretical Model

The origin of the large macroscopic crystal distortions which occur in the compounds $[(1-x)La, xCa]MnO_3$ lies in a co-operative ordering phenomenon of local distortions caused by the Jahn–Teller effect. According to the theorem of Jahn and Teller, molecules or complexes with orbitally degenerate electronic ground states are unstable in the symmetric configuration. The molecule will always find at least one vibrational co-ordinate along which it can distort to split the degeneracy and lower its energy. For $3d^4$ configurations ($Mn^{3+}$) in octahedral co-ordination the distortions are large and of the prolate tetragonal type; for $3d^3$ configurations ($Mn^{4+}$) there is no distortion.[7]

In the ideal cubic $ABO_3$ perovskite structure the A cations and the anions together form a close-packed cubic lattice with the B cations occupying the octahedral interstices of the anion sub-lattice. The simple cubic lattice formed by the B cations may be subdivided into two interpenetrating face-centred cubic lattices; a B cation in one sub-lattice has six B nearest neighbours of the other sub-lattice. It is convenient for further investigation to consider the perovskite structure as built up by $BO_6$ octahedral complexes, centred on the B ion sites, the apexes falling on the cubic close-packed oxygen positions. Each $BO_6$ complex of the first sub-lattice shares one oxygen ion with each of the six neighbouring $BO_6$ complexes.

According to the Jahn–Teller theorem, each $Mn^{3+}O_6$ complex is distorted; the distortions in neighbouring complexes interfere with each other. At high temperature, this interaction is small with respect to the thermal vibration: each complex will distort independently. At any instant an equal number of complexes are distorted in the three cubic directions; the structure will appear as statistically cubic. Upon lowering the temperature, the interactions between distortions become more and more important. Finally, below a certain critical temperature, long-range ordering occurs and as a consequence a macroscopically distorted phase results.

For the purpose of further analysis it is convenient to define a set of pseudo-cubic axes with respect to which the orientation of the long axes of the octahedra can be specified. We also assume that the magnitude of the distortion of the individual octahedra is a constant, independent of the temperature and the orientations of neighbouring distortions. Furthermore we assume that only pairwise interaction between nearest neighbouring Jahn–Teller distortions contribute appreciably to the configuration energy of the system, and that all the B sites are occupied by $Mn^{3+}$ ions, sensitive to Jahn–Teller distortion. In this manner there are four different pair potentials, depending on the relative orientation of the octahedra. The four possibilities are schematically represented in Figure 3, in which we have drawn the three mutually perpendicular axes

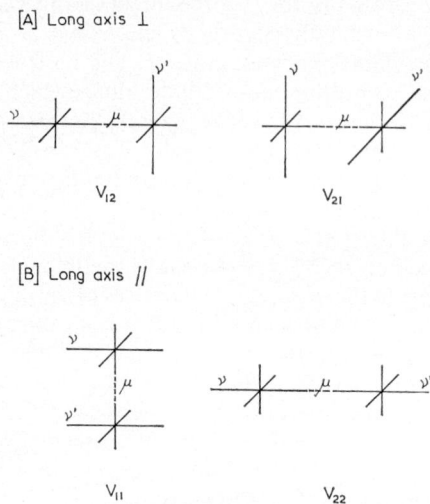

FIGURE 3.—The four possible pair interactions between octahedra in the perovskite structure.

of the distorted octahedra, one long axis (heavy lines) and two short axes (in full lines). Since the long axes of the octahedra are mutually perpendicular, we have the two potentials $V_{12}$ (a long axis and a short axis collinear) and $V_{21}$ (two short axes collinear); in addition, there exist two pair potentials when the long axes of the octahedra involved are parallel: $V_{11}$ (the short axes co-planar) and $V_{22}$ (the long axes collinear). Taking into account the strain induced in the lattice by the distortions, it is reasonable to compare the magnitude of the pair potentials as follows:

$$V_{12} < V_{11} \sim V_{21} \ll V_{22} \qquad . \qquad . \qquad . \qquad . \qquad . \qquad (1)$$

Finally, we assume these pair potentials to be independent of the temperature and of the orientations of other octahedra in the neighbourhood of the pair in question. This theoretical model for the perovskite structure is adapted from the one developed by WOJTOWICZ[8] for the study of the spinel structure and leads to the same conclusions as the model based on the effects of covalence on bond length, formerly proposed by J. B. GOODENOUGH.[9]

### 3.2 The Hamiltonian Configuration

Let us consider a particular configuration of $N$ octahedra in the crystal ($N=$ Avogadro's number). The Hamiltonian configuration is the sum of all the nearest-neighbour pair-interaction potentials. For the purpose of simplification, we divide the system of octahedra into the two sub-lattices mentioned above (let us call these $s$ and $t$); in each sub-lattice there will be $N/2$ sites. To describe all the possible configurations of the system, we define as in reference 8 a set of occupation variables $\rho_i^s$ and nearest-neighbour selector factors $\lambda_{ij}^{st}$ as follows:

$\rho_i^s(\nu) = 1$, if site $i$ on sub-lattice $s$ is occupied by an octahedron distorted in the $\nu$ direction.

$\rho_i^s(\nu) = 0$, if site $i$ on sub-lattice $s$ is occupied by an octahedron distorted in any other direction.

$\lambda_{ij}^{st} = 1$, if site $i$ on sub-lattice $s$ is a nearest neighbour to site $j$ on sub-lattice $t$.

$\lambda_{ij}^{st} = 0$, otherwise

where $i, j = 1, \ldots, N/2$ and $\nu = x, y, z$.

The occupation variables, which specify the orientation of the long axis of the distorted octahedra, satisfy the following relations:

$$\sum \rho_i^s(\nu) = 1 \quad ; \quad \sum_i \rho_i^s(\nu) = N_s^\nu \quad ; \quad \sum_\nu N_s^\nu = \frac{N}{2} \qquad . \quad (2)$$

where $N_s^\nu$ is the number of octahedra on sub-lattice $s$ with the long axis in the $\nu$ direction. The nearest-neighbour selector factors, which indicate whether two arbitrary sites $i$ and $j$ respectively of sublattice $s$ and $t$ are nearest neighbours or not, are related by:

$$\sum \lambda_{ij}^{st} = \sum \lambda_{ij}^{st} = z = 6 \quad . \qquad . \qquad . \qquad . \quad (3)$$

where $z$ is the total number of nearest neighbours for each site. The configurational potential energy for any pair of sites $i, j$ in the crystal can now be written as

$$\rho_i^s(\nu)\lambda_{ij}^{st}\rho_j^t(\nu')V(\nu,\nu';\mu) \qquad . \qquad . \qquad . \qquad (4)$$

$V(\nu, \nu'; \mu)$ is the pair interaction potential resulting from a pair of octahedra distorted in the $\nu$ and $\nu'$ directions respectively, and having their line of centres along the $\mu$ direction. The possible values of the potentials in this notation are easily identified with the potentials in the notation of 3.1, which leads to Table 2.

**Table 2**
**Identification of the Pair Potentials**

| Configuration | $V(\nu, \nu'; \mu)$ |
|---|---|
| $\nu \neq \nu' = \mu$ | $V_{12}$ |
| $\nu = \mu \neq \nu'$ | $V_{12}$ |
| $\nu = \nu' \neq \mu$ | $V_{21}$ |
| $\nu \neq \nu' = \mu$ | $V_{11}$ |
| $\nu = \nu' \neq \mu$ | $V_{22}$ |

For each neighbouring pair $i, j$ in the crystal, there will be nine terms of the form (4), due to the three allowed directions for distortions: only one term will be non-vanishing. The Hamiltonian function for the whole system can now be obtained by carrying out the summation

$$H = \sum_{i,j} \sum_{\nu,\nu'} \rho_i^s(\nu)\,\lambda_{ij}^{st}\,\rho_j^t(\nu')\,V(\nu,\nu';\mu) \qquad . \qquad . \qquad (5)$$

Taking into account the relations (2) and (3) and using Table 2, Equation (5) can be rewritten as:

$$H = \frac{N}{2}zV_{21} + p(2V_{11} + V_{22} - 3V_{21}) + q(V_{12} - V_{21}) . \qquad (6)$$

wherein

$$p = \sum_{i,j} \sum_{\mu} \rho_i^s(\mu)\,\lambda_{ij}^{st}\,\rho_j^t(\mu)$$

is the total number of nearest neighbour pair contacts of the kind contributing $V_{22}$ to the configurational energy, and where

$$q = \sum_{i,j} \lambda_{ij}{}^{st} \sum_{\mu < \mu'} [\rho_i{}^s(\mu) \, \rho_j{}^t(\mu') + \rho_i{}^s(\mu') \, \rho_j{}^t(\mu)]$$

is the total number of the kind contributing $V_{12}$ to this same energy. The Hamiltonian function can be used to deduce the structure of the low-temperature distorted phase, resulting from the spatial ordering of the distorted octahedra. The completely ordered arrangement of the octahedra, which minimized $H$ at $T=0°K$, will be the one that makes $p$ minimum and $q$ maximum simultaneously. This is accomplished when the long axes of the nearest-neighbour distorted octahedra are mutually perpendicular: $p$ takes then its absolute minimum, zero, while $q$ reaches its absolute maximum $(2/3).(N/2).z$. The configurational energy of this phase is

$$H = \frac{1}{3}\frac{N}{2}zV_{21} + \frac{2}{3}\frac{N}{2}zV_{12},$$

one-third of the neighbour contacts contributing $V_{21}$, two-thirds contributing $V_{12}$ (Figure 4).

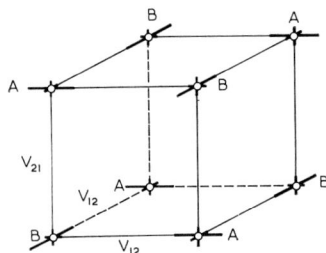

FIGURE 4.—The completely ordered arrangement of the octahedra at absolute zero.

Such arrangement results necessarily in a macroscopic ortho-rhombic distortion at $T=0°K$, maintained but progressively weakened as temperature rises. However, when a certain critical temperature is attained, we can argue that the thermal agitation will dominate the co-operative interaction, resulting in an orthorhombic to cubic phase transformation, in agreement with the experimental evidence.

Based on the Hamiltonian obtained above and using methods of statistical mechanics, the thermodynamic behaviour of our model can now be derived.

As the efficiency of the co-operative interaction is reduced by the presence of ions such as $Mn^{4+}$, it is desirable to extend the theory in such a way that the composition dependence is also included, which will make a direct comparison between theory and experiment possible. This generalization is now in progress.

## ACKNOWLEDGMENTS

The authors thank Prof. Dr J. L. Verhaeghe for his stimulating interest in this work and Prof. Dr W. Dekeyser, in whose laboratory the X-ray diffraction patterns were taken. One of us (C.M.I.) is much indebted to the Nationaal Fonds voor Wetenschappelijk Onderzoek.

## REFERENCES

1. ROBBRECHT, G. G., and JACOBS, E. I., *Rev. Sci. Instr.*, **33**, (3), 376, 1962.
2. VERHAEGHE, J. L., ROBBRECHT, G. G., ISERENTANT, C. M., VAN OUTRYVE, E. J., DOCLO, R. J., Med. v. d. Kon. Vl. Akad., Klasse der Wet. XXVI, 7, (1964).
3. WOLLAN, E., and KOEHLER W., *Phys. Rev.*, **100**, 545, 1955.
4. WOLD, A., and ARNOTT, R., *J. Phys. Chem. Solids*, **9**, 176, 1959.
5. JONKER, G., *Physica*, **22**, 707, 1956.
6. GELLER, S., and WOOD, E., *Acta Cryst.* **9**, 563, 1956.
7. DUNITZ, J. D., and ORGEL, L. E., *J. Phys. Chem. Solids*, **3**, 20, 1957.
8. WOJTOWICZ, P. J., *Phys. Rev.*, **116**, 32, 1959.
9. GOODENOUGH, J. B., *Phys. Rev.* **100**, 564, 1955.

# 31.—Origin of Lattice Defects in Titanate Materials and their Influence on Behaviour

By A. KOLLER

*Výzkumný Ústav Elektrotechniké Keramiky,
Hradec Králové, Czechoslovakia*

*ABSTRACT* A51/E732

*When titanate materials are fired, the atmosphere affects not only the perfection of the crystals that develop but also their properties, particularly the electrical ones. Photosensitive materials with characteristic behaviour in an electric field originate when an oxidizing atmosphere is changed to moderately reducing. If the atmosphere is highly reducing, a blue body occurs that is characterized by considerable electrical conductivity. Defect bodies that develop while being fired in a moderately reducing atmosphere are dealt with briefly. Lattice defects were studied chiefly on materials of the $BaTiO_3$ type; they generally occur in furnaces having SiC elements. Crystal-lattice defects develop in almost all titanate materials. By replacing Ba ions by Sr ions in $BaTiO_3$, the formation of defects increases and the defects produce ionic conductivity which causes colour centres to develop. In other titanate structures (e.g. $TiO_2$) the formation of colour centres in less evident, and in materials of the $BaTi_4O_9$ type, colour centres do not occur at all. When these materials are irradiated, e.g. by X-rays, unstable colour centres develop which can be studied by electron emission. The composition of materials in a moderately reducing atmosphere is assumed to be: $R\ Ti^{4+}(KV)_x(AV)_{2x}O^{2-}{}_{3-2x}$. Colour centres also develop in these materials when they are annealed in oxygen and during rapid cooling. From changes in weight during the formation of colour centres, the number of oxygen vacancies taking part in their development can be deduced.*

***Origine des défauts reticulaires dans les matériaux à base de titanate.
Influence de ces défauts sur le comportement de ces matériaux***

*Au cours de la cuisson des matériaux à base de titanate, l'atmosphère a une influence non seulement sur la perfection des cristaux qui se développent mais également sur leurs propriétés, particulièrement les propriétés électriques. Des matériaux sensibles à la lumière et manifestant un comportement caractéristique dans un champ électrique prennent naissance quand on change une atmosphère oxydante en une atmosphère modérément réductrice. Si l'atmosphère est fortement* *réductrice, il se forme un corps bleu caractérisé par une conductivité électrique très élevée. Les produits présentant des défauts qui se développent dans une atmosphère modérément réductrice sont brièvement étudiés. Les défauts réticulaires ont été étudiés principalement sur des matériaux du type $BaTiO_3$; ils apparaissent généralement dans les fours à éléments en SiC. Quand on remplace, dans $BaTiO_3$, les ions Ba par des ions Sr, la formation de défauts s'accentue et ces derniers produisent une conductivité ionique donnant naissance au développement de centres colorés. Dans d'autres structures de titanate (par exemple $TiO_2$), la formation de centres colorés est moins évidente et dans les matériaux du type $BaTi_4O_9$, il n'y a aucune apparition de centres colorés. Quand ces matériaux sont irradiés, aux rayons X par exemple, il se forme des centres colorés instables qui peuvent être étudiés par émission électronique. On admet que dans une atmosphère modérément réductrice les matériaux ont la composition suivante: $RTi^{4+}(KV)_x(AV)_{2x}O^{2-}_{3-2x}$. Des centres colorés se développent également dans ces matériaux quand ils subissent un recuit dans l'oxygène et au cours d'un refroidissement rapide. Il est possible de déduire le nombre de lacunes d'oxygène qui participent au développement des centres colorés à partir des variations de poids qui se produisent pendant leur formation.*

## Ursprung von Gitterfehlern in Titanatmaterialien und ihr Einfluss auf das Verhalten

*Wenn Titanatmaterialien gebrannt werden, beeinflußt die Atmosphäre nicht nur die Vollkommenheit der entstehenden Kristalle, sondern auch ihre Eigenschaften, besonders die elektrischen. Lichtempfindliche Materialien mit einem charakteristischen Verhalten in einem elektrischen Feld entstehen, wenn eine oxydierende Atmosphäre geringfügig reduzierend gemacht wird. Wenn die Atmosphäre stark reduzierend ist, entsteht ein blauer Körper, der durch eine beträchtliche elektrische Leitfähigkeit gekennzeichnet ist. Die Defekte in den Körpern, die beim Brennen in mäßig reduzierender Atmosphäre entstehen, werden kurz behandelt. Gitterfehler werden hauptsächlich an Materialien von $BaTiO_3$-Typ untersucht; sie entstehen vornehmlich in Öfen mit SiC-Heizelementen. Gitterfehler entstehen in fast allen Titanatmaterialien. Beim Ersatz von Ba-Ionen durch Sr-Ionen im $BaTiO_3$ nimmt die Bildung von Defekten zu, und letztere erzeugen Ionenleitfähigkeit, die das Auftreten von Farbzentren im Gefolge hat. In anderen Titanatstrukturen (z.B. $TiO_2$) ist die Bildung von Farbzentren weniger auffällig, und in Materialien vom $BaTi_4O_9$-Typ entstehen überhaupt keine Farbzentren. Wenn diese Materialien bestrahlt werden, z.B mit Röntgenstrahlen, entstehen instabile Farbzentren, die* *durch Elektronenemission untersucht werden können. Für die Zusammensetzung der Materialien aus mäßig reduzierender Atmosphäre wird folgende Formel angenommen:* $R\,Ti^{4+}(KV)_x(AV)_{2x}O^{2-}{}_{3-2x}$. *In diesen Materialien entstehen ebenfalls Farbzentren beim Tempern in Sauerstoff und beim raschen Abkühlen. Aus den Gewichtsänderungen während der Farbzentrenentstehung läßt sich die Anzahl beteiligter Sauerstoffleerstellen ableiten.*

## 1. INTRODUCTION

When ceramic capacitors age, their properties change in several ways. Permittivity changes with the time below the Curie point. In an electric field, Ag from the electrodes may penetrate into the body, sometimes the ions are separated by electrolysis, and for example the electrical resistance increases, etc.

Under certain conditions, when direct current is used we can observe interesting colour changes which cause strong deterioration of the dielectric properties. This effect does not generally occur in normal industrial materials, but it can often be observed when the properties of pure titanates are studied. W. A. WEYL[1] was one of the first to describe this effect and he called it degradation. BURSIAN and KOSMAN[2] and BŘEZINA and JANOVEC[3] have described similar behaviour in BaTiO$_3$ monocrystals, and the degradation of ceramics based on TiO$_2$ has been described by KUNIN and CIKIN.[4]

The changes in the capacitor materials are caused by defects in the lattice and this paper describes the origin of these defects, the results of studying the defects by exo-electron emission and the effects caused by them.

Titanate materials have different electrical properties according to the medium in which they are fired—oxidizing or reducing. A reducing firing gives the characteristic blue colour of materials possessing considerable electrical conductivity. The oxidizing firing of pure perovskite-type ferroelectrics is generally characterized by light colours, high capacitor resistance and low dielectric losses. Between these two typical firing conditions there exists a very narrow region which, however, is of great importance in connection with the origin of the defects. In this region, reduction does not occur and the fired bodies exhibit characteristics very similar to those of materials fired in an oxidizing atmosphere, but always somewhat less favourable. This is observed particularly in capacitor resistance at higher temperature and dielectric loss. The dielectric itself exhibits rather high photosensitivity, owing to which the ceramic darkens when exposed to sunlight. Such a moderately reducing atmosphere is to be found for example in electric furnaces with silicon carbide heating-elements.

480     KOLLER:

The conditions for the formation of the lattice defects in individual materials vary and depend on the composition, the firing temperature, the firing curve and the water vapour concentration in the firing atmosphere.

## 2. EXPERIMENTAL

With bodies of the $BaTiO_3$ type in particular, the formation of such defects is very characteristic. By substituting Sr ions for Ba ions, we obtained a material that was very suitable for studying defects and their influence on the electrical properties. The defects formed and the consequent degradation were compared with those of ceramics made from almost pure $TiO_2$ or $BaTi_4O_9$.

We used exo-electron emission to study the defects.[5] X-irradiated samples were put under a point electron counter capable of indicating negative particles released—due to heating of the sample—from the unstable colour centres.

Curves (Figure 1) of exo-emission of materials containing defects show several peaks up to 300°C and even thermoluminescence could be observed. According to the position of the emission peaks and thermoluminescence it is possible to estimate the thermal dissociation energy of the individual defect types as approximately 1·05, 1·3 and 1·45 eV. These indicate at least three kinds of defect in the anion lattice, in which the electrons released from F centres had been trapped after irradiation. These materials with defects can also form stable colour centres, which form owing to constant anion conductivity or to diffusion of oxygen ions into the ceramic at high temperature.

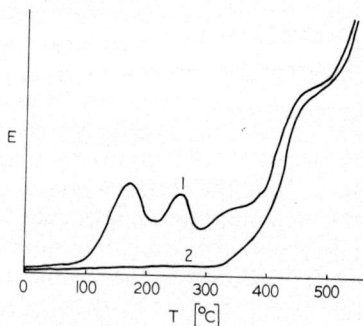

FIGURE 1.—Exo-electron emission curve:

E = Emission intensity
T = Temperature (°C)
1 = Temperature curve of exo-emission of a degrading sample
2 = Similar curve of a non-degrading sample

The first type of these additive centres forms during the degradation of the capacitor. The formation of these additive centres was followed on disc capacitors. Formation does not depend on the electrode material and can be followed by measuring the variation in current through the capacitor at a constant temperature. Up to 320°C the process is accelerated with increasing temperature but with further rise in temperature the changes in current become smaller, even though the bulk conductivity is increasing, and at 450°C additive colour centres are no longer formed. At this temperature all the colour centres have already dissociated.

In oxygen or in air, dark brown colour centres of mostly one type are found. They originate from the anode and eventually spread through the whole capacitor. In vacuum, in nitrogen, or in oil, two layers of different colours are found between the electrodes.

On the basis of these facts, we can say that stable colour centres of both V type and F type are formed: the former when an electron hole is trapped at a Schottky defect in the cation lattice, the latter when electrons are trapped at the anion vacancies, which compensate the loss of charge in the lattice originating from the movement of oxygen ions.

The second type of additive V centre is formed when the material containing defects is heated in an oxidizing atmosphere to 1200°C and quenched. The colour centres formed have the same colouring as the V centres originating in the electric field. Both ways of forming additive colour centres increase the weight of the samples, which indicates absorption of oxygen. From this increase the number of oxygen ions which take part in the formation of the colour centres could be determined, and amounted to at least $1\text{--}3 \cdot 10^{17}/\text{cm}^3$.

Because additive and unstable colour centres form after irradiation, we can assume that, when titanate materials are fired in a moderately reducing atmosphere, a large quantity of Schottky defects occur in both the anion lattice and the cation lattice. The schematic composition of these bodies is

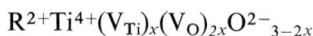

$$R^{2+}Ti^{4+}(V_{Ti})_x(V_O)_{2x}O^{2-}_{3-2x}$$

where $V_{Ti}$ and $V_O$ represent the cation and anion vacancies.

## REFERENCES

1. WEYL, W. A., and TERHUNE, N. A., *Ceramic Age*, **62**, (2), 23, 1953.
2. KOSMAN, M. S., and BURSIAN, J. V., *DAN U.S.S.R.*, **115**, (3), 483, 1957.
3. BŘEZINA, B., and JANOVEC, V., *Silikaty*, **3**, 189, 1961.
4. KUNIN, V. JA., and CIKIN, A. N., *Solid State Physics* (Leningrad U.S.S.R.), **2**, (10), 2359, 1961.
5. KRAMER, J., *Ber. dtsch. keram. Ges.*, **30**, 203, 1953.

SBit
105/WB.